Antenna Designs for 5G/IoT and Space Applications

Antenna Designs for 5G/IoT and Space Applications

Editors

Faisel Tubbal
Ladislau Matekovits
Raad Raad

MDPI • Basel • Beijing • Wuhan • Barcelona • Belgrade • Manchester • Tokyo • Cluj • Tianjin

Editors
Faisel Tubbal
University of Wollongong
Australia

Ladislau Matekovits
Politecnico di Torino
Italy

Raad Raad
University of Wollongong
Australia

Editorial Office
MDPI
St. Alban-Anlage 66
4052 Basel, Switzerland

This is a reprint of articles from the Special Issue published online in the open access journal *Electronics* (ISSN 2079-9292) (available at: https://www.mdpi.com/journal/electronics/special_issues/Antenna_Space).

For citation purposes, cite each article independently as indicated on the article page online and as indicated below:

LastName, A.A.; LastName, B.B.; LastName, C.C. Article Title. *Journal Name* **Year**, *Volume Number*, Page Range.

ISBN 978-3-0365-5151-7 (Hbk)
ISBN 978-3-0365-5152-4 (PDF)

Contents

About the Editors

Faisel Tubbal

Faisel Tubbal (Senior Member, IEEE) received the B.E. degree from the college of Electronic Technology, Tripoli, Libya, in 2004, the M.S. degrees in engineering management and Telecommunication Engineering both from the University of Wollongong, Wollongong, Australia, in 2012 and 2013, respectively. He has also received a PhD in telecommunication engineering from the University of Wollongong, Wollongong, Australia in 2017. His Ph.D. thesis was entitled S-band Planar Antenna Designs for CubeSat Communications. Dr. Faisel has been working as a researcher with Libyan Centre for Remote Sensing and Space Science (LCRSSC), Tripoli, Libya. In 2012, he joined the University of Wollongong as an academic assistant at school of electrical, computer and telecommunication engineering. In 2019, Faisel was promoted to the position of labs manager at School of Computer, Electrical and Telecommunication Engineering (SECTE), University of Wollongong, Australia. Since 2017, he has been a unit Convenor with school of Computing, Engineering and Mathematics, Western Sydney University, Australia. Faisel Is currently the Labs manger and Work Integrated Learning Coordinator at School of Computer, Electrical and Telecommunication Engineering (SECTE), University of Wollongong. He is the author of three chapters and more than 55 conference and journal articles. His research interests include antenna designs for CubeSat applications, wearable antennas, antenna designs using metamaterials, Metasurface antennas and metamaterials. Dr. Faisel was the recipient of the Vice-Chancellor's Awards for Outstanding Contribution to Teaching and Learning (OCTAL) in 2016 and 2021. He is a Fellow at Wollongong Academy for Tertiary Teaching and Learning Excellence (WATTLE), and a member of Australasian Association for Engineering Education (AAEE).

Ladislau Matekovits

Ladislau Matekovits, received the degree in electronic engineering from Institutul Politehnic din Bucureşti, Bucureşti, Romania, and the Ph.D. degree (Dottorato di Ricerca) in electronic engineering from Politecnico di Torino, Torino, Italy, in 1992 and 1995, respectively. Since 1995, he has been with the Department of Electronics and Telecommunications, Politecnico di Torino, where in 2014 he was appointed as Associate Professor. Beginning 1 July 2009, for two years he has been a Marie Curie Fellow at Macquarie University, Sydney, NSW, Australia. Since 2020 he is Honorary Professor at Polytechnic University of Timisoara, Romania, and Associate of the Italian National Research Council. His main research activities concern numerical analysis of printed antennas and in particular development of new, numerically efficient full-wave techniques to analyse large arrays, and active and passive metamaterials for cloaking applications. Material parameter retrieval of these structures by inverse methods and different optimization techniques has also been considered. In the last years, bio-electromagnetic aspects have also been contemplated, as for example design of implantable antennas or development of nano-antennas for example for drug delivery applications. He has published 400+ papers, including 110+ journal contributions, and delivered seminars on these topics all around the world: Europe, USA (AFRL/MIT-Boston), Australia, China, Russia, etc. Prof. Matekovits has been invited to serve as Research Grant Assessor for government funding calls (Romania, Italy, Croatia, Kazakhstan, and Iceland) and as International Expert in PhD thesis evaluation by several Universities from Australia, India, Pakistan, Spain, etc. Prof. Matekovits has been a recipient of various awards in international conferences, including the 1998 URSI Young Scientist Award (Thessaloniki, Greece), the Barzilai Award 1998 (young Scientist Award, granted every two years by the Italian National Electromagnetic Group), and the Best AP2000 Oral Paper on Antennas, ESA-EUREL Millennium Conference on Antennas and Propagation (Davos, Switzerland). He is recipient of the Motohisa Kanda Award 2018, for the most cited paper of the IEEE Transactions on EMC in the past five years, and more recently he has been awarded with the 2019 American Romanian Academy of Arts and Sciences (ARA) Medal of Excellence in Science and by the Ad Astra Award 2020, Senior researcher, for Excellence in Research. He serves as Associated Editor of the IEEE ACCESS, IEEE Antennas and Wireless Propagation Letters and IET MAP and reviewer for different journals.

Raad Raad

Raad Raad received the Bachelor of Engineering degree (Hons.) in electrical engineering and the master's degree from the Switched Networks Research Centre, University of Wollongong, Wollongong, NSW, Australia, in 1997, and the Ph.D. degree in neuro-fuzzy logic admission control in cellular mobile networks, in 2006. Since 2004, he has been with the School of Electrical, Computer and Telecommunications Engineering, University of Wollongong, where he works as the Deputy Head of the School. His current research interests include wireless communications, CubeSat, the IoT, and antenna design. He received an Australian Postgraduate Award that was matched by Telstra Research Laboratories. He received a scholarship from the Motorola Australian Research Centre in the later part of his degree.

electronics

Editorial

Antenna Designs for 5G/IoT and Space Applications

Faisel Tubbal [1,2,*], Ladislau Matekovits [3,4,5] and Raad Raad [1]

1 School of Electrical, Computer and Telecommunication Engineering, University of Wollongong, Wollongong 2500, Australia
2 Technological Projects Department, Libyan Center for Remote Sensing and Space Science, Tripoli 21821, Libya
3 Dipartimento di Department, Elettronica e Telecomunicazioni, Politecnico di Torino, 10129 Torino, Italy
4 Istituto di Elettronica e di Ingegneria dell'Informazione e delle Telecomunicazioni, National Research Council, 10129 Turin, Italy
5 Department of Measurements and Optical Electronics, Politehnica University Timisoara, 300006 Timisoara, Romania
* Correspondence: faisel@uow.edu.au

Citation: Tubbal, F.; Matekovits, L.; Raad, R. Antenna Designs for 5G/IoT and Space Applications. *Electronics* **2022**, *11*, 2484. https://doi.org/10.3390/electronics11162484

Received: 2 August 2022
Accepted: 8 August 2022
Published: 10 August 2022

1. Introduction

Antenna design has received renewed attention in the last few years. This is thanks to an explosion of interest in a range of applications, from the Internet of Things, low frequency long-range applications to high frequency mmWave 5G mobile technologies. There has also been renewed interest in wearable antennas that form body area networks. These include wearable garments, as well as materials that directly attach themselves to skin, such as e-skin. In addition to this, a renewed interest in space and space exploration has renewed interest in satellite technologies and applications, such as CubeSats, intersatellite communications and deep space exploration. All these emerging applications bring a renewed interest in looking at special materials and new designs for antenna systems. This will bring new challenges in designing such antennas.

For this purpose, this Special Issue is intended to shed some light on recent advances in antenna design for these new emerging applications and identify further research areas in this exciting field of communications technologies. We invite researchers and practicing engineers to contribute original research articles that discuss issues related but not limited to:

- Antenna design for Internet of Things;
- Beamforming and smart antennas for 5G;
- Antenna design for wearable applications;
- Antenna design for body area networks;
- Antenna design for chipless RFID;
- Metamaterial-based antennas;
- Smart antennas, beamforming and MIMO;
- Aeronautical and space applications;
- Antenna design for CubeSat;
- Antenna design for deep space communication
- Antenna design for biomedical systems and applications;
- Implanted antennas;
- UWB and multispectral technologies and systems;
- MM-wave and THz antennas.

2. Short Presentation of the Papers

Liu et al. [1] presented a comprehensive survey of antennas used for 120 CubeSat missions from the period from 2003 to 2022, as well as their techniques and approaches. The aim of this paper is to provide an introductory guide on CubeSats antennas for CubeSat enthusiasts and a state of the art for CubeSat designers in this ever-growing field. It presents a background on CubeSats and their subsystems to provide newcomers with the fundamental knowledge on CubeSat technologies. In addition, the authors investigated different designs for CubeSats, including single-element antennas and antenna arrays, as well as the aim of capturing the current and future CubeSat trends from an antenna point of view. They also proposed a pictorial representation of how to select an antenna for different types of CubeSat missions.

Farasat et al. [2] surveyed the recent base station antenna (BSA) designs for 5G Sub-6 GHz and their challenges. The authors provided a comprehensive summary of BSAs with particular interest placed on lower microwave bands in sub-6 GHz range. They also provided a brief discussion on the development of the BSA technologies through mobile generations, as well as a detailed discussion on the challenges associated with the BSAs. These challenges include the achieving wide impedance bandwidth while meeting the port-to-port isolation requirements, making the radiating elements of one band transparent to another band and vice versa without causing scattering and undesirable effects on the radiation pattern and meeting the size constrains. The authors have also provided a cohesive list of design techniques adopted to overcome the aforementioned challenges.

Khaleel et al. [3] proposed a graphene plasmonic two port MIMO antenna for 6G/IoT applications. The proposed antenna operates in Terahertz frequency band (e.g., 3.2–3.8 THz) and provides good radiation performance. A technique that the authors used to reduce the mutual coupling between radiating elements is the etching of the dumbbell-structure metamaterial (MTM) unit cell in the ground plane. They reported a high isolation of −55 dB between radiating elements, simulated gain of 7.23 dB and low channel capacity loss of 0.006.

Khan et al. [4] presented an ultra-wideband pentagonal fractal antenna for new-field microwave imaging applications. The proposed compact antenna has a size of 24 mm × 30 mm × 0.787 mm and used the Rogers RT/Duroid 5880 dielectric substrate. The authors used the co-planar waveguide (CPW) feed technique and trapezoidal ground plane to achieve low losses and to enhance the impedance matching, respectively. The proposed antenna provides a fractional bandwidth of 123.56%, and a measured −10 dBi impedance bandwidth of 9.7 GHz (3–12.7 GHz) and a total gain of 3.6 dBi at 8.5 GHz.

Nuñez et al. [5] designed and proposed a compact single-input single-output antenna with a slotted decahedral patch for 5G applications. The main idea is the use of decahedral patch with the eight-pointed star-shaped slot geometry and two rectangular grooves to enhance the antenna's performance and to achieve the stability of the radiation pattern across the wide −10 dB impedance bandwidth (e.g., 23–29.9 GHz). The proposed antenna has a total size of 13 mm × 13 mm × 0.787 mm and achieves a wide −10 dB impedance matching of 6.84 GHz (23.1–29.9 GHz), 89.4% radiation efficiency, a gain of 6.56 dBi and reflection coefficient of −21.5 dB at 29 GHz.

Munir et al. [6] presented a multi-circular loop planar antenna array for next generation mm-wave communication system. The proposed antenna consists of multi circular loop rings, uses Rogers 5880 substrate and has a small size of 18.5 mm × 12.5 mm × 0.254 mm. To enhance the performance (e.g., −10 dB bandwidth, gain and reflection coefficient) of the proposed antenna, the authors used a partial ground plane with a square slot, and four elements. The measure and simulated results are in good agreement and the proposed antenna array provides a dual band with a narrow beamwidth and a small reflection coefficient, wide −10 dB impedance bandwidth (e.g., 26–38.5 GHz) and radiation efficiency of 95%.

Hussain et al. [7] proposed a low-profile meandered loop slot-line antenna for the Internet of Things (IoT) application. The proposed antenna operates in wideband tuning, 0.758–1.034 GHz, has a size 60 mm × 27 mm × 0.76 mm and operates over a wide band of sub-GHz. The miniaturization of the antenna is achieved by using the meandered structured loop slot line loaded on a varactor diode. The varactor diode was used to achieve the wideband frequency reconfigurability. The authors reported an efficiency ranging from 54–67% and total gain ranging from 0.86–1.8 dBi.

Faouri et al. [8] presented a microstrip meandered bowtie antenna with multi-resonant and rejection bands for 5G applications. The antenna has a total size of 30 mm × 30 mm × 1.6 mm and was built on double-sided FR-4 substrate. The shorting via is used to connect one portion of the bowtie to the ground plane, hence enhancing the resonances. Moreover, length, width and spacing of the meander line arms are optimized to produce more resonances with an acceptable bandwidth. The proposed antenna covers S, C and X-bands, has an efficiency of 90.3%, achieved a gain of 4.46 dB and resonates at 2.7, 4.05, 5.05, 6.04, 7.15, 7.9 and 11.55 GHz.

Sadananda et al. [9] presented a compact three-port wide angular coverage patch antenna array for mmWave 5G smartphones. The proposed multi-port electrically connected patch antenna system operates at 28 GHz (Ka-band) and has a size of 24 mm × 6.2 mm. The multiport radiator and ground plane are shared and electrically connected. To achieve the pattern diversity with angular coverage, the authors integrated the stepped impedance transformers with the corner-most elements. Moreover, the feed line is connected to a two-way non-Wilkinson-based power divider. The authors reported a wide angular coverage of 100°, antenna efficiency of 97%, a mutual coupling less than 10 dB and a gain variation of 6 to 11 dBi across the ports.

Alharbi et al. [10] investigated and proposed a compact two antenna MIMO with a dual wideband operation for wireless and satellite applications. The antenna has a total size of 32 mm × 20 mm × 0.8 mm and achieves a dual wideband operation of 3.3–7.8 GHz and 8–12 GHz. The key idea was to load the monopole antenna with number of narrow slots and multiple slotted stubs to achieve multiband characteristics. Furthermore, to achieve an isolation of less than 20 dB, five concentric ring elements were etched between the two adjacent antenna elements. The proposed two antenna MIMO antenna provides a low band operation of 81.08% (3.3–7.8 GHz) and a high band operation of 40% (8–12 GHz), which makes it suitable for the Sub-6 GHz 5G new radio (NR) n77/78/79, IEEE 802.11ac/ax, X-band/C-band wireless and satellite applications. The authors reported a gain of 3–4 dBi and a radiation efficiency of 69–80%.

Khan et al. [11] presented a high efficiency circular-shaped patch antenna array for Sub-6 GHz 5G applications. The proposed antenna operates at 5.57 GHz and has a total size of 160 mm × 70 mm. The main idea was the use of transmission line and quarter wavelength transformation techniques for impedance and phase matching. To achieve the desired operating frequency band of 5.6–5.67 GHz, a circular shape of radiating element with an inner-circular slots and rectangular at its right edge is used. The four-element array configuration of the proposed antenna provides a total gain of 12.4 dB. The authors also reported an isolation between the two used ports of more than −30 dB and a total efficiency of 85.1%.

Kulkarni et al. [12] proposed a planar four-port and four element smartphone flexible antenna for Sub-6 5G and WLAN smart terminals. Its features, including flexibility, bi/omnidirectional radiation pattern, planar structure, wide bandwidth covering Sub6 GHz and WLAN bands, and high inter-elemental isolation make it a good candidate for use on smartphone applications. The proposed antenna consists of one contacting ground at the center placed on the top of a flexible polyamide substrate and four conducting MIMO elements (Radiators) at the four corners on the top of the substrate. The proposed antenna has a total size of 70 mm × 145 mm × 0.2 mm, a wide impedance bandwidth ranging from 2.37 to 5.85 GHz, gain of 4–5.5 dBi, efficiency of 85% and self-isolation of about 17.5 dB.

Azizi et al. [13] presented a wide-band dual concentric phase gradient modulated surface (PGMS) for surface radar cross-section (RCS) reduction at wideband frequency range, 20.9–45.7 GHz (75%). The main idea is using the two single band concentric modulated surfaces, e.g., MS1 and MS2 to reduce the RCS from 20.5–32 GHz (45%) for MS1 and 31.8–46.5 GHz (40%) for MS2, consequently. To enhance the bandwidth, the authors placed MS2 in the central part of MS1 to obtain a concentric configuration and hence achieve a wideband 75% (20.9–45.7 GHz) RCSR performance.

Huang et al. [14] designed a quad-port antenna MIMO antenna array for 5G smartphone applications. The proposed antenna array consists of four elements printed on two side boards, which are positioned vertically to the 150 mm × 75 mm main board. The total size of the main board is 150 mm × 75 mm × 0.8 mm, and the size of the side boards is 150 mm × 6.2 mm × 0.8 mm. Each single element has a size of 14.9 mm × 7 mm × 0.8 mm and includes an L-shaped feed, a strip, a parasitic rectangular strip and a Z-shaped radiation strip that is connected to ground plane. The main idea of printing the quad-port antenna array along two long frames of the smartphones is to reserve space for 2G/3G/4G antennas. Parametric analysis had been completed by the authors to obtain the required results and the optimal dimensions of the proposed antenna array design. The proposed design achieved a dual band operation at 3.5 GHz (3.4–3.6 GHz) and at 5 GHz (4.8–5 GHz). The authors reported a measured efficiency of 82%, isolation of 16.5 dB, and total measured gains of 4.7 and 5 dBi at 3.6 and 5 GHz, respectively.

Author Contributions: Conceptualization, F.T., L.M. and R.R.; methodology, F.T.; validation, L.M. and R.R.; investigation, R.R., F.T. and R.R.; resources, F.T., L.M. and R.R.; writing—original draft preparation, F.T.; writing—review and editing, L.M. and R.R.; visualization, F.T., L.M. and R.R.; supervision, F.T.; project administration, F.T., L.M. and R.R. All authors have read and agreed to the published version of the manuscript.

Funding: This research received no external funding.

Conflicts of Interest: The authors declare no conflict of interest.

References

1. Lu, S.; Theoharis, P.I.; Raad, R.; Tubbal, F.; Theoharis, A.; Iranmanesh, S.; Abulgasem, S.; Khan, M.U.A.; Matekovits, L. A Survey on CubeSat Missions and Their Antenna Designs. *Electronics* **2022**, *11*, 2021. [CrossRef]
2. Farasat, M.; Thalakotuna, D.N.; Hu, Z.; Yang, Y. A review on 5G sub-6 GHz base station antenna design challenges. *Electronics* **2021**, *10*, 2000. [CrossRef]
3. Khaleel, S.A.; Hamad, E.K.; Parchin, N.O.; Saleh, M.B. MTM-Inspired Graphene-Based THz MIMO Antenna Configurations Using Characteristic Mode Analysis for 6G/IoT Applications. *Electronics* **2022**, *11*, 2152. [CrossRef]
4. Khan, M.A.; Rafique, U.; Savci, H.Ş.; Nordin, A.N.; Kiani, S.H.; Abbas, S.M. Ultra-Wideband Pentagonal Fractal Antenna with Stable Radiation Characteristics for Microwave Imaging Applications. *Electronics* **2022**, *11*, 2061. [CrossRef]
5. Olan Nuñez, K.N.; Murphy Arteaga, R.S. A Novel and Compact Slotted-Decahedral Antenna for 5G Devices. *Electronics* **2022**, *11*, 1813. [CrossRef]
6. Munir, M.E.; Al Harbi, A.G.; Kiani, S.H.; Marey, M.; Parchin, N.O.; Khan, J.; Mostafa, H.; Iqbal, J.; Khan, M.A.; See, C.H. A New mm-Wave Antenna Array with Wideband Characteristics for Next Generation Communication Systems. *Electronics* **2022**, *11*, 1560. [CrossRef]
7. Hussain, R.; Alhuwaimel, S.I.; Algarni, A.M.; Aljaloud, K.; Hussain, N. A Compact Sub-GHz Wide Tunable Antenna Design for IoT Applications. *Electronics* **2022**, *11*, 1074. [CrossRef]
8. Faouri, Y.S.; Ahmad, S.; Parchin, N.O.; See, C.H.; Abd-Alhameed, R. A novel meander bowtie-shaped antenna with multi-resonant and rejection bands for modern 5G communications. *Electronics* **2022**, *11*, 821. [CrossRef]
9. Sadananda, K.G.; Elfergani, I.; Zebiri, C.; Rodriguez, J.; Koul, S.K.; Abd-Alhameed, R.A. A Wide-Angle Pattern Diversity Antenna System for mmWave 5G Mobile Terminals. *Electronics* **2022**, *11*, 571. [CrossRef]
10. Alharbi, A.G.; Kulkarni, J.; Desai, A.; Sim, C.-Y.-D.; Poddar, A. A Multi-Slot Two-Antenna MIMO with High Isolation for Sub-6 GHz 5G/IEEE802. 11ac/ax/C-Band/X-Band Wireless and Satellite Applications. *Electronics* **2022**, *11*, 473. [CrossRef]
11. Khan, J.; Ullah, S.; Tahir, F.A.; Tubbal, F.; Raad, R. A sub-6 GHz MIMO antenna array for 5G wireless terminals. *Electronics* **2021**, *10*, 3062. [CrossRef]
12. Kulkarni, J.; Alharbi, A.G.; Desai, A.; Sim, C.-Y.-D.; Poddar, A. Design and analysis of wideband flexible self-isolating MIMO antennas for sub-6 GHz 5G and WLAN smartphone terminals. *Electronics* **2021**, *10*, 3031. [CrossRef]
13. Azizi, Y.; Soleimani, M.; Sedighy, S.H.; Matekovits, L. Low-Cost, Low-Profile Wide-Band Radar Cross Section Reduction Using Dual-Concentric Phase Gradient Modulated Surface. *Electronics* **2021**, *10*, 1552. [CrossRef]
14. Huang, J.; Dong, G.; Cai, J.; Li, H.; Liu, G. A quad-port dual-band MIMO antenna array for 5G smartphone applications. *Electronics* **2021**, *10*, 542. [CrossRef]

electronics

MDPI

Review

A Survey on CubeSat Missions and Their Antenna Designs

Sining Liu [1], Panagiotis Ioannis Theoharis [1], Raad Raad [1], Faisel Tubbal [1,2,*], Angelos Theoharis [1,3], Saeid Iranmanesh [1], Suhila Abulgasem [1], Muhammad Usman Ali Khan [1,4] and Ladislau Matekovits [5,6,7]

1 School of Electrical, Computer and Telecommunications Engineering, University of Wollongong, Wollongong, NSW 2522, Australia; sining@zzuli.edu.cn (S.L.); pit289@uowmail.edu.au (P.I.T.); raad@uow.edu.au (R.R.); at925@uowmail.edu.au (A.T.); saeidim@uow.edu (S.I.); sgsa450@uowmail.edu.au (S.A.); muak803@uowmail.edu.au (M.U.A.K.)
2 Technological Projects Department, The Libyan Center for Remote Sensing and Space Science, Tripoli 21218, Libya
3 Department of Electronics, Information and Bioengineering, Politechnico di Milano, 20125 Milan, Italy
4 Department of Electronic Engineering, The Islamia University of Bahawalpur, Bahawalpur 63100, Punjab, Pakistan
5 Department of Electronics and Telecommunications, Politecnico di Torino, 10129 Turin, Italy; ladislau.matekovits@polito.it
6 Istituto di Elettronica e di Ingegneria dell'Informazione e delle Telecomunicazioni, National Research Council, 10129 Turin, Italy
7 Department of Measurements and Optical Electronics, Politehnica University Timisoara, 300006 Timisoara, Romania
* Correspondence: faisel@uow.edu.au

Citation: Liu, S.; Theoharis, P.I.; Raad, R.; Tubbal, F.; Theoharis, A.; Iranmanesh, S.; Abulgasem, S.; Khan, M.U.A.; Matekovits, L. A Survey on CubeSat Missions and Their Antenna Designs. *Electronics* 2022, *11*, 2021. https://doi.org/10.3390/electronics11132021

Academic Editor: Raed A. Abd-Alhameed

Received: 1 June 2022
Accepted: 22 June 2022
Published: 27 June 2022

Abstract: CubeSats are a class of miniaturized satellites that have become increasingly popular in academia and among hobbyists due to their short development time and low fabrication cost. Their compact size, lightweight characteristics, and ability to form a swarm enables them to communicate directly with one another to inspire new ideas on space exploration, space-based measurements, and implementation of the latest technology. CubeSat missions require specific antenna designs in order to achieve optimal performance and ensure mission success. Over the past two decades, a plethora of antenna designs have been proposed and implemented on CubeSat missions. Several challenges arise when designing CubeSat antennas such as gain, polarization, frequency selection, pointing accuracy, coverage, and deployment mechanisms. While these challenges are strongly related to the restrictions posed by the CubeSat standards, recently, researchers have turned their attention from the reliable and proven whip antenna to more sophisticated antenna designs such as antenna arrays to allow for higher gain and reconfigurable and steerable radiation patterns. This paper provides a comprehensive survey of the antennas used in 120 CubeSat missions from 2003 to 2022 as well as a collection of single-element antennas and antenna arrays that have been proposed in the literature. In addition, we propose a pictorial representation of how to select an antenna for different types of CubeSat missions. To this end, this paper aims is to serve both as an introductory guide on CubeSats antennas for CubeSat enthusiasts and a state of the art for CubeSat designers in this ever-growing field.

Keywords: antenna arrays; CubeSat missions; CubeSat antennas; miniaturized satellites; single element antennas; CubeSat subsystems; antenna selection; antenna designs

1. Introduction

CubeSats are a type of spacecraft called miniaturized satellites that are categorized based on their size, namely minisatellites, microsatellites, nanosatellites, picosatellites, and femtosatellites [1]. CubeSats belong to the class of nanosatellites having the smallest dimensions of 10 cm × 10 cm × 10 cm (1U) and a mass of 1.3 kg. As shown in Figure 1, other available sizes range from 2U up to 12U with a mass of 15.6 kg. In addition, Table 1 shows the weight and dimensions of the CubeSats in each category. However, nanosatellites of less than 1 kg have yet to be designed for commercial applications [2].

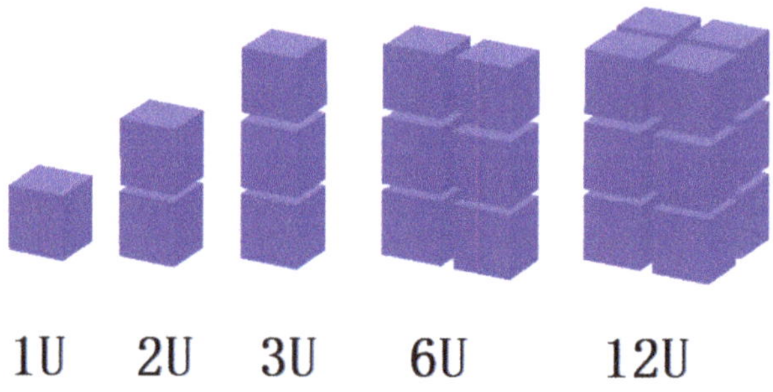

Figure 1. Different CubeSat standards.

Table 1. Size and mass of CubeSats.

Size	Dimensions	Wet Mass
1U	10 cm × 10 cm × 10 cm	1.3 kg
2U	10 cm × 10 cm × 20 cm	2.6 kg
3U	10 cm × 10 cm × 30 cm	3.9 kg
6U	10 cm × 20 cm × 30 cm	7.8 kg
12U	10 cm × 10 cm × 60 cm	15.6 kg

Having such small and light features enables developers to provide a time- and cost-effective solution to designing fully functional satellites. The development cost is significantly less than standard satellite missions, as CubeSats can be launched as a secondary payload from standardized ejection modules such as the Poly-Picosatellite Orbital Deployer (P-POD) [3] or being deployed from the International Space Station [4]. The original reference of CubeSat design specifications dates back to 1999, when it was proposed by professors Jordi Puig-Suari from California Polytechnic State University and Bob Twiggs from Stanford University. These specifications, including mechanical, electrical, operational, and testing requirements, set the foundations for CubeSat design [5,6]. By referring to the maturing standards gradually developed over the years, a CubeSat mission can be finished within a few years.

The primary goal of introducing this new class of satellites was to educate graduate students on developing skills for designing, building, testing, and operating small satellites in low Earth orbit (LEO) and developing new scientific research methods, as well as advancing new space technologies. Meanwhile, the first CubeSat mission was launched in 2003, and only a handful of CubeSats were launched every year before 2013. As mentioned, this is because most CubeSats were developed by universities or research institutions. The number of CubeSat missions increased rapidly only when commercial applications joined the field [7]. To date, many CubeSats have been designed, launched, and operated successfully at low Earth orbit (LEO). Examples include CanX-1, CUTE-1, and AUU [8]. Very few CubeSats are designed, launched, and operated for deep space communications, such as those in [9,10]. On the other hand, CubeSats have limited functionality compared with other larger in size satellites. For instance, fitting an on-board propulsion system [11], large solar panels [12], radiators [13], as well as high-gain antennas [14] has proven to be a series of challenging tasks due to the limited room on the CubeSat [15]. However, the advancements in printed circuit board (PCB) technology [16,17] and the availability of off-the-shelf components (COTS) alongside the development of more powerful processors such as Field Programmable Gate Arrays (FPGAs) [18] have enabled researchers to develop cost-effective CubeSats for various challenging missions.

One of the key challenges is an antenna design that achieves a high gain while having a small size. This is because higher gain leads to a higher data rate. However, the higher the gain, the more highly directional it is. This means the CubeSat's antenna must be pointed with high accuracy toward the ground station to communicate at a high rate over the highly focused radio signal. According to the recent surveys in [19–21], most CubeSat missions use UHF bands typically at 438 MHz, with possible data rates of up to 9.6 kbps. While only CubeSats employ higher frequency bands such as S-band to increase the data rates, considering the data rates for a standard mobile phone, 9.6 kbps is indeed a throwback to another era. This issue has also been raised by NASA's investigator Doug Rowland [22]. In addition, the orbital lifetime of CubeSat is another limitation, which is 0–100 days when it is orbiting at lower than 300 km. Higher altitudes might lead to an extended lifetime of up to 2 years at 400 km [23].

This paper is organized as follows. CubeSats and their subsystems are introduced in Section 2. The background on CubeSat antennas and several restrictions posed by the CubeSat standards are presented in Section 3. Furthermore, Section 3 reviews 120 CubeSat missions dating from 2003 to 2022 and draws conclusions on the mission type, frequency bands, and antenna type popularity among the CubeSat community. In Section 4, single-element antennas for CubeSat are presented, while in Section 5, antenna arrays and arraying techniques for CubeSats are reviewed. Section 6 presents a comparison between single-element antennas and antenna arrays. The paper concludes with Section 7, which includes the proposed pictorial representation of how to select an antenna for different types of CubeSat missions.

2. Background on CubeSats and Their Subsystems

Compared with conventional large and medium satellites, CubeSats weigh less and require less time for development (e.g., 1 year) [24]. Because of their limited size and weight, several restrictions should be considered in the design of a CubeSat system. The development of CubeSats has enabled the study and testing of novel ideas in the field of low-power microelectronics, digital signal processing, and communication protocols in space without spending millions of dollars. This section provides an overview of CubeSats.

2.1. Mass and Volume

A small satellite often has limited room for the installation of electronics, antennas, payload, and solar panels. More specifically, large-volume antennas that generally have complex deployable systems are more likely to be excluded from a CubeSat mission. Thus, smaller antennas are preferable as they occupy small space on the CubeSat. Another important factor that influences the antenna's choice and design for CubeSat missions is the mass. A typical CubeSat should weigh around 1.3 kg. Therefore, the weight of the antenna should be considered along with the different major weight contributions from the payloads, solar panels, core processors, batteries, and the chassis of the CubeSat. Failure to meet the CubeSat standards in terms of mass limitations would result in failure to launch.

2.2. Low Earth Orbit (LEO)

Most CubeSat missions take place on LEO, which ranges from about 150 km up to approximately 600 km and is below the ionosphere. Within this region, there are many science satellites and the International Space Station (ISS) [25]. When orbiting in LEO, a CubeSat undergoes different heat inputs and passes through different light ranges. The antenna might be designed to radiate certain amounts of power, but it should also be designed accordingly to reject any power received from unwanted sources. The noise temperature is a parameter which should be considered and modelled properly when designing an antenna for space missions where the thermal environment is harsh. For example, Earth can be considered an ideal blackbody in equilibrium that absorbs all the electromagnetic energy and emits energy at the same rate, which implies that Earth is an unwanted power source for the antenna. These fluctuations in the external space environment should be considered when designing a CubeSat. Figure 2 shows the altitude classifications for geocentric orbits.

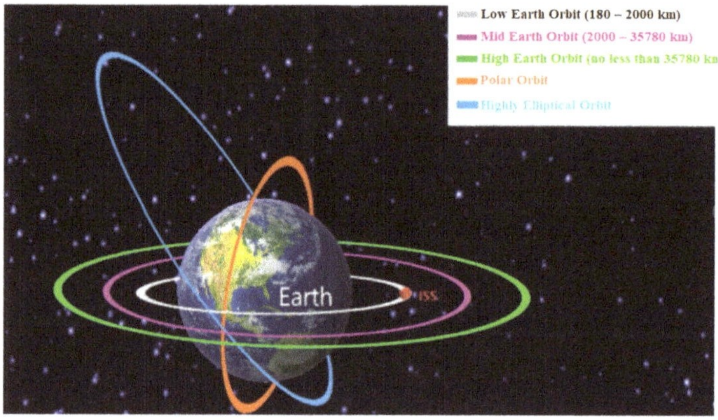

Figure 2. Altitude classifications for geocentric orbits.

2.3. Electrical Power Subsystem (EPS)

Another essential part for the satellite subsystems is the power subsystem, which constantly supplies the required power during the operation of the CubeSat. Power budgeting is one of the primary objectives of the CubeSat's design, and it must follow accurate calculations and obey certain limitations. Solar power is the main source of CubeSat power, since sunlight is the only available energy source in space. Other extra energy sources such as fuel cells and radioactive decay units are not practical for CubeSats. The expected power usage of the CPU, the radio, and the sensors need to be predictable and tightly defined. An optimized scheme of power budgeting needs to be developed accordingly.

Many CubeSats use 5-V buses as their core components of the power system, as a 5-V microcontroller is common and popular. A voltage of 3.3 V is becoming increasingly popular for CubeSats as well. According to [26], the capacity of space-grade lithium polymer batteries in CubeSat applications can range from 1.1 Ah to 1.4 Ah among different suppliers. The authors of [27] presented a design for a CubeSat bus that can provide 1.3 W of power to support some kinds of Earth observation missions and makes an allocation of the power budget to each subsystem.

2.4. Command and Data Handling Subsystem (C&DH)

The main processor subsystem, also known as Command and Data Handling (D&DH), coordinates complex actions taken by different parts within the system and provides stable and synchronized operation. The main processor of a CubeSat needs to be small and consume a low amount of power. Most current CubeSat missions use microcontrollers, but microprocessors are being considered for future missions. Some common processors such as ARMA, PC-104, and H8S-2674R have been chosen for some CubeSat missions [28–30]. Processors such as the PIC series and AT91SAM series are also available from providers, namely Pumpkin and Tyvak, and have also been applied in this kind of small satellite project. Additionally, applications of BasicX-24 and Arduino as the main processors of CubeSats are compared in [31]. FPGAs are also an attractive solution for CubeSat C&DHs as they offer in-obit reconfiguration, and their fabric can be designed to be tailored to specific mission requirements [17].

2.5. Propulsion Subsystem

CubeSats may use propulsion systems to realize active attitude control, reaction wheel desaturation, drag recovery, orbit changes and proximity operations [32]. In addition, a proper propulsion system can help a CubeSat slow down orbital decay and extend its lifetime. Because of the restrictions on CubeSats, their propulsion systems can only be used on specific occasions. The following technologies have been applied on current CubeSat missions: solar sail, cold gas, electric propulsion systems, and chemical propulsion systems [33]. The design of propulsion systems is still being developed to push the capabilities of CubeSats even further.

2.6. Attitude Determination and Control Subsystem (ADCS)

The ADCS for CubeSats is responsible for controlling the orientation of the spacecraft from the current attitude to a desired one. Various sensors or gyroscopes are used to record the orientation. For example, magnetic coils and reaction wheels are applied to provide necessary torques to reorient the satellite. Generally, two types of stabilization methods are utilized, namely spin stabilization and three-axis stabilization. Spin-stabilization requires one of the satellite axes to be fixed toward a specific direction, and then the body of the satellite is rotated accordingly. For this method, an initial force will be applied to the body of the satellite around an axis, and then the satellite will keep rotating because of the moment of inertia in space. Secondly, with three-axis stabilization, a satellite can be reoriented and stabilized in three different orthogonal axes instead of spinning around one axis, which results in maintaining a fixed attitude relationship with Earth or a successful inter-satellite link, respectively [34].

3. Background on Antennas and CubeSat Missions

An antenna or an aerial is defined as an element that is capable of sending and receiving radio waves (IEEE Std 124-1983). Technically, antennas may be resonant or non-resonant devices and operate efficiently when their geometry and impedance characteristics are tuned to a specific frequency. Antennas direct energy in a specific direction (or in all directions) and do not add or subtract power to a transmitted or received signal. An ideal transmitting antenna is one that radiates power without reflecting energy back to the feeding circuit. An ideal receiving antenna is one that absorbs the entire incidence electromagnetic wave without reflection.

Antennas can be classified by their radiation pattern as isotropic, omnidirectional, or directional. An isotropic antenna is an ideal reference and not physically feasible since it has equal radiation in all directions. Omnidirectional antennas, or whip antennas, have equal radiation in a given plane (e.g., the horizontal plane), and radiation is reduced outside that plane. Usually, single-element antennas have a wide radiation pattern and relatively low gain, making them suitable for general communication purposes, where the location of the receiver or transmitter is unknown, while at the same time, they are less efficient for applications that require long-distance, reliable, and effective communication links. The most common antenna used on a CubeSat is a whip antenna, which is a single-element monopole antenna. In [30], a CubeSat with two whip antennas of different lengths is shown. It consists of a flexible wire where the bottom part is connected, usually through a coaxial feed to the receiver or transmitter, and typically mounted above a ground or metal plane. A popular whip antenna is a $\frac{\lambda}{4}$ monopole, where λ is referred to as the resonant wavelength. The length of the antenna L in meters can be calculated with Equation (1):

$$L = \frac{\lambda}{4} = \frac{300}{4f} \tag{1}$$

where λ is the wavelength and f is the operating frequency in MHz. In a typical CubeSat application where the operating frequency is 433 MHz, a quarter-wavelength whip antenna should be 17.3 cm. However, this length exceeds the available CubeSat surface area. In this case, the antenna needs to be folded or rolled, stowed within the CubeSat body, and then deployed to its functional size after launch. This adds to the design complexity, as it brings in a potential point of failure for an antenna deployment mechanism. If the size is reduced, the antenna will be less efficient at the operating frequency. This simple example illustrates at least one design consideration when looking at the communication and antenna subsystem of a CubeSat.

On the other hand, directional antennas could radiate a relatively higher amount of power in specific directions and thus are more suitable than omnidirectional antennas when a higher gain is desirable. Realizing a directional radiation pattern can be accomplished by employing antenna arrays, reflectarrays, parabolic reflector antennas, horn antennas, or inflatable antennas. For example, in the case of antenna or phased arrays, the radiation fields from the elements of the array interfere constructively in the desired direction while cancelling each other out in other directions. Therefore, antenna arrays can synthesize a variety of radiation patterns which can be accurately pointed in the desired direction. As a result, higher data rate links can be established through higher signal-to noise-ratios (SNRs) and larger bandwidths (B). In short, antenna arrays have the following advantages. First, they have higher gain compared with single-element antennas and a better trade-off between antenna size and performance. Secondly, arrays can enable an electronically steerable beam with appropriate feeding of the elements. This can make satellites more versatile. Indeed, instead of physically reorienting a CubeSat to establish a link, its antenna beam can be electronically steered. Finally, they are more flexible in terms of mission-specific synthesis, as additional elements can be added when the mission needs a larger aperture or finer beam control, and the failure of a single element may not significantly impact the overall system's performance.

Other methods that have been proposed to increase the overall CubeSat downlink capacity are known as CubeSat swarms or constellations [35]. As demonstrated in [34], a CubeSat swarm is a group of individual CubeSats working cooperatively and sharing resources with each other. Interestingly, a subset of CubeSats can form a virtual antenna array to realize a larger aperture area with higher reliability. As a result, the power, memory, and bandwidth can be shared and distributed inside the swarm. However, a swarm of CubeSats requires reliable and efficient inter-CubeSat communication links [36,37]. Specifically, high-speed inter-satellite links will help to facilitate formation-flying missions, where CubeSats maintain the desired relative separations, positions, and orientations. Omni-directional antennas are the first choice for inter-satellite links because they can easily respond to the constant reorienting or repositioning of CubeSats.

3.1. CubeSat Antenna Specifications

Antenna design is a complex task and must be tailored to the CubeSat mission under consideration. A CubeSat's typical operation is performed in the VHF, UHF, or S-band, and the antennas are placed on one of the CubeSat surfaces. The size of a monopole or dipole antenna designed to operate in the VHF or UHF band usually exceeds a CubeSat's surface, which is 10 cm $\times$ 10 cm. From this point of view, when designing antennas for CubeSat missions, the following concept should be followed: *reduce the physical size of the antenna but maintain the desired radiation performance to meet the mission requirements*. Telemetry, tracking and command, a high-speed downlink for payload data, GPS and GNSS signal reception, and inter-satellite communication links are some of the fundamental functions performed by the antenna system. In order to design CubeSat antennas for high-level missions, such as inter-satellite communications for distributed CubeSat swarms, it is necessary to properly define the specifications based on the communications requirements and the platform or mission aspects [38]. A summary of the restrictions posed by the CubeSat platform and their corresponding descriptions are presented in Table 2. These restrictions must be considered by the CubeSat antenna designer during the design and integration phase of the antenna with the CubeSat.

Table 2. Restrictions imposed by the CubeSat platform on antenna designs [17,39–42].

Restrictions	Description
Size and Mass	Light weight and compact to fit the 10 cm × 10 cm size of a CubeSat surface in the case of 1U (without considering deployment volume).
Deployment	Deployment mechanism must be chosen or designed to minimize risk of deployment failure.
Attitude Control	Choice between active or passive control systems will determine the antenna pointing accuracy, as well as the choice of fixed or steering beam antenna.
Frequency Band and Bandwidth	Set by the mission specifications and allocated by the ITU or FCC. Dictates the uplink and downlink.
Loss	Antenna radiation or aperture efficiency must be higher than 50%. Antenna must match well with a reflection coefficient less than −10 dB.
Orbit (Communication Range)	Low Earth orbit: 400–2000 km Inter-satellite: Depends on swarm architecture Deep space: >2 × 10^6 km
Gain	Choice between low gain (LG), medium gain (MG), or high gain (HG), according to the available RF power budget and orbit.
Link Budget	High enough gain to provide the required SNR according to the modulation used.
Footprint	Dictate the beam width or the shape of the radiation pattern, as well as the size of the aperture.
Space Environment and Durability	Withstand thermal variations from −40 to +85 Celsius. Must pass thermal-vacuum cycling test (TVCT) and vibration test.
Cost	Off-the-shelf materials to reduce budget.
Polarization	Circular polarization to reduce losses due to polarization mismatch. Satisfy the cross-polarization levels set by the mission specifications.

3.2. CubeSat Missions and Their Antenna Designs

Based on the specifications defined in Table 2, a type of antenna may be selected for a specific CubeSat mission according to the communications' requirements for the mission and the limitations set by the CubeSat standards. The simplest and most used structures of CubeSat antennas are wire and patch antennas. Furthermore, planar antennas have become increasingly attractive for small satellite missions due to their low profile and compatibility with RF and microwave circuits. Microstrip patch antennas and slot antennas are two types of popular planar antenna designs for CubeSats. A literature survey based on planar antenna designs and their potential for picosatellite applications is presented in [21]. To better understand how the aforementioned restrictions are considered in current CubeSat missions, in the remaining part of this section, a comprehensive survey of the current antennas used in CubeSat missions is presented. Table 3 shows the well know frequency bands. Table 4 lists 120 CubeSat missions from 2003 to 2022 with the aim of informing the reader about the different antenna types, frequency bands, and CubeSat sizes used for different mission types. The mission selection is not exhaustive but illustrative. The choice of the missions depends on the availability of mission data related to the communication subsystem of each mission. The frequency bands used in Table 4 correspond to the IEEE standards shown in Table 3.

Table 3. IEEE standard letter designations for different frequency bands [43].

Band Designator	Dimensions
HF	3–30 MHz
VHF	30–300 MHz
UHF	300–1000 MHz
L	1–2 GHz
S	2–4 GHz
Ku	12–18 GHz
Ka	26.5–40 GHz
V	40–75 GHz
W	75–110 GHz

Table 4. Collection of CubeSat missions (nonexhaustive).

CubeSat Mission Name	Antenna Type	Mission Type	Size	Frequency Band	Year
XI-IV (CO-57) [44]	Monopole and Dipole	Educational	1U	VHF/UHF	2003
DTUSat-1 [45]	Monopole and Dipole	Educational	1U	UHF	2003
QuakeSat [30]	Four Monopoles	Earth Observation Technology Demonstration	3U	UHF	2003
CUTE-1 [46]	Three Monopoles	Technology Demonstration	1U	UHF	2003
Cute-1.7 + APD [47]	Dipole	Technology Demonstration	2U	UHF and L-Band	2003
XI-V (CO-58) [44]	Dipole	Technology Demonstration	1U	UHF	2005
NCube [48]	Monopole and Patch	Educational	1U	VHF, UHF, and S-Band	2005
Cute-1.7 + APD II [49]	Three Monopoles	Educational Technology Demonstration	2U	UHF and L-Band	2006
CP1 [16,50]	Dipole	Scientific	1U	UHF	2006
GeneSat-1 [51]	Monopole	Scientific	3U	UHF and S-Band	2006
Mea Huaka [52]	Monopole	N/A	1U	UHF	2006
MEROPE [53]	Dipole	Scientific	1U	UHF	2006
KUTESat-2 [54]	Dipoles	Technology Demonstration	1U	UHF and VHF	2006
ION [55]	Dipole	Technology Demonstration	3U	UHF	2006
CP2/CP4 [50,56]	Dipole	Technology Demonstration	1U	UHF	2007
CAPE-1 [57]	Monopole	Educational	1U	UHF	2007
CSTB1 [58]	Dipole	Technology Demonstration	1U	UHF	2007
COMPASS-1 [59]	Monopole and Dipole	Earth Observation	1U	UHF and VHF	2008
CanX-1 [60]	Monopoles	Technology Demonstration	1U	UHF	2008
Delfi-C3 [61]	Monopole and Dipole	Educational	3U	UHF and VHF	2008
CanX-2 [62]	Monopole and Patch	Technology Demonstration	3U	UHF and S-Band	2008
AAU [63]	Dipoles	Technology Demonstration	1U	UHF	2008
PSSCT [64,65]	Patch	Technology Demonstration	12.5 cm × 12.5 cm × 25 cm	UHF	2008
SEED-2 [66]	Monopoles	Technology Demonstration	1U	UHF	2008
SwissCube [67]	Monopoles	Scientific	1U	UHF	2009
BeeSat (Known as DRAGON SAT with AggieSat2) [46]	Monopoles	Technology Demonstration	1U	UHF	2009
CP3/CP6 [50]	Dipole	Technology Demonstration	1U	UHF	2009
HAUSAT-2 [68]	Dipoles	Educational	1U	UHF	2009
ITUpSat-I [69]	Four Monopoles	Technology Demonstration	1U	UHF	2009
Pharmasat- [70]	Patch	Scientific	3U	S-Band	2009
AubieSat-1 [71]	Dipole	Educational Technology Demonstration	1U	UHF	2011
CP5 [50]	Dipole	Technology Demonstration	1U	UHF	2011
Hermes [72]	Monopole	Technology Demonstration	1U	UHF and S-Band	2011
KySat [45]	Three Monopoles	Educational Technology Demonstration	1U	UHF, VHF, and S-Band	2011

Table 4. *Cont.*

CubeSat Mission Name	Antenna Type	Mission Type	Size	Frequency Band	Year
M-Cubed [73]	Monopole and Dipole	Educational Technology Demonstration	1U	UHF and VHF	2011
E1P-2 [74]	Monopole	Scientific	1U	UHF	2011
AtmoCube [74]	Dipole	Scientific	1U	UHF	2012
Aeneas [75]	Parabolic Meshed Reflector	Surveillance	3U	UHF	2012
Goliat [76]	Monopole	Earth Observation	1U	UHF and S-Band	2012
UNICUBESAT [77]	Monopole	Technology Demonstration	1U	UHF	2012
PWSat [78]	Two Monopoles	Technology Demonstration	1U	UHF	2012
XaTcobeo [79]	Four Monopoles	Technology Demonstration	1U	UHF	2012
e-st@r [80]	Dipole	Educational	1U	UHF	2012
CAPE-2 [57]	Monopole	Educational	1U	UHF and VHF	2013
CP8(IPEX) [50,81]	Monopole	Technology Demonstration	1U	UHF	2013
Delfin3Xt [82]	Monopole and Patch	Technology Demonstration	3U	UHF, VHF, and S-Band	2013
ExoplanetSat [83]	Patch	Technology Demonstration	3U	S-Band	2013
FireFly [22]	Monopole	Scientific	3U	UHF	2013
ZACUBE-1 [84]	Dipole	Scientific	1U	HF	2013
MOVE I [85]	Dipole	Educational	1U	UHF and VHF	2013
UWE-3 [86]	Monopoles	Technology Demonstration	1U	UHF	2013
FunCube [78]	Monopoles	Educational	1U	VHF and UHF	2013
TJ3Sat [87]	Monopoles	Educational	1U	UHF	2013
ALL-STAR [88]	Cavity-Backed Antenna	Educational	3U	UHF and S-Band	2014
CanX-4&5 [89]	Monopole, Dipole, and Patch	Technology Demonstration	1U	VHF, UHF, and S-Band	2014
DTUSat-2 [73]	Dipole	Technology Demonstration	1U	S-Band and L-Band	2014
MicroMAS [90]	Parabolic Reflector and Monopole	Earth Observation	3U	UHF	2014
OPUSat [91]	Monopole	Technology Demonstration	1U	UHF and VHF	2014
VELOX-PII [85]	Dipole	Educational	1U	UHF and VHF	2014
AeroCube-OCSD [44,92]	Patch	Technology Demonstration	1.5U	UHF	2015
Firebird [93]	Dipole	Scientific	1.5U+1.5U	UHF and VHF	2015
GOMX-3 [94]	Four Monopoles, Patch, and Helical	Technology Demonstration	3U	UHF, S-Band, and L-Band	2015
AggieSat2 (Known as DRAGON SAT with Bevo-1) [95]	Dipole and Patch	Educational Technology Demonstration	1U	S-Band	2016
CP10(ExoCube) [50,81]	Parabolic Reflector	Scientific	3U	UHF	2016
OUFTI-1 [94]	Monopoles	Educational	1U	UHF	2016
BEVO-1 [95]	Dipole and Patch	Educational Technology Demonstration	1U	S-Band	2016
Aalto-I [96,97]	Crossed-Dipole and Patch	Technology Demonstration	3U	VHF, UHF, and S-band	2017
CXBN-2 [70]	Quadrature Spring Steel Array	Technology Demonstration	2U	UHF and S-Band	2017

Table 4. *Cont.*

CubeSat Mission Name	Antenna Type	Mission Type	Size	Frequency Band	Year
EC0 (UNSW-EC0) [98]	Monopole	Education	2U	UHF	2017
ICECube [99]	Dipole and Patch	Technology Demonstration	3U	UHF	2017
QBITO [100]	Four Monopoles	Education purpose	2U	UHF	2017
ISARA [101]	Reflectarray Integrated with Solar Panels	Technology Demonstration Communications	3U	UHF and Ka-Band	2017
RadSat [99]	Monopole	Technology Demonstration	3U	UHF	2018
SPATIUM [102]	Monopole	Scientific	2U	UHF	2018
UWE-4 [103]	Dipole	Technology Demonstration	1U	UHF	2018
CANYCAL-X [102]	Patch and Monopole	Technology Demonstration	1U+2U	UHF and S-Band	2018
AeroCube-11R3 [104]	Patch	Technology Demonstration	3U	UHF	2018
KNACKSA [105]	Two Dipoles	Technology Demonstration Earth observation	1U	VHF and UHF	2018
CHOMPTT [62]	Monopole	Technology Demonstration	3U	UHF	2018
MarCO [9]	Reflectarray, Patch Array, and Loop	Interplanetary Exploration	6U	UHF and X-Band	2018
RainCube [106]	Parabolic Mesh Reflector	Earth Observation Technology Demonstration	6U	Ka-Band	2018
AzTechSat-1 [107]	Patch	Educational	1U	UHF and VHF	2019
ANGELS [108]	Square Array Inverted F	Technology Demonstration	12U	UHF and L-Band	2019
ARMADILLO [51]	Monopole	Technology Demonstration	3U	UHF	2019
EyeSat [109]	Patch	Educational	3U	S-Band and X-Band	2019
OPS-SAT [92]	Dipole	Technology Demonstration	3U	UHF and S-Band and X-Band	2019
SORTIE [103]	Dipole	Technology Demonstration	6U	UHF	2019
Artemis [110]	Horn	Technology Demonstration	1U	L-Band and Ku-Band	2020
SERB [111]	Patch	Technology Demonstration	3U	S-Band	2020
Landmapper-BC5 [112]	Monopole and Horn	Earth observation	6U	UHF and Ka-Band	2020
Lemur-2 [113]	Monopoles and Patch Array	Earth observation	3U	UHF and S-Band	2020
Flock [114]	N/A	Earth observation	3U	UHF and X-Band	2020 and 2022
Kepler [115]	Phased Array	Communications IoT	6U	S-Band and Ku-Band	2018 and 2020–2022
TTU100 [116]	Dipole and Patch Array	Technology Demonstration Earth Observation	1U	UHF and X-Band	2020
NetSat [117]	Dipoles	Technology Demonstration Educational	3U	UHF	2020
TRISAT [118]	Patch and Dipoles	Technology Demonstration	3U	UHF and S-Band	2020
Quetzal	Dipoles	Educational	1U	UHF	2020
OSM1-CICERO [119]	Phased Array	Earth Observation	6U	UHF and X-Band	2020
PICASSO [120]	Patch and Dipoles	Earth Observation	3U	VHF, UHF, and S-Band	2020
AMICal Sat [121]	Patch and Dipoles	Demonstration Earth Observation	2U	VHF, UHF, and S-Band	2020
Astrocast [122]	Patch and Patch Array	Communications IoT	3U	L-band	2021
BEESAT 5–8 [123]	N/A	Technology Demonstration	0.25	UHF	2021

Table 4. *Cont.*

CubeSat Mission Name	Antenna Type	Mission Type	Size	Frequency Band	Year
RADCUBE [124]	Dipoles	Scientific	3U	UHF	2021
ExoCube-2 [125]	Monopoles	Scientific	3U	UHF	2021
Cesium Satellite [126]	Active Phased Array	Technology Demonstration	6U	Ka-Band	2021
CAS-9 [127]	Monopole	Communications	6U	VHF and UHF	2021
SOMP2b [128]	Monopoles	Communications	2U	UHF	2021
W-Cube [129]	Array of Concentric Ring Antennas (Bull's Eye)	Scientific Technology Demonstration	3U	W-Band	2021
Centauri [130]	Active Phased Array	Communications IoT	6U	S-Band	2018, 2021, and 2022
ELO Alpha [131]	Helical, Patch Array, and Dipoles	Communications IoT	3U	ISM	2021
IDEASSat [132]	Monopole and Patch	Technology Demonstration Earth Observation	3U	UHF and S-Band	2021
KSF1 [133]	Monopoles, Patch Array, and Helical	Surveillance	6U	VHF and S-Band	2021
D2/AtlaCom-1 [134]	Monopoles and Patch Array	Educational Earth Observation	6U	UHF and X-Band	2021
FORESAIL-1 [135]	Monopoles	Scientific Technology Demonstration	3U	UHF	2022
IRIS-A [136]	Monopoles	Communications IoT	2U	UHF	2022
HYPSO [137]	Patch, Monopole, and Dipoles	Technology Demonstration Earth Observation	6U	UHF and S-Band	2022
Spark-2 [138]	Patch and Monopoles	Communications IoT (5G)	12U	S-Band	2022
SanoSat-1 [139]	Dipole	Educational	1U	UHF	2022
Planetum-1 [140]	Monopole	Educational	1U	UHF	2022
SpaceBEE [141]	Dipole	Communications IoT	0.25U	VHF	2018–2022

Figure 3 shows a breakdown of the mission types corresponding to the missions presented in Table 4. Nearly half of the missions were focused on technology demonstration, where CubeSats were used as a cost-effective way to test and validate innovative hardware or software such as novel propulsion systems, attitude control systems, or inter-satellite links, as in the case of CP6 or Cesium. The second biggest portion (21% of the missions) was devoted to educational purposes, where CubeSat were used by different academic institutions around the world to motivate students and familiarize them with satellite subsystems. It is worth mentioning that offering undergraduate students opportunities to get involved in a space-related project like CubeSats has led to a rapid increase in CubeSat launches lately. Earth observation and scientific missions each contributed 12% of all the investigated missions. In these missions, CubeSats may be used to study radiation levels in outer space, the ionosphere, and space as well as Earth weather or even monitor crops from space. Lately, CubeSat missions related to communications such as high-speed downlinks, IoT, M2M, and 5G from space have started to emerge, which corresponded to 7% of all the missions. Finally, the smallest portions of the CubeSat missions were found in surveillance with 2% and interplanetary exploration led by NASA, corresponding to 1%.

In this survey, the sizes of existing and previous CubeSats were also recorded in an effort to illustrate the overall trend in CubeSat designs. Figure 4 shows the popularity of CubeSat sizes over the 120 missions under investigation. The smallest size in the literature corresponded to 0.25U, with only one mission found. The most popular CubeSat size was found to be 1U, which is defined as the basic unit for CubeSat design. The next most popular size was 3U with 36 CubeSats in total, which showcases the potential of using 3U to accommodate more complex functions and carry more advanced technology. With the development of CubeSat missions, some larger sizes such as 6U and 12U emerged, which are associated with missions with increased power and RF budgets or payloads that require more integration real estate.

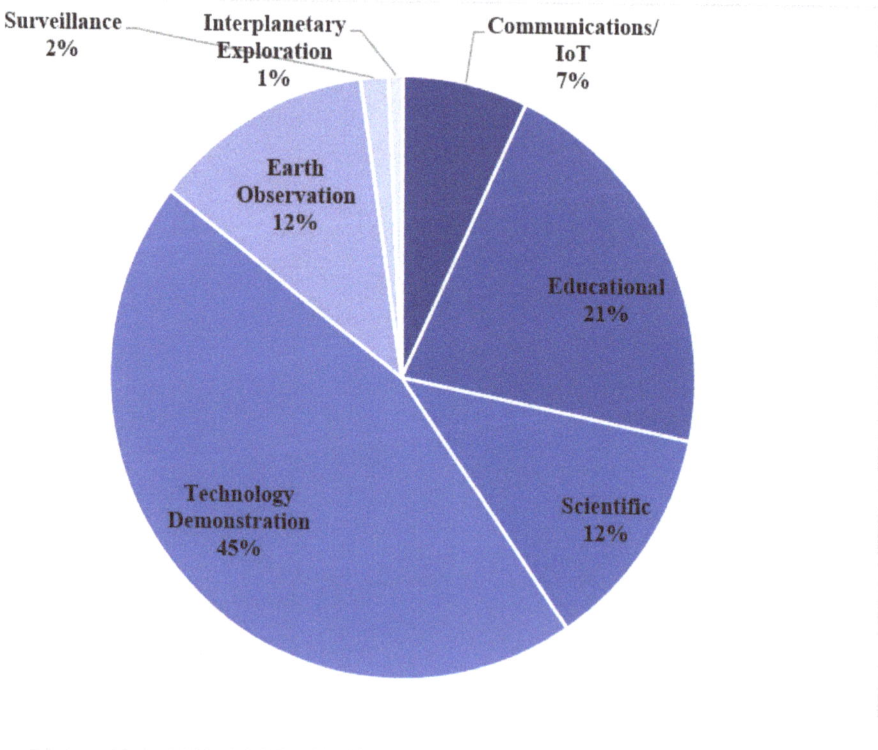

Figure 3. Popularity of CubeSat mission types from 2003 to 2022.

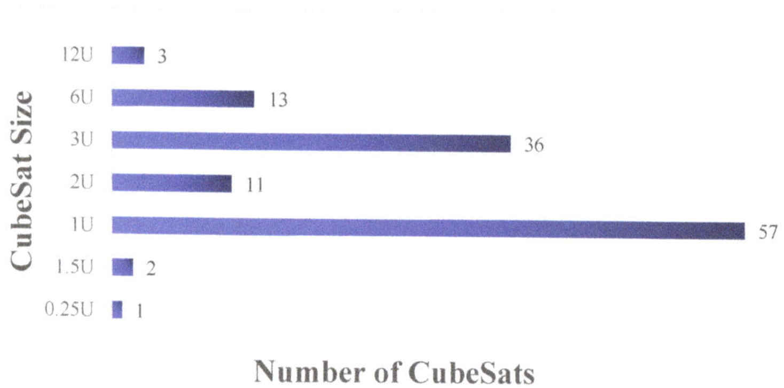

Figure 4. Popular CubeSat sizes used in the 120 CubeSat missions presented in Table 4.

Figure 5 shows the antenna types used in the investigated CubeSat missions. We can see that monopole and dipole antennas were widely used as the primary uplink or downlink band, which normally operate at a VHF or UHF. Patch antennas operating on the

S-band or L-band were the second most popular antenna type since they have a low profile, are lightweight, and usually do not require any deployment mechanism. In addition, patch antennas can provide a considerable gain improvement over monopoles and dipoles. For that reason, monopoles or dipoles and patch antennas are usually used in the same mission, where the monopole or dipole is used for low-data rate telemetry functions and the patch antennas can be used for higher data rate communications. A helical antenna is another antenna candidate that has been used in CubeSat missions. They can provide a medium directive pattern and can be easily stowed and deployed in a similar way to monopole and dipole antennas. If higher gain and high-speed downlinks are required, a higher frequency must be used (e.g., X-band, Ku-band, or Ka-band), combined with antenna arrays such as microstrip patch arrays, reflectarrays, or meshed reflectors. In more recent missions, phased arrays operating at the Ka-band have been used, where the antenna beam can be steered electronically to realize intersatellite links or accommodate varying traffic as in the case of IoT missions. Horn or guided wave antennas have also been used in CubeSat missions, operating at the Ka-band and W-band to realize high-speed downlinks and push the current CubeSat communication capabilities to their limits. Last but not least, X-band reflectarray antennas have been used in MARS CubeSat missions launched by NASA, proving that CubeSats can be used as a cost-effective way to explore different planets.

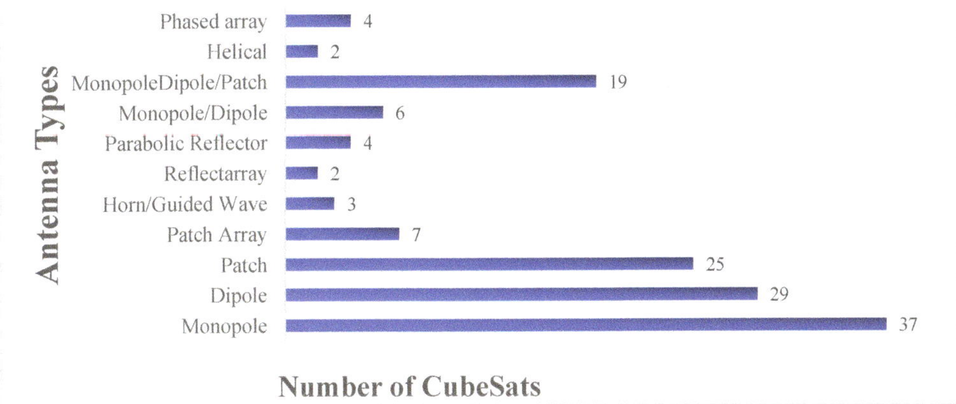

Figure 5. Antenna types used in the 120 CubeSat missions presented in Table 4.

As shown in Figure 6, the most commonly used frequency band among the CubeSat missions is UHF and the combination of UHF with VHF, L-band, X-band, Ku-band, or Ka-band. Reagrdless of the mission objectives, the UHF band is mostly found in CubeSat missions, even as the primary downlink or as the backup secondary radio frequency. Moreover, S-band has also been widely used either by itself or combined with higher frequencies such as X-band or Ku-band. There are very few CubeSats that employ only the VHF, L, Ka- or W-band. This highlights the importrance of using various frequency bands in a single CubeSat mission to satisfy the variety of the mission's requirements.

Lastly, it would be interesting to observe when each antenna type apeared in CubeSat missions through the years, starting from 2003 amd leading up to 2022. Initially, around 2003, monopole and dipole antennas were solely used in CubeSat missions due to their simplicity of design, low cost, and ease of deployment. As we move forward to 2005–2011, patch antennas made their appearance in the CubeSat community. Around 2012, meshed reflector antennas started emerging in CubeSat missions, while during 2014 and 2015, helical and reflectarray antennas were adopted. In 2017 and 2018, CubeSat missions were equiped with antenna arrays, followed by phased arrays during 2018 and 2019. Finally, during the last 3 years, horn and guided wave antennas appeared in CubeSat missions. It is important to note that once a specific antenna type appeared at a point in time, it was being used for the upcoming years as well. For example, during the last 3 years, a combination of antenna types can be found in a single CubeSat mission, which was not the case before 2011. Through the years, the complexity of CubeSat missions has increased. Therefore, the CubeSat capabilities must increase accordingly to meet the mission requirments. This is reflected in the variety of antenna types and frequency bands used by recent CubeSat missions. More advanced missions require higher data rates, higher gains, and reduced antenna sizes while keeping the satellite power budget as low as possible. As a result, from Table 5 we can draw the following conclusion: *The evolution of CubeSat antennas is dictated by the requirements and complexity of the missions.*

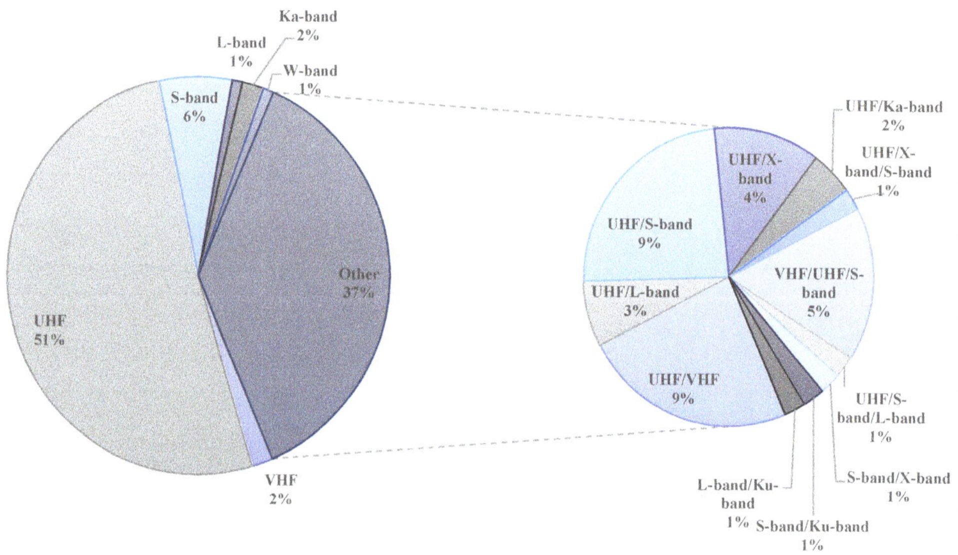

Figure 6. Frequency bands used in the 120 CubeSat missions presented in Table 4.

Table 5. The appearance of different antenna types in CubeSat missions from 2003 to 2022.

	2003	2005–2011	2012–2013	2014–2015	2016–2017	2018–2019	2020–2022
Monopole and Dipole	✓	✓	✓	✓	✓	✓	✓
Patch		✓	✓	✓	✓	✓	✓
Reflector			✓	✓	✓	✓	✓
Helical				✓	✓	✓	✓
Reflectarrays				✓	✓	✓	✓
Arrays					✓	✓	✓
Phased Arrays						✓	✓
Horn							✓

4. Single-Element CubeSat Antenna Designs

Single-element antennas vary from monopole and dipole antennas to planar, conical, and helical antennas, as well as guided wave structures like the bull's eye antenna and metasurfaces. They are easier to construct than the antenna arrays but do not achieve such a high radiation performance compared with the antenna arrays. In 2001, monopole and dipole antennas were initially chosen and used for the communication subsystem of the CubeSat [142]. As research on CubeSats communication systems drew more scientific interest, more complex antennas were introduced. Figure 7 presents four popular antenna types after deployment on a 3U CubeSat.

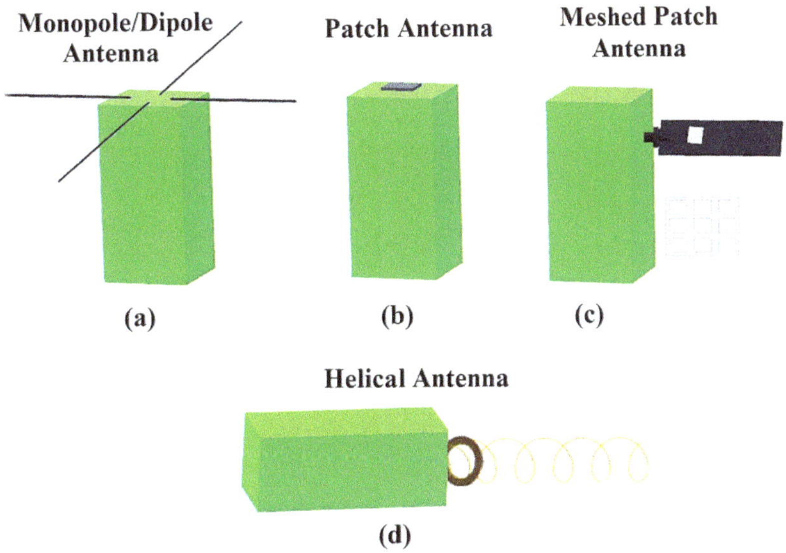

Figure 7. Popular antenna designs proposed in the literature. (**a**) Monopole/Dipole; (**b**) Patch; (**c**) Meshed Patch and (**d**) Helical Patch.

4.1. Monopole and Dipole Antennas

CubeSats operating at low frequencies such as VHF or UHF are equipped with wire antennas. In most cases, a wire antenna needs to be deployed after a CubeSat is launched into space. Deployment usually involves a composite tape spring [143–145]. In addition, wire antennas in dual-band operation can be used to transmit and receive data at the same time, similar to the 3 $\lambda/4$ dual band monopole antenna in [146]. Instead of simply employing the monopole antenna in the form of a straight wire, the authors in [147] proposed a design for a monopole G-shaped antenna that is mounted on a CubeSat. It consists of two rectangular wire loops. The size of the antenna was smaller than the size of the CubeSat's surface, so it could be mounted and stowed within the CubeSat body and did not need a deployment mechanism. Three structures were available based on three different frequencies of 150 MHz, 180 MHz, and 330 MHz. In [94], the authors proposed four monopoles, which were combined to form a circular polarized isotropic antenna that established communication at the initial stages of CubeSat deployment. Similar to the monopole antenna, in many instances, the length of dipole antenna is a restriction for CubeSat missions to be solved. For example, in [143], a deployable dipole antenna using a curved, bi-stable composite tape spring was outlined. The dipole antenna operated at 250 MHz, and its total length was 55.88 cm, around five times larger than a CubeSat's edge.

In some designs, monopole and dipole antennas can also be combined and used together to achieve the goals of a mission. In 2002, one of the first CubeSats, namely "XI-IV", was launched and presented, where a 56-cm monopole antenna was used for the uplink at 144 MHz and placed vertical to a 35-cm (tip to tip) dipole antenna used for downlink communication (telemetry and beacon) at 430 MHz. Crucial health data were broadcasted constantly via a beacon signal at 430 MHz [142]. The authors in [148] proposed a combined VHF/UHF (144/435 MHz) system. Both a dual-band dipole as well as a dual-band monopole were presented with the monopole configuration, acting as a fallback solution. An LC circuit was used as a notch filter in the UHF band to decide whether the dipole or monopole configuration was to be activated. Moreover, the authors in [149] presented a communication system consisting of an S-band transmitter along with a UHF dipole as a backup transmitter. The downlink operated at 437 MHz and the uplink at 145 MHz, and these were implemented using a VHF monopole antenna. A dual-band folded-end dipole antenna was proposed in [150] for plastic CubeSats. The folded dipole does not require any deployment mechanism as it has a low profile and wraps around the CubeSat's body. The dipole operates at 2.5 and 4.7 GHz and is fed by a microstrip line, which is placed inside the CubeSat's body. The proposed antenna achieved one of the highest gains that could be found on wire antennas of 7.55 and 4.9 dBi at 4.7 and 2.5 GHz, respectively. Table 6 summarizes the findings related to monopole and dipole CubeSat antennas.

Table 6. Performance of monopole and dipole antennas.

Type of Antenna	Frequency (GHz)	Gain (dBi)	−10 dB BW (MHz)	Deployable
Monopole and Dipole [142,148,149]	0.146 (uplink) 0.438 (downlink)	2–4	N/A	Yes
G-Shaped Monopole (3 structures) [147]	Structure I: 0.150 Structure II: 0.180 Structure III: 0.330	3.757 2.671 2.774	58 77 147	No
Dual-Band Folded-End Dipole [150]	2.5 4.7	4.9 7.55	500 250	No

4.2. Planar Antennas

One of the main objectives of wireless communications and especially satellite communications is to use multiband or wideband low-profile antennas [151]. Patch antennas are good alternatives to wire antennas, as they have a low profile, do not require a deployment mechanism, are easy to fabricate, and have relatively low costs. Patch antennas are ideal for S-band communication, which is one of the international amateur satellite frequency ranges for high bit rates. They are characterized by higher gain than the wire antennas but suffer from narrow bandwidths. An interesting use of patch antennas can be found in [152]. A standard patch antenna was used as a feed for a parabolic reflector placed inside an inflatable volume. The patch antenna had dimensions of 9 cm × 9 cm, which complied with the CubeSat standards, and its gain was 8 dBi at 2.4 GHz. By adding the parabolic reflector, the antenna gain was increased to 16 dBi at 2.4 GHz. The authors showed that a patch antenna can be used as a feed to a parabolic reflector instead of the traditional choice of a horn feed.

Another novel patch antenna design is proposed in [153], where an F-shaped patch antenna was under investigation. The results show that the patch antenna could achieve a gain of 8.5 dBi at 2.45 GHz. Furthermore, in [154] a dual-feed, L/S dual-band-stack patch antenna design is presented. This antenna operated in the L-band at 1.57 GHz for receiving the position signals from GPS satellites and in the S-band at 2.2 GHz for downlink transmission to the ground station. Even though the antenna consisted of 3 layers, it weighed less than 120 g and maintained a low profile of 11 mm, which conforms to the CubeSat standards. A Koch curve microstrip fractal antenna was presented which efficiently utilized the available space by maintaining a wide bandwidth [155]. The antenna was attached on a FR-4 substrate with dimensions of 3.5 cm × 4.5 cm and had an operating frequency range from 2.25 GHz to 2.45 GHz. The planar antennas that have been proposed for CubeSats are shown in Table 7.

Table 7. Performance of planar antennas.

Type of Antenna	Frequency (GHz)	Gain (dBi)	−10 dB BW (MHz)	Deployable
Inflatable parabolic reflector with patch feed [152]	2.4	16	N/A	Yes
F-shaped patch [153]	2.45	8	1200	No
Dual-feed, L/S dual-band stack patch [154]	1.57 2.2	5.4 6	N/A	No
Koch curve fractal microstrip [155]	2.3	4.18	290	No
Shorted patch [156]	2.45	2.52	550	No
Metal only patch [157]	2.45	8.5	1100	No
Tapered line feeder patch [158]	2.46	9.6	Narrow	No
V-shaped asymmetrical slits [159]	2.285	6	0.4	No
Tapered peripheral slits [160]	0.436	0.7–1.4	4	No
L-band patch with frequency tuning slots [94]	1.54	5.5	40	No

4.3. Antenna Integrated with Solar Panels

Solar panels represent the main source of power for a CubeSat, and thus it is important to reserve available space for their installation on the satellite's body. On the other hand, the antenna is another subsystem of the CubeSat that requires instalment space and is also of great importance when it comes to mission success in terms of communication. Consequently, integrating antennas with solar panels was proven to be a very efficient approach to using a CubeSat's available space while at the same time avoiding the requirement of a deployment mechanism [161]. An integrated solar panel-antenna system must keep the received solar energy loss to low levels. Four types of integration between solar panels and antennas are presented in [162]:

- Place patch antennas under the solar cells;
- Create slot antennas and deposit solar cells directly on top of them [161];
- Place transparent antennas directly on top of solar cells [161];
- Integrate transparent antennas on solar cells [163].

Thus, solar cells with patch antennas and slot antennas where the transparency is kept high are two possible solutions for space missions.

In Table 8, examples of meshed patch antennas on a transparent borosilicate glass substrate suitable for integration with solar panels are presented as in [164]. A meshed patch antenna is similar to a microstrip patch antenna, but instead of continuous solid patch shapes, some metallic areas are removed to form a mesh. Along similar lines, the design philosophy of slot antennas takes advantage of the gaps between solar cells by creating radiating cavity-backed slots in these gaps. A slot antenna integrated with solar cells is proposed in [165], where a closed-loop meander-shaped slot antenna is wrapped around the top and bottom halves (1.5U) of a 3U CubeSat for the downlink and uplink, respectively. The way the slot antenna is wrapped around the chassis allows effective installment for solar cells. Finally, another cavity-baked slot antenna design can be found in [166]. The design can switch between left-handed circular polarization (LHCP) and right-handed circular polarization (RHCP) by changing the on and off states of the installed PIN diodes.

Table 8. Performance of meshed planar antennas.

Type of Antenna	Frequency (GHz)	Gain (dBi)	−10 dB BW (MHz)	Deployable
Transparent meshed [167]	2.4	6.16	40	No
Dual-feed meshed patch [164]	2.4	6.25	N/A	No
Dual patch meshed (81% transparency) [168]	2.5	5.09	N/A	No
L-shaped slots meshed (89% transparency) [168]	2.4	4.4	140	No
Meander shaped slot [165]	0.485 0.500	4	15	No
Polarization reconfigurable slot [166]	2.3	7	N/A	No
Transparent mesh patch [169]	2.4	N/A	80	No
Transparent meshed [167]	2.4	6.16	40	No
Dual-feed meshed patch [164]	2.4	6.25	N/A	No

4.4. Conical Spiral Helix Antenna

Conical spiral and helix antennas can easily change from a 2D structure while stowed to a 3D structure during deployment. When compressing the conical spiral into a panel, it is possible to mount the antenna on one of the CubeSat's surfaces. Once the CubeSat is launched into space, the conical spiral antenna can be easily released and flick up to its functional 3D shape. The challenges for this kind of antenna are usually related to feeding mechanisms. Designers need to avoid any impedance mismatch and provide a stable deployable mechanism. Furthermore, feeding from the top or bottom of the conical spiral shows different radiation performance. The maximum gain direction of this kind of antenna is from the smallest element to the biggest one. As shown in Table 9, in [170], a deployable helical UHF antenna is presented. The antenna can be stowed in a 10 cm × 10 cm × 5 cm package, and when deployed, it can reach up to 137.16 cm in length and 35 cm in diameter. A 5-turn helix contributes to a gain of 10 dBi, while a 2-turn taper helps to improve the axial ratio to less than 2 dB. The conductive element of the antenna is made of copper adhesive tape. Another design that employs a similar deployment concept is described in [171].

Table 9. Performance of conical spiral and helical antennas.

Type of Antenna	Frequency (GHz)	Gain (dBi)	−10 dB BW (MHz)	Deployable
Helical [170]	0.400	13	N/A	Yes
Hemispherical Helical [171]	2.45	13.2	N/A	Yes
Bottom-Fed [172]	2.2–3.1	11.2	900	Yes
Conical [173]	0.300–0.600	5	300	Yes
Quadrifilar Helix [174]	0.250–0.500	5.41	Various	Yes

4.5. Other Antenna Designs

As shown in Table 10, the "bull's eye" antenna, as demonstrated in [175], has grooves in the shape of a bullseye. It is useful for inter-CubeSat communications, and it is possible to be employed and used for a CubeSat swarm. In addition, it also has a low profile and high gain. Another special design presented in [176] is that of a dual-band cylindrical dielectric resonator antenna (DRA). In general, a DRA presents some interesting features, such as a small size, wide bandwidth, high radiation efficiency, and low loss, making them a good alternative for CubeSat integration. The proposed DRA utilizes a microstrip line as a feed and a rotated plus-shaped slot placed on an FR4 epoxy sheet to excite the cylindrical dielectric resonator. The resonator has a height of 5 mm and a diameter of 10 mm. The modes of operation of the resonator determine the resonant frequency of the antenna, which in this case is at 7.7 GHz and at 11.4 GHz. The simulated and measured results of this design verify the feasibility of using DRAs for CubeSat missions. The authors of [177] presented a novel S-band antenna design that provides good radiation performance and occupies a small area on a CubeSat. The antenna consists of 4 balanced inductive exciters (BIEs) placed at the top surface of a 1U CubeSat. The main idea is to transform the 1U CubeSat platform into an efficient radiator by employing characteristic mode analysis. The proposed antenna has the capability to steer the radiation beam at both principle planes as well as the capability of changing its polarization between CP and LP. Recently, metasurface antennas have been considered for deep space CubeSat missions [178]. More specifically, a metal-only modulated metasurface antenna excited by surface waves from a circular waveguide was proposed. The antenna operates at 32 GHz and achieves a gain of 24.4 dBi without needing a deployment mechanism.

Table 10. Performance of other antennas.

Type of Antenna	Frequency (GHz)	Gain (dBi)	−10 dB BW (GHz)	Deployable
Bull's Eye [175]	60.08	19.1	5.06	No
Cylindrical DRA Antenna [176]	7.4 11.1	5.2 4.8	0.6 0.8	No
Balance Inductive Exciters (BIEs) [177]	2.425	6.3	0.5	No
Metal-Only Metasurface [178]	32	24.4	2	No

4.6. Recommendation for Single-Element Antennas and CubeSat Missions

To summarize from the previous section, most of the current existing CubeSats employed single-element antennas. Dipole, monopole, and patch antennas are widely chosen as they are easy to design and deploy. The deployment mechanisms of whip antennas have been successfully implemented on many CubeSat missions. It is recommended to operate whip antennas at a lower frequency band such as UHF or VHF, as their wide signal coverage enables tracking and telemetry on these bands. However, considering the lengths of whip antennas might be too long, a CubeSat mission can deploy multiple dipole or monopole antennas to avoid complex deployment issues. Whip antennas operating on frequencies higher than S-band will have lengths several times smaller than those operating at UHF and VHF. An interesting alternative is to utilize the CubeSat body as an efficient radiator, which may provide beam steering and polarization agility benefits. More challenging applications of CubeSats where the data rates are considerably higher than 9.6 kbps require the use of higher frequency bands, such as L-, S-, Ka-, or V-band. In these cases, more sophisticated antenna structures are needed, such as the metasurface or bull's eye antenna.

The higher potential of single-element antennas is still being explored, and many new structures other than patch and whip antennas have been developed for several CubeSat missions. However, they all aim at optimizing the antenna characteristics within the available space and weight limitations. In other words, the design of single-element antennas becomes a task of balancing the antenna performance and the limitations imposed by the

CubeSat standard. Thus, flexible materials for supporting the deployment mechanism need to be investigated for specific antenna designs. A typical example is a design of membrane antennas [179] that employs a foldable membrane plated with conductors.

5. Antenna Arrays for CubeSat

There are two ways to increase the directivity and gain of an antenna. The first approach involves increasing the electrical size of the antenna, and the second approach involves combining single elements under a specific geometrical configuration. It is obvious that the first approach is more impractical for CubeSat-specific applications, as the available area is limited by the CubeSat size standards. The second approach is commonly referred to as antenna arrays and represents a strong candidate for CubeSat missions when a high gain and beam steering are required. Antenna arrays can be classified according to the spacing between the elements, the excitation phase, and the amplitude of each element, as well as the radiation pattern of each element [180]. The most common implementations of antenna arrays are the linear and two-dimensional planar arrays.

A subcategory of planar arrays is called phased arrays, which are characterized by their electronic beam scanning and the beam forming capabilities. This can be accomplished by electronically controlling the phase and amplitude at the element or subarray level. In practice, varying the phase differences and excitations of each element in an array is often achieved by phase shifters and power amplifiers. Phase shifters assign complex weights to each element, and the output signal from an array is the combination of signals from each element multiplied by their complex weights. When choosing a phase shifter, the distortion caused by group delay, inter-element interference, as well as beam squinting needs to be taken into consideration. Other effects on array performance related to phase shifters are caused by the insertion loss envelope and phase accuracy [181]. Furthermore, the operation point of the power amplifiers may influence the bit error rate (BER), considering the modulation type and the number of carriers. Therefore, the receiver array employs a low-noise amplifier, while the transmit array employs a linear power amplifier. Finally, phase array antennas offer one of the most versatile and powerful antenna candidates for CubeSat missions when beam steering or beamforming capabilities are required. Despite their attractive features, phased arrays present high power consumption, which must be taken into account in the CubeSat power budget calculations [182].

To date, most CubeSat missions have employed standard antenna designs such as wire and patch antennas (see Table 4). The dimensions and the geometry of single-element antennas impose limitations related to gain, directivity, and beam steering capability, resulting in a limited attainable data rate as well as restricted mission capabilities. For example, in IoT CubeSat missions, there are thousands of Earth terminals to which the CubeSat constellation should establish reliable links. In addition, the CubeSat swarm can form a distributed array, which can compensate for any gain variations or satellite orientation errors [183]. In such cases, antenna arrays or phased arrays are promising CubeSat antenna candidates. A collection of antenna array designs suitable for CubeSat missions is discussed in the following sections.

5.1. Linear Arrays for CubeSats

A sophisticated Yagi-Uda antenna for CubeSat attached to the eXtendable Solar Array System (XSAS) can be found in [184]. Given that the typical dipole configuration results in a gain of around 5 dBi, the authors tried to achieve a high gain with a 6-element Yagi-Uda antenna array incorporated into the deployable solar panels. The design comprises one reflector, one driven element, and four directors. Experiments showed that the length and spacing of the directors were critical for maximizing the gain of the antenna. The proposed antenna, when deployed, reaches 1.2 m in length, being attached to the 30° tilted solar panels. As shown in Table 11, at a 435-MHz operating frequency, the gain obtained was 11.5 dBi, exhibiting superior performance compared with the 5-dBi gain of a dipole configuration. Achieving a higher gain in satellite antennas would also allow for higher altitude deployment of the small satellites, relieving them from suffering high atmospheric drag.

Table 11. Performance of linear antenna arrays.

Type of Antenna	Frequency (GHz)	Gain (dBi)	−10 dB BW (MHz)	Deployable
Yagi-Uda on solar panels [184]	0.435	11.5	53	Yes
Quad 4-monopole array [185]	0.436	2d	N/A	Yes
W-Band 16 × 32 series fed phased antenna array [186]	85.4	30	800	No
Printed Yagi [187]	1.3 2.4 3.0	5.28 6.12 8.17	100 190 250	No

In addition, a quad monopole antenna array was proposed with a communication system consisting of a UHF transceiver, RF splitter, and a quad four-monople antenna array. The RF splits the RF signal into a phasing network to form a single circular polarized antenna. The gain of the antenna array was 2 dBi at 436 MHz. LituanicaSAT-2 is part of the QB50 mission which in 2017 launched a 36-CubeSat swarm to explore the temporal and spatial distributions of the parameters measured in the thermosphere [185]. In [186], a W-band, 16 × 32 element, circular, polarized, phased antenna array on a single layer was presented. The antenna operated at 85 GHz with 30 dBic. The antenna also achieved beam scanning of +/−30° using a 4-bit phase shifter at the cost of rapid gain drops, especially in the range from −30° to −10°. Overall, the antenna efficiency was more than 56% for the investigated scanning range. A printed Yagi antenna with multi-frequency operation was proposed in [187]. This Yagi antenna has four printed elements and an integrated balun. Moreover, by adjusting the angle between the antenna and the CubeSat, the antenna can operate at different L- and S-band frequencies and achieve beam steering.

5.2. Planar Arrays for CubeSats

First of all, it is important to note that when incorporating an antenna array on a CubeSat, a major limiting factor will be the available surface. This factor limits the operating frequency range, as an antenna array requires a wavelength-specific spacing between its elements. Another issue with integrating antenna arrays on CubeSats is the power required for phase shifting techniques [188]. Table 12 lists the performance of planar antenna arrays. The design in [188] is an active phased planar array that complies with the CubeSat size standards. As the design was tested in the anechoic chamber, it was found that it could deliver a 5-dBi average gain at 2.5 GHz, and it might be possible to expand the design to a 16-element deployable antenna array able to produce a gain of 11 dBi on a 2U CubeSat. In [189], a reconfigurable S-band patch antenna was proposed as an improvement to the previous work performed in [190]. The antenna consists of four rectangular patches which can generate three different radiation patterns as well as three different polarisations when excited in different ways, in terms of the excitation phase.

Another design relevant to planar phased arrays on CubeSats can be found in [38]. In this design, the antenna array is aimed at inter-CubeSat communications and can enable beam scanning, as the antenna's beam can be steered up to 40°. The array consists of several subarrays. Each subarray contains 4 patches and has a size of 30 mm × 30 mm. At a frequency of 5.8 GHz, the subarray has a gain of 5.1 dBi, while the complete array achieves a gain around 5.8 dBi. This planar phased array is placed on one of the 1U CubeSat surfaces, and all of its subarrays share the same substrate panel and are designed under the same frequency. Furthermore, another planar array design compatible with CubeSats is presented at [191]. More specifically, a 2 × 2 antenna array is proposed, utilizing annular patches as the elements which are connected with strips to a ring resonator functioning as a feeding network. The array presents circular polarization at 8.25 GHz with a gain of 13 dBi, in contrast to a single annular patch where the gain is only 7 dBi.

Table 12. Performance of planar antenna arrays.

Type of Antenna	Frequency (GHz)	Gain (dBi)	−10 dB BW (GHz)	Deployable
2 × 2 Planar Active Phased Array [188]	2.4	5	0.1	No
4-Element Patch Array [189,190]	2.45	8.3	1500	No
Planar Phased Array [38]	5.8	5.8	N/A	No
2 × 2 Annular Planar Array [191]	8.25	13	0.7	No
4 × 4 Dual-Frequency, Dual-Polarization Stack Array [192]	14 35	15.82 14.84	0.3 1.7	No
6 × 6 Planar Patch Array [193]	10.4	20.1	2	No

Moreover, in [192], a 4 × 4 planar antenna array design is proposed as a feed for parabolic reflector antennas for satellite remote sensing applications and especially the global water cycle that affects the Earth's climate. The design has dimensions of 8.7 cm × 7.2 cm and can fit onto one of the CubeSat surfaces. The proposed array is characterized by dual frequencies at 14 GHz and 35 GHz and dual linear polarization, namely vertical and horizontal. The design utilizes the concept of aperture-coupled patch antennas. It is important to note that for lower coupling and lower undesired radiation, a thin substrate with a high dielectric constant is used for the feed network (RT/Duroid 6010). On the other hand, a thick substrate with a low dielectric constant (RO4003C) is used for the antennas, as this allows for a larger bandwidth. The design has three layers, where the 35-GHz array lies inside the empty space of a 2 × 2 14 GHz subarray. This eliminates the need for independent substrates for the different arrays. Furthermore, a 36-element RHCP patch antenna array that operates at 10.4 GHz was proposed in [193]. The array has a size of 0.9 cm × 0.9 cm and can be mounted on any 1U CubeSat surface without the need for any deployment mechanism. The array achieves a high gain of 20.1 dBi at 10.4 GHz with a low side lobe level of −14 dB while keeping the cross-polarized LHCP radiation at −18 dB.

5.3. Reflectarrays, Reflectors, and Transmitarrays for CubeSats

Deployable reflectors and reflectarrays are some of the most popular solutions for satellite missions requiring high gain in high-frequency bands. However, the option of employing deployable reflector-based antennas for CubeSat applications is still under investigation, and some novel concepts such as foldable reflectarrays, transmitarrays, and mesh deployable antennas have been proposed to solve the problems such as scaling and deployable mechanisms. As shown in Table 13, the reflectarray described in [194] consists of three flat rectangular panels that will deploy perpendicular to the side of feeding bus. They are stacked on one side of the CubeSat before flipping out, which brings design challenges when considering the thickness of the substrate. Their deployment mechanism is controlled by simple spring-loaded hinges. Reflectarrays are usually lightweight and inexpensive, but they are characterized by narrow bandwidths. A novel concept for deployable mesh reflector antennas was first proposed in [195]. The folded size of the mesh reflector can be stored in a 1.5U volume satellite body, and its functional dimension can support a 6U class CubeSat. Its physical function is similar, with an umbrella that deploys as a parabolic reflector. Compared with reflectarrays, this design can provide a higher gain and larger bandwidth but also has a larger stowage volume.

There are some designs where a reflectarray is combined with a solar array. In [101], the Integrated Solar Array and Reflectarray Antenna (ISARA) is presented as a deployable antenna operating on the Ka-band which is compatible with a 3U CubeSat and can be used for radar applications in space. Radars applications on CubeSats require a satisfactory SNR performance and have their power limited to a few watts. Thus, the antenna gain should not be lower than 35 dBi. The design consists of three 33.9 cm × 8.26 cm reflectarray panels and a microstrip feed. More specifically, square reflectarray patches are printed on a 15-mil substrate (ε_r = 3.00) underneath the solar panels, while the feed is composed of a 4 × 4 element microstrip patch array facing the bottom surface solar arrays at a distance of 14.67 cm. A major advantage of the ISARA against other deployable mesh reflectors or inflatable reflector antennas is that it does not occupy any payload space and is extremely lightweight. The ISARA mission was at Technological Readiness Level (TRL) 5, and it flew for 5 months to reach TRL 7.

Following the ISARA mission, the same reflectarray antenna concept was utilized in the first CubeSat mission to Mars, which was called Mars Cube One (MarCo) [196]. The CubeSats in this mission would be used as a twin communication relay for the InSight mission. More specifically, during the entry descent and landing (EDL) phase, InSight would transmit spacecraft status data at a UHF band. Each MarCo CubeSat would receive this data using a circular polarized loop antenna, and then each CubeSat would transmit at an X-band link to a 70-m Deep Space Network antenna at a distance of 160 million kilometers. The downlink antenna design was modified from the one used in the ISARA mission. First, the antenna has a small stowage volume of 0.1U, and the deployment mechanism utilizes hinges for the reflectarray and a flip-out feed. The design consists of three 19.9 cm × 33.5 cm × 1.25 cm reflectarray panels and a 4 × 2 element microstrip patch feed. The reflectarray panels are designed on a Rogers RO4003 woven glass reinforced hydrocarbon ceramic material with a thickness of 0.812 mm and a constant of ε_r = 3.55, while the spacing of the elements is 1.168 cm and 1.189 cm in the x and y directions, respectively. The spacing of the patches and the thickness of the substrate were chosen to provide a sufficient bandwidth [197]. The MarCO 6U CubeSats demonstrated the suitability of CubeSats for deep space missions and also the importance of high-gain antennas and folded panel reflectarrays for high-data rate communications.

Another reflectarray design suitable for a 6U CubeSat can be found on [196]. The design consists of 15 reflectarray deployable panels forming an array of 255 × 212 elements that are 81.8 cm × 98.4 cm in size, a feed horn, 3 telescoping waveguides, a rectangular hyperboloid subreflector, a rectangular-to-circular waveguide, and 3 struts to align the subreflector with respect to the feed. The simulated gain was found to be 48 dBi at 35.75 GHZ with an aperture efficiency of 44%. Along similar lines, a Large-Area Deployable Reflectarray (LADeR) concept was proposed by the authors of [198]. In this design, the array elements are etched on a polyimide sheet which is supported by a flexible substrate made of a quartz-epoxy composite. The reflectarray, when stowed, occupies 4U, and when deployed, it has a surface area of 1.5 m × 1.5 m, which provides a high X-band gain of 39.6 dBi. The proposed LADeR concept can be used in future deep space missions or to establish high-speed downlinks. A similar concept to reflectarrays is transmitarray antennas, where the transmission characteristics of the impinging wave are studied. A circular polarized transmitarray was proposed for CubeSat inter-satellite links in [199]. The transmitarray is fed by an offset, 3-D printed, corrugated CP horn, which would be placed and deployed from the side of the 3U CubeSat. A transmitarray utilizes the variable rotation technique to provide the desired phase shift and collimate the beam toward the direction $(\varphi, \theta) = (0^\circ, -20^\circ)$. The array achieves an LHCP gain of 31.6 dBi at 24.6 GHz and a 1-dB gain bandwidth of 5.7%.

Table 13. Performance of reflectors, reflectarrays, and transmitarrays.

Type of Antenna	Frequency (GHz)	Gain (dBi)	−10 dB BW (MHz)	Deployable
Reflectarray [194]	26G 8.425	33.5 >28	>100 >100	Yes Yes
Mesh Reflector [195,200]	35.75	42.6	N/A	Yes
Integrated Solar Array and Reflectarray [101]	36	33.5	>100	Yes
MarCO [196]	8.425	29.2	50	Yes
Cassegrain Reflectarray [196]	35.75	48	N/A	Yes
LADeR Reflectarray [198]	8.4	39.6	N/A	Yes
Transmitarray [199]	24.6	31.6	N/A	Yes

5.4. Log-Periodic Crossed-Dipole Arrays

Dipoles and wire antennas have been two of the most popular designs used in CubeSats. A disadvantage of employing dipole antennas is the fact that they do not exhibit circular polarization unless crossed together, which makes the antenna more flexible in terms of capturing electromagnetic waves and also more insensitive to signal degradation due to harsh weather conditions. Thus, in [201], a log-periodic crossed-dipole array was proposed. This design could be a major candidate for CubeSat space communications due to its directive radiation pattern, high gain, and wide bandwidth. The antenna is fabricated using a curved composite bi-stable tape spring, which will allow for a very compact and small stowed volume as well as a simple deployment roll-unroll mechanism. Moreover, each dipole element should be crossed with one another while the differences between the dipole pair lengths are calculated on a log-periodic scale. Finally, the multiband operation of the antenna elements results in an antenna array with wideband operation.

5.5. Slotted Waveguide Antenna Arrays

An antenna design that can be used as a synthetic-aperture radar (SAR) for remote sensing applications can be found in [202]. More specifically, a low profile, high efficiency, and high-power capacity parallel plate slotted waveguide antenna array design is proposed. First of all, the radiating slots are located on the wall of a waveguide. The antenna system consists of two layers. The first layer is an RF feeder panel, and the second is an aluminium parallel plate with a slotted array. There are six square antenna panels that can be folded into three adjacent CubeSat surfaces and be deployed as a large rectangular panel. Each single panel weighs around 1 kg, has dimensions of 0.7 m × 0.7 m, and has multiple coupling slots that are not parallel to each other. The antenna array when deployed is 4.9 m in length, and it can achieve a gain of 34.9 dBic at 9.56 GHz. To summarize, the proposed design is suitable for small satellite applications, but improvements can be made for reducing the system loss and weight due to its complex structure.

5.6. Inflatable Antenna Arrays

Inflatable antennas can realize CubeSat interplanetary missions as they can increase the achievable data rate as well as the antenna gain [152]. The use of inflatable antennas as cooperative communication techniques to form antenna arrays is investigated in [203]. Such arrays could be used to relay information from different places in the solar system, enabling CubeSat missions in geostationary orbit (GEO). The technique of forming a beam from several inflatable antennas installed on different CubeSats is investigated. The gain of the array with n inflatable antennas is increased by a factor of n^2 minus the losses of the system. The gain of the array is also increased by increasing the diameter of the individual inflatable antennas constituting the array. On the other hand, when employing the beam forming technique, atomic clocks on each CubeSat may be required to synchronize the transmission and the inter-satellite communication.

5.7. Retrodirective or Self-Steering Antenna Arrays

CubeSat is an emerging and rapid-growing technology that might replace conventional big satellites. Several CubeSat missions involve the deployment of CubeSat swarms which can overcome the capabilities and functionalities of larger satellites. To accomplish this, an efficient crosslink is required for inter-CubeSat communication. An interesting solution to this can be given by employing retrodirective arrays (RDAs) which make the CubeSat network dynamically reconfigurable [204]. An RDA can be considered an alternative beam-steering antenna design for nanosatellite applications. The potential of a self-steering antenna array application for distributed CubeSat networks is investigated in [205]. A self-steering antenna, also named a retrodirective antenna, can sense and record the direction of the incoming signal and then send an outgoing signal along that same direction. An RDA has the advantage of not requiring prior knowledge of the position of the intended receiver, and the steering is also performed at the hardware level, which eliminates the need for complex digital signal processing. The main challenge that comes with the integration of RDA into CubeSat is the power limitation imposed by the nanosatellite platform. In [204], the design is divided into three modules: the detection of arrival (DOA), the communication array, and the tracking and steering control. First of all, the antenna array consists of 4 patch elements spaced at half-wavelength distances and fabricated on Rogers RT/Duroid 6002 substrate operating at 9.67 GHz for receiving mode and at 9.59 GHz for transmitting mode. The DOA array utilizes the null scanning technique for power detection, and this information is used by the control module to retrodirect a signal back to the interrogator. The overall design is composed by two four-layer PCBs: one for communication purposes and one for power detection. Finally, the 4-element 1-D RDA was designed to fit a 1.5U CubeSat and consumes 1 W of power, which complies to the CubeSat standards.

Moreover, in [206], another RDA design was proposed that eliminates phase shifters. This design consists of a cross-shaped patch array with a total of eight elements printed on a Rogers TMM3 substrate and a quadruple subharmonic mixer utilizing anti-parallel diodes as a mixer. The reason why quadruple subharmonic mixing is used is to achieve phase conjugation. Phase conjugation is a technique used to achieve retrodirectivity without any use of phase shifters, appearing on phased antenna arrays, as well as to relax the requirement of a high-frequency local oscillator (LO). The retrodirective array operates at 10.5 GHz, and the array spacing is at 1.38 cm, demonstrating circular polarization and two-dimensional steering [207].

5.8. Interferometer (Large Antenna Arrays)

The research about antenna array technology in the Deep Space Network (DSN) has been being undertaken by NASA since the 1960s [208]. The antenna array for deep space consists of elements located in different positions far away from each other forming a very large antenna aperture. Generally, a large antenna array plays the role of the receiver that receives the signals from a deep space source. By combining the signals received by each element according to their coherence and incoherence among the noise signal, a high SNR can be achieved. As a result, by improving the SNR of the receiver, the deep space network could support an increased data load with higher efficiency.

Compared with an antenna array installed in a single CubeSat, as mentioned in previous sections, this large antenna array is formed by combining many CubeSats located in different locations. This may introduce synchronization problems, as each CubeSat runs on a different clock. Therefore, when performing signal synthesis, it is necessary to solve the problems of phase correlation, time synchronization, and delay compensation. An example of a CubeSat swarm network was realized in the QB50 project, a collaboration among

50 different multi-national partners. The resulting cooperative QB50 CubeSat network is envisaged to have a higher degree of functionalities than conventional satellites. Another example of an interferometer antenna array is the Very Large Array (VLA) located in New Mexico in the United States [209]. It consists of 27 steerable antennas forming a 3-arm array in the shape of the letter "Y". The largest distance between antenna elements to the center of the "Y" is up to 21 km. A various range of operating bandwidths is available in this large receiving system from 50 MHz to 97 kHz.

5.9. Arraying Techniques and Correlation Algorithms

There are five basic arraying techniques, namely full-spectrum combining (FSC), complex-symbol combining (CSC), symbol-stream combining (SSC), baseband combining (BC), and carrier arraying (CA). Among these techniques, FSC can achieve an optimized remote sensing performance [210]. It can be used in the case of a weak carrier signal that is hard to track with a single antenna. The gross signal delay and phase offset between antennas are adjusted before signal combination from geometry calculations. The residual relative delay and phase can be estimated from the signal cross-correlation of each individual antenna.

There are several correlation algorithms that can be employed in antenna arrays. The Eigen algorithm can maximize the SNR from complex weights, but the computational complexity is proportional to the number of antennas squared [211]. Furthermore, there is the SUMPLE algorithm that can be applied to weak signals through the cross-correlation of each antenna element. The number of iterations of SUMPLE is proportional to the number of antennas [212]. To decrease the combining loss introduced in SUMPLE, a matrix-free (MF) algorithm can be used [213]. Based on the SUMPLE and MF algorithms, a new method named the variable step size matrix-free power (SVS-MF) method was proposed by the National Laboratory of Science and Technology on Antennas and Microwaves at Xidian University [214]. Aside from using the cross-correlation of each antenna, this method computes the variance of weights and updates data for each iteration, thus achieving a low combining loss for very weak signals with a high convergence rate. To conclude, a high-performance correlation algorithm is equally significant to the geometry and aperture of the antenna array for a successful deep space CubeSat swarm. Based on the existing algorithms, improvements could be made in terms of reducing the computational complexity and further improving the SNR at the receiver.

5.10. Recommendations for Antenna Arrays

As CubeSat applications are becoming more popular, CubeSat mission communication requirements are becoming more demanding. Researchers and developers around the world design CubeSats with increased capabilities in terms of achievable data rate, gain, and bandwidth. To meet such requirements, the migration from single-element antennas to antenna arrays for CubeSats is imminent. Antenna arrays are an effective way to increase the gain of single-element antennas by combining single antennas under a specific geometrical configuration. The integration of antenna arrays on the CubeSat platform can be approached using two different methods. In the first method, an antenna array along with a corresponding deployment mechanism is installed on a single CubeSat, while in the second method, many CubeSats, namely a swarm, are required to form the overall antenna array, where each satellite has a single-element antenna installed on it.

To date, most CubeSat missions employ single-element antennas for their communication. Missions that utilize antenna arrays are few, as the integration of antenna arrays on the CubeSat platform is still under research. The reason for this is that antenna arrays might require a complex deployment mechanism as in the case of linear arrays. Furthermore, antenna arrays may require phase shifters for beam steerability, which in turn demands board processing capabilities and an increased power budget. The choice of arraying techniques and correlation algorithms plays a vital role in the effective function of an antenna array. Lastly, when using CubeSat swarms to form an antenna array, clock synchronization between the different CubeSats is of high importance.

Despite the current difficulties of combining CubeSats with antenna arrays, there are many advantages that encourage their use for CubeSat missions. First, CubeSats equipped with a high-gain slotted waveguide antenna array might be used for radar and remote sensing applications such as SAR. Moreover, from the recent NASA mission called MarCo, it can be observed that CubeSats with reflectarrays could be used for deep space applications to relay data between the Earth and the main mission spacecraft. In addition, inflatable antennas installed on several CubeSats in different locations can form an array of inflatable antennas with beam forming capabilities, realizing geostationary Earth orbit or even interplanetary CubeSat missions. Finally, by using retrodirective antenna arrays, effective intersatellite links (ISLs) can be established inside a CubeSat swarm. A swarm equipped with RDAs can even outrun the capabilities of conventional big satellites. The current research on different types of antenna arrays for CubeSats for applications that exceed their current capabilities is of considerable importance. The typical operating frequency of antenna arrays is higher than that of single-element antennas spanning from S-band to W-band. Most of the research is focused on the X- or Ka-band frequency bands by using active phased planar arrays or reflectarrays.

6. Comparison of Single-Element Antennas with Antenna Arrays

6.1. Operating Frequency Bands

When examining the aforementioned antenna designs, we conclude that most of the single-element antennas including whip antennas, patch antennas, helix antennas, and other special antennas are operating on UHF, VHF, or S-band. More specifically, monopole and dipole antennas are more likely to be designed on the UHF and VHF bands, while patch antennas and microstrip antennas usually operate on S-band. Very few antennas such as the bull's eye antenna might aim for a much larger frequency of up to 60 GHz. On the other hand, for the antenna arrays, a large diversity in operating frequencies can be observed. As presented in Figure 8, although a number of antenna arrays operate on UHF or VHF, most of the investigated designs operate on much higher frequency bands such as X-band, C-band, or Ka-band. As far as the bandwidth is concerned, single-element antennas and antenna arrays can offer a bandwidth within range from 400 kHz to 1.5 GHz and from 50 MHz to 1.7 GHz, respectively, even though there are some antenna designs that turned out to be extremely wideband, with bandwidths as high as 10 GHz. Nonetheless, there is no unique characterization of bandwidths that can be applied to different antennas, as the antenna specifications are highly associated with the operating frequency.

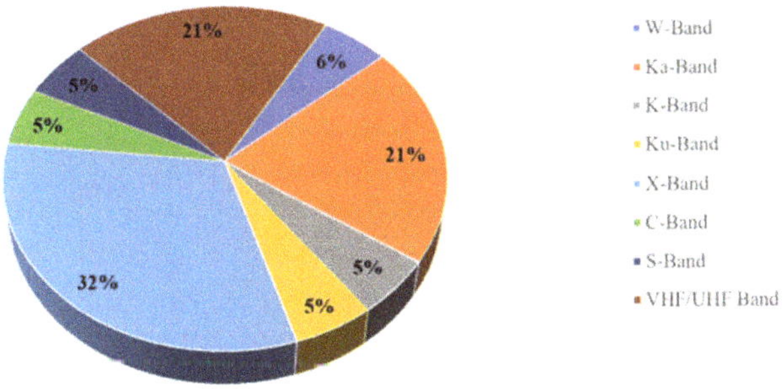

Figure 8. Operating frequency bands used in antenna arrays.

6.2. Gain

Single-element antennas and antenna array designs for CubeSats show different gain characteristics. The use of single-element antennas on a space-restricted platform like the CubeSat greatly limits the mission's high gain potential. Designs such as monopole, dipole, and planar antennas present a gain between 2–9 dBi. Furthermore, when a deployment mechanism is included in the antenna design, the gain of single-element antennas can be increased. Example designs include conical spiral, helical, and inflatable antennas. In such designs, the antenna size, when deployed, can be much larger than the CubeSat itself, and thus the gain can reach a value of 16 dBi. In addition, the gain can be increased by using antenna arrays. For instance, in the case of reflectarrays, the gain can reach up to 48 dBi, which enables long-distance deep space CubeSat missions. A comparison between the achievable gain of single-element antennas and antenna arrays is illustrated in Figure 9. As expected, the antenna arrays offer higher gain compared with single-element antennas by efficiently combining the radiation patterns of different elements in a desired direction.

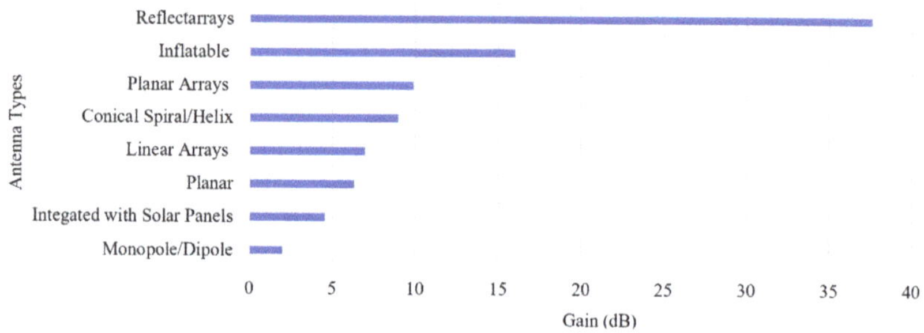

Figure 9. Gain of single-element antennas and antenna arrays used in CubeSats.

7. Conclusions

7.1. Lessons Learned

The mission requirement is the one of the most crucial parameters that a CubeSat designer must consider when deciding upon the type of antenna. Single-element antennas such as monopoles or dipoles can be found in all types of missions as either as the primary or secondary low data rate radios. On the other hand, the gain superiority and beam steering capabilities of arrays renders them suitable for deep space exploration, Earth observation, and communication or IoT missions. That aside, their beam steering ability and flexible radiation patterns make them suitable candidates for CubeSat swarms. In the flowchart in Figure 10, a guide on antenna selection for different CubeSat missions is presented. The antenna designer first chooses the type of mission under development, and then several recommended frequency bands are proposed. The next condition relates to the gain requirements, namely low or high gain, and the operating frequency. Following the gain condition, and considering the cost of the antenna, the designer is given a pool of different antennas from which the desired antenna type can be easily selected. Finally, after selecting a specific antenna type, the designer can refer to the related section of the text in which the selected antenna type is being discussed and analyzed more thoroughly.

7.2. Concluding Remarks

This paper investigated different antenna designs for CubeSat, which are categorized as single-element antennas and antenna arrays. The background on CubeSats and their subsystems was presented to provide newcomers with the fundamental knowledge on CubeSat technologies. Furthermore, the specifications and restrictions posed by the CubeSat platform were analyzed. Following that, 120 CubeSat missions dating from 2003 to 2022 were reviewed, and their antenna specifications and characteristics were extracted and tabulated. In addition, a survey on single-element antennas and antenna array designs was conducted with the aim of capturing the current as well as future CubeSat trends from an antenna point of view.

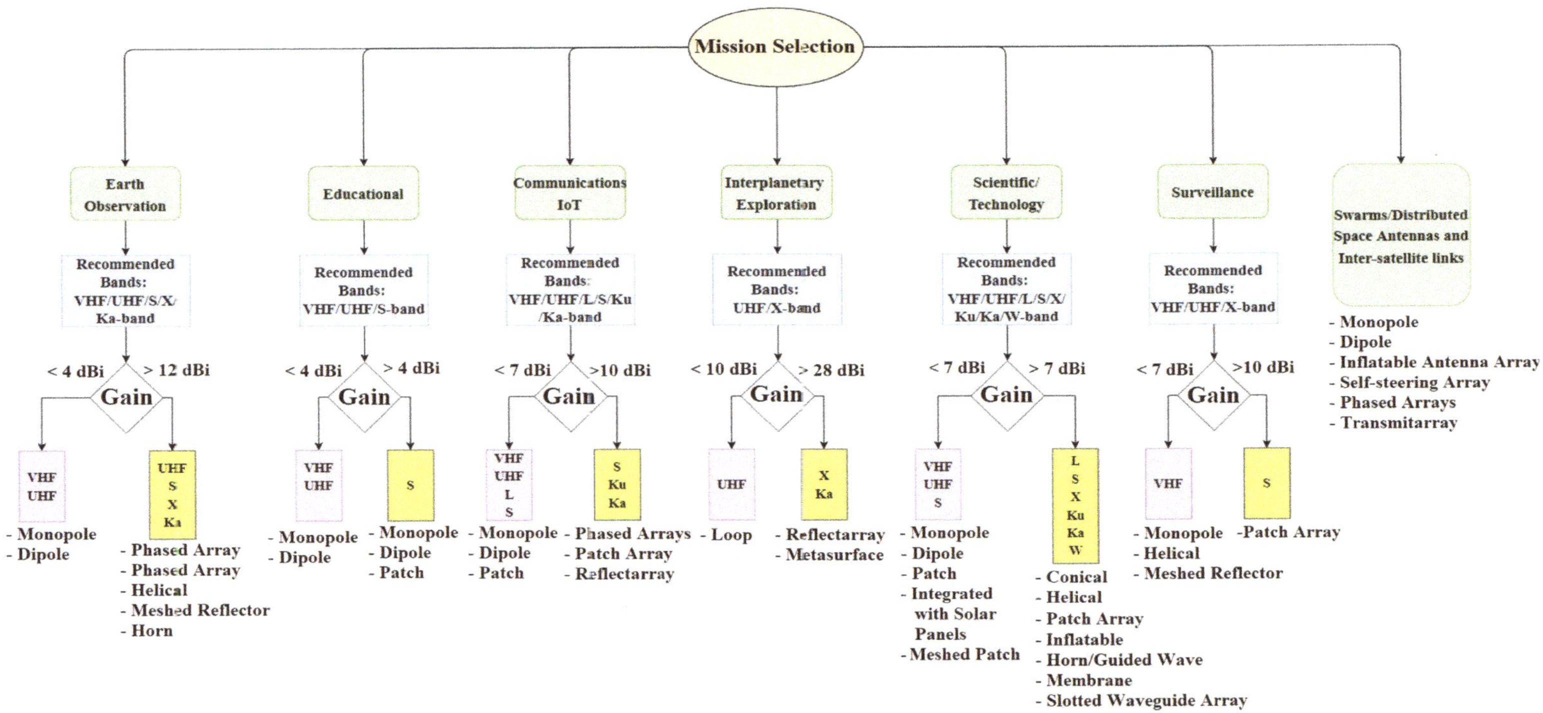

Figure 10. Antenna selection guide for different CubeSat missions.

Single-element antennas can be classified as whip antennas (monopole or dipole), patch antennas, antennas integrated with solar panels, inflatable antennas, conical spiral antennas, helical antennas, and some specially designed antennas. Whip antennas are good candidates for downlink antennas as they are relatively easy and cheap to construct. Monopole and dipole antennas along with a deployment mechanism are popular among current educational and scientific CubeSat missions. On the other hand, planar antennas present a number of advantages against their monopole and dipole counterparts. First of all, they have low profiles, higher gain, and eliminate the need for a deployment mechanism. Planar antennas can be placed on any of the CubeSat's surfaces and operate mostly on S-band. Moreover, the printed antennas on the solar panels have similar characteristics to patch antennas, but their advantage is that they do not occupy any payload or chassis space. Planar antennas are currently being used in test and demonstration CubeSat missions, which makes them very likely to replace their monopole and dipole counterparts soon. Inflatable, helical, and conical spiral antennas are currently under research, where only the helical antenna was found to be used in a CubeSat mission. Such antenna designs require a deployment mechanism, but they present superior gain characteristics to the whip or the planar antennas, so they are recommended for technology demonstration missions. As far as the antenna arrays are concerned, current research is more focused on reflectarrays, transmitarrays, and planar arrays, as well as different arraying techniques and correlation algorithms. Antenna arrays are more suitable for deep space, communications or the IoT, and Earth observation missions.

From the 120 investigated CubeSat missions, it can be concluded that the interest of the space community lies in new technology demonstration, where the CubeSat serves as a cost-effective way of launching and evaluating experiments and technologies in space. Aside from that, a big portion of CubeSat missions is devoted to educational purposes. This plays an important role in the advent of CubeSat technology in recent years, as CubeSats are now widely designed and built by academic institutions and start-up companies. Exploring the universe and learning more about the composition of the weather in space and on Earth is also another potential use of CubeSats. Finally, the most recent missions have been focused on high-speed communications, the IoT, and 5G from space. These missions will be the main driver for the future development of CubeSat technology for the years to come. It is important to note that most of the IoT and 5G from space missions are based on CubeSat constellations or swarms. In such cases, the tiny form of a CubeSat is not a disadvantage but rather an advantage for scalability. The coordinated function of each CubeSat in the swarm will result in unprecedented capabilities by redefining space missions. All these technological advances would be impossible without tiny, but at the same time mighty, CubeSats.

Author Contributions: Writing—original draft preparation, S.L., P.I.T. and A.T.; methodology, S.L., P.I.T., R.R., F.T. and S.A.; formal analysis, S.L., P.I.T., R.R., F.T. and S.A.; investigation, L.M. and F.T.; resources, S.L., P.I.T. and A.T.; data curation, S.L.; writing—original draft preparation, S.L., P.I.T., R.R. and A.T.; writing—review and editing, L.M., M.U.A.K., S.I., F.T. and S.A.; visualization, P.I.T.; supervision, R.R., F.T. and L.M.; project administration, S.L. All authors have read and agreed to the published version of the manuscript.

Funding: This research received no external funding.

Conflicts of Interest: The authors declare no conflict of interest.

References

1. Shimmin, R.; Agasid, E.; Burton, R.; Carlino, R.; Defouw, G.; Perez, A.; Karacaliglu, A.; Klamm, B.; Rademacher, A.; Schalkwyck, J. *Small Spacecraft Technology State of the Art*; NASA Mission Design Division Ames Research Center, Moffett Field: Santa Clara, CA, USA, 2015; Volume NASA/TP–2015–216648/REV.
2. Xue, Y.; Li, Y.; Guang, J.; Zhang, X.; Guo, J. Small satellite remote sensing and applications—History, current and future. *Int. J. Remote Sens.* **2008**, *29*, 4339–4372. [CrossRef]

3. Puig-Suari, J.; Turner, C.; Ahlgren, W. Development of the standard CubeSat deployer and a CubeSat class PicoSatellite. In Proceedings of the 2001 IEEE Aerospace Conference Proceedings (Cat. No.01TH8542), Big Sky, MT, USA, 10–17 March 2001; Volume 1, pp. 347–353.
4. Wuerl, A.; Wuerl, M. Lessons learned for deploying a microsatellite from the International Space Station. In Proceedings of the 2015 IEEE Aerospace Conference, Big Sky, MT, USA, 7–14 March 2015; pp. 1–12.
5. Twiggs, B.; Puig-Suari, J. *CUBESAT Design Specifications Document*; Stanford University: Stanford, CA, USA; California Polytechnical Institute: San Luis Obispo, CA, USA, 2003.
6. ESA. First P-Pod Integration. Available online: http://www.esa.int/spaceinimages/Images/2012/02/First_P-POD_integration2 (accessed on 30 December 2021).
7. Swartwout, M. The first one hundred CubeSats: A statistical look. *J. Small Satell.* **2013**, *2*, 213–233.
8. Klofas, B.; Anderson, J.; Leveque, K. A Survey of Cubesat Communication Systems. In Proceedings of the 5th Annual CubeSat Developers Workshop, San Luis Obispo, CA, USA, 9–11 April 2008.
9. Chahat, N.; Decrossas, E.; Gonzalez-Ovejero, D.; Yurduseven, O.; Radway, M.J.; Hodges, R.E.; Estabrook, P.; Baker, J.D.; Bell, D.J.; Cwik, T.A.; et al. Advanced cubesat antennas for deep space and earth science missions: A review. *IEEE Antennas Propag. Mag.* **2019**, *61*, 37–46. [CrossRef]
10. Malphrus, B.K.; Freeman, A.; Staehle, R.; Klesh, A.T.; Walker, R. 4—Interplanetary CubeSat missions. In *Cubesat Handbook*; Cappelletti, C., Battistini, S., Malphrus, B.K., Eds.; Academic Press: Cambridge, MA, USA, 2021; pp. 85–121.
11. Tummala, A.R.; Dutta, A. An overview of cube-satellite propulsion technologies and trends. *Aerospace* **2017**, *4*, 58. [CrossRef]
12. Santoni, F.; Piergentili, F.; Donati, S.; Perelli, M.; Negri, A.; Marino, M. An innovative deployable solar panel system for Cubesats. *Acta Astronaut.* **2014**, *95*, 210–217. [CrossRef]
13. Bunce, D.T.; Bassett, K.P.; Ghosh, A.R.; Barnett, P.R.; Haken, D.M.; Coverstone, V.L.; Yost, B.D.; Feller, J.R.; Agasid, E.F. Microvascular Composite Radiators for Small Spacecraft Thermal Management Systems. In Proceedings of the 30th Annual AIAA/USU Conference on Small Satellites, Logan, UT, USA, 6–11 August 2016.
14. Gao, S.; Brenchley, M.; Unwin, M.; Underwood, C.I.; Clark, K.; Maynard, K.; Boland, L.; Sweeting, M.N. Antennas for small satellites. In Proceedings of the 2008 Loughborough Antennas and Propagation Conference, Loughborough, UK, 17–18 March 2008; pp. 66–69.
15. Selva, D. A survey and assessment of the capabilities of Cubesats for Earth observation. *Acta Astronaut.* **2012**, *74*, 50–68. [CrossRef]
16. Schaffner, J. The Electronic System Design, Analysis, Integration, and Construction of the Cal Poly State University CP1 CubeSat. In Proceedings of the 16th Annual AIAA/USU Conference on Small Satellites, Logan, UT, USA, 12–15 August 2002.
17. Theoharis, P.I.; Raad, R.; Tubbal, F.; Khan, M.U.A.; Liu, S. Software-Defined Radios for CubeSat Applications: A Brief Review and Methodology. *IEEE J. Miniat. Air Space Syst.* **2020**, *2*, 10–16. [CrossRef]
18. Olivieri, S.J.; Aarestad, J.; Pollard, L.H.; Wyglinski, A.M.; Kief, C.; Erwin, R.S. Modular FPGA-based software defined radio for CubeSats. In Proceedings of the 2012 IEEE International Conference on Communications (ICC), Ottawa, ON, Canada, 10–15 June 2012; pp. 3229–3233.
19. Bouwmeester, J.; Guo, J. Survey of worldwide pico-and nanosatellite missions, distributions and subsystem technology. *Acta Astronaut.* **2010**, *67*, 854–862. [CrossRef]
20. Abulgasem, S.; Tubbal, F.; Raad, R.; Theoharis, P.I.; Lu, S.; Iranmanesh, S. Antenna Designs for CubeSats: A Review. *IEEE Access* **2021**, *9*, 45289–45324. [CrossRef]
21. Tubbal, F.E.; Raad, R.; Chin, K.W. A Survey and Study of Planar Antennas for Pico-Satellites. *IEEE Access* **2015**, *3*, 2590–2612. [CrossRef]
22. Rowland, D.E.; Hill, J.; Uribe, P.; Klenzing, J.; Hunsaker, F.; Fowle, M.; Simms, K.; Hancock, H.; Saulino, M.; Guzman, D.; et al. The NSF Firefly CubeSat mission: Rideshare mission to study energetic electrons produced by lightning. In Proceedings of the 2011 Aerospace Conference, Big Sky, MT, USA, 5–12 March 2011; pp. 1–12.
23. Qiao, L.; Rizos, C.; Dempster, A.G. Analysis and comparison of CubeSat lifetime. In Proceedings of the 12th Australian Space Development Conference, Adelaide, Australia, 8–10 July 2013.
24. Sweeting, M.N.; Underwood, C.I. Small Satellite Engineering and Applications. In *Spacecraft Systems Engineering*; John Wiley & Sons, Ltd.: Hoboken, NJ, USA, 2011; pp. 575–605.
25. Wilson, J.; Cucinotta, F.; Golightly, M.; Nealy, J.; Qualls, G.; Badavi, F.; De Angelis, G.; Anderson, B.; Clowdsley, M.; Luetke, N.; et al. International space station: A testbed for experimental and computational dosimetry. *Adv. Space Res.* **2005**, *37*, 1656–1663. [CrossRef]
26. Navarathinam, N.; Lee, R.; Chesser, H. Characterization of Lithium-Polymer batteries for CubeSat applications. *Acta Astronaut.* **2011**, *68*, 1752–1760. [CrossRef]
27. Waydo, S. CubeSat design for LEO-based Earth science missions. In Proceedings of the IEEE Aerospace Conference, Big Sky, MT, USA, 9–16 March 2002; pp. 435–445.
28. Wells, J.; Stras, L.; Jeans, T. Canada's Smallest Satellite: The Canadian Advanced Nanospace Experiment (CanX-1). In Proceedings of the 16th Annual AIAA/USU Conference on Small Satellite, Logan, UT, USA, 12–15 August 2002.
29. Schmidt, M.; Zeiger, F.; Schilling, K. Design and implementation of in-orbit experiments on the pico-satellite UWE-1. In Proceedings of the 57th International Astronautical Congress, IAC-06-E2, Valencia, Spain, 2–6 October 2006; Volume 1.

30. Long, M.; Lorenz, A.; Rodgers, G.; Tapio, E.; Tran, G.; Jackson, K.; Twiggs, R.; Bleier, T.; Solutions, S. A cubesat derived design for a unique academic research mission in earthquake signature detection. In Proceedings of the 16th Annual AIAA/USU Conference on Small Satellite, Logan, UT, USA, 12–15 August 2002.
31. Antunes, S. *DIY Satellite Platforms: Building a Space-Ready General Base Picosatellite for Any Mission*; O'Reilly Media, Inc.: Sebastopol, CA, USA, 2012.
32. Lemmer, K. Propulsion for CubeSats. *Acta Astronaut.* **2017**, *134*, 231–243. [CrossRef]
33. Mueller, J.; Hofer, R.; Ziemer, J. *Survey of Propulsion Technologies Applicable to Cubesats*; Jet Propulsion Laboratory, National Aeronautics and Space Administration: Pasadena, CA, USA, 2010.
34. Falbel, G.; Puig-Suari, J.; Peczalski, A. Sun oriented and powered, 3 axis and spin stabilized CubeSats. In Proceedings of the IEEE Aerospace Conference, Big Sky, MT, USA, 9–16 March 2002; Volume 1, pp. 447–455.
35. Cockrell, J.; Alena, R.; Mayer, D.; Sanchez, H.; Luzod, T.; Yost, B.; Klumpar, D. EDSN: A Large Swarm of Advanced Yet Very Affordable, COTS-Based Nanosats that Enable Multipoint Physics and Open Source Apps. In Proceedings of the 26th Annual AIAA/USU Conference on Small Satellite, Logan, UT, USA, 13–16 August 2012.
36. Budianu, A.; Castro, T.J.W.; Meijerink, A.; Bentum, M.J. Inter-satellite links for cubesats. In Proceedings of the 2013 IEEE Aerospace Conference, Big Sky, MT, USA, 2–9 March 2013; pp. 1–10.
37. Radhakrishnan, R.; Edmonson, W.W.; Afghah, F.; Rodriguez-Osorio, R.M.; Pinto, F.; Burleigh, S.C. Survey of Inter-Satellite Communication for Small Satellite Systems: Physical Layer to Network Layer View. *IEEE Commun. Surv. Tutor.* **2016**, *18*, 2442–2473. [CrossRef]
38. Martínez, R.-O.R.; Fueyo, R.E. A Hands-On Education Project: Antenna Design for Inter-CubeSat Communications [Education Column]. *IEEE Antennas Propag. Mag.* **2012**, *54*, 211–224.
39. Taylor, J. *Deep Space Communications*; John Wiley & Sons, Inc.: Hoboken, NJ, USA, 2016; The Deep Space Network; pp. 15–35. [CrossRef]
40. Palo, S.E. High rate communications systems for CubeSats. In Proceedings of the 2015 IEEE MTT-S International Microwave Symposium, Phoenix, AZ, USA, 17–22 May 2015; pp. 1–4.
41. Popescu, O. Power Budgets for CubeSat Radios to Support Ground Communications and Inter-Satellite Links. *IEEE Access* **2017**, *5*, 12618–12625. [CrossRef]
42. Park, Y.-K.; Kim, G.-N.; Park, S.-Y. Novel Structure and Thermal Design and Analysis for CubeSats in Formation Flying. *Aerospace* **2021**, *8*, 150. [CrossRef]
43. National Academies of Sciences and Medicine. *Handbook of Frequency Allocations and Spectrum Protection for Scientific Uses*; National Academies Press: Washington, DC, USA, 2015.
44. Funase, R.; Takei, E.; Nakamura, Y.; Nagai, M.; Enokuchi, A.; Yuliang, C.; Nakada, K.; Nojiri, Y.; Sasaki, F.; Funane, T.; et al. Technology demonstration on University of Tokyo's pico-satellite "XI-V" and its effective operation result using ground station network. *Acta Astronaut.* **2007**, *61*, 707–711. [CrossRef]
45. Doering, T.J. Development of a Reusable Cubesat Satellite Bus Architecture for the KYSAT-1 Spacecraft. Master's Thesis, University of Kentucky, Lexington, KY, USA, 2009.
46. Kayal, H.; Baumann, F.; Briess, K.; Montenegro, S. BEESAT: A pico satellite for the on orbit verification of micro wheels. In Proceedings of the Recent Advances in Space Technologies, Istanbul, Turkey, 14–16 June 2007; pp. 497–502.
47. Konda, Y.; Usuda, T.; Sagami, T.; Omagari, K.; Kashiwa, M.; Matunaga, S. Development of attitude determination and control system for pico-satellite cute-1.7+ APD. In Proceedings of the 16th Workshop on JAXA Astrodynamics and Flight Mechanics, Sagamihara, Japan, 1–2 August 2006; pp. 242–247.
48. Eide, E.; Ilstad, J. NCUBE-1, the first Norwegian CUBESAT student satellite. In Proceedings of the 16th ESA Symposium on European Rocket and Balloon Programmes and Related Research, St. Gallen, Switzerland, 2–5 June 2003; pp. 85–88.
49. Maeno, M.; Omagari, K.; Iljic, T.; Masumoto, S.; Fujiwara, K.; Konda, Y.; Yamanaka, T.; Tanaka, Y.; Ueno, T.; Ashida, H.; et al. Development of Tokyo Tech Nano-Satellite Cute-1.7+ APD II. In Proceedings of the 17th Workshop on JAXA Astrodynamics and Flight Mechanics, Sagamihara, Japan, 23–24 July 2007.
50. PSSCT (Pico Satellite Solar Cell Testbed). eoPortal Directory. Available online: https://directory.eoportal.org/web/eoportal/satellite-missions/p/pssct (accessed on 11 January 2022).
51. Kitts, C.; Hines, J.; Agasid, E.; Ricco, A.; Yost, B.; Ronzano, K.; Puig-Suari, J. The GeneSat-1 Microsatellite MissionA Challenge in Small Satellite Design. In Proceedings of the 20th Annual AIAA/USU Conference on Small Satellites, Logan, UT, USA, 14–17 August 2006.
52. Ichikawa, D. CubeSat-to-Ground Communication and Mobile Modular Ground-Station Development. *HSGC Rep. Number* **2006**, *7–15*, 34.
53. Hunyadi, G.; Klumpar, D.M.; Jepsen, S.; Larsen, B.; Obland, M. A commercial off the shelf (COTS) packet communications subsystem for the Montana EaRth-Orbiting Pico-Explorer (MEROPE) CubeSat. In Proceedings of the IEEE Aerospace Conference, Big Sky, MT, USA, 9–16 March 2002; Volume 1, pp. 473–478.
54. Sorensen, T.; Prescott, G.; Villa, M.; Brown, D.; Hicks, J.; Edwards, A.; Lyke, J.; George, T.; Mobasser, S.; Yee, K.; et al. KUTESAT-2, a Student Nanosatellite Mission for Testing Rapid-Response Small Satellite Technologies in Low Earth Orbit. In Proceedings of the AIAA 3rd Responsive Space Conference, Los Angeles, CA, USA, 25–28 April 2005.

55. Tag Archives: First-MOVE. AMSAT-UK. Available online: https://amsat-uk.org/tag/first-move/ (accessed on 11 January 2022).
56. Noe, C. *Design and Implementation of the Communications Subsystem for the Cal Poly CP2 Cubesat Project*; California Polytechnic State University: San Luis Obispo, CA, USA, 2004.
57. LaBerteaux, J.; Moesta, J.; Bernard, B. Cajun advanced picosatellite experiment. In Proceedings of the 2007 IEEE/AIAA 26th Digital Avionics Systems Conference, Dallas, TX, USA, 21–25 October 2007; p. 5-E.
58. Taraba, M.; Rayburn, C.; Tsuda, A.; MacGillivray, C. Boeing's CubeSat TestBed 1 Attitude Determination Design and on-Orbit Experience. In Proceedings of the 23rd Annual AIAA/USU Conference on Small Satellite, Logan, UT, USA, 10–13 August 2009.
59. Scholz, A.; Ley, W.; Dachwald, B.; Miau, J.; Juang, J. Flight results of the COMPASS-1 picosatellite mission. *Acta Astronaut.* **2010**, *67*, 1289–1298. [CrossRef]
60. Stras, L.N.; Kekez, D.D.; Wells, G.J.; Jeans, T.; Zee, R.E.; Pranajaya, F.M.; Foisy, D.G. The design and operation of the Canadian advanced nanospace eXperiment (CanX-1). In Proceedings of the AMSAT-NA 21st Space Symposium, Toronto, ON, Canada, 17–19 October 2003; pp. 150–160.
61. Bouwmeester, J.; Aalbers, G.; Ubbels, W. Preliminary mission results and project evaluation of the delfi-c3 nano-satellite. In *4S Symposium Small Satellites Systems and Services*; European Space Agency: Rhodes, Greece, 2008.
62. Rankin, D.; Kekez, D.D.; Zee, R.E.; Pranajaya, F.M.; Foisy, D.G.; Beattie, A.M. The CanX-2 nanosatellite: Expanding the science abilities of nanosatellites. *Acta Astronaut.* **2005**, *57*, 167–174. [CrossRef]
63. Alminde, L.; Bisgaard, M.; Vinther, D.; Viscor, T.; Østergard, K.Z. The AAU-Cubesat Student Satellite Project: Architectural Overview and Lessons Learned. *IFAC Proc. Vol.* **2004**, *37*, 949–954. [CrossRef]
64. OUFTI-1 (Orbital Utility for Telecommunicaion Innovation). eoPortal Directory. Available online: https://directory.eoportal.org/web/eoportal/satellite-missions/o/oufti-1 (accessed on 12 January 2022).
65. Janson, S.; Hinkley, D. Spin Dynamics of the Pico Satellite Solar Cell Testbed Spacecraft. In Proceedings of the 23rd Annual AIAA/USU Conference on Small Satellite, Logan, UT, USA, 10–13 August 2009.
66. RadSat (Radiation-tolerant SmallSat Computer System). Available online: https://directory.eoportal.org/web/eoportal/satellite-missions/r/radsat (accessed on 13 January 2022).
67. Noca, M.; Jordan, F.; Steiner, N.; Choueiri, T.; George, F.; Roethlisberger, G.; Scheidegger, N.; Peter-Contesse, H.; Borgeaud, M.; Krpoun, R.; et al. Lessons Learned from the First Swiss Pico-Satellite: SwissCube. In Proceedings of the 23rd Annual AIAA/USU Conference on Small Satellite, Logan, UT, USA, 10–13 August 2009.
68. Young-Keun, C.; Byoung-Young, M.; Ki-Lyong, H.; Soo-Jung, K.; Suk-Jin, K. Development of the HAUSAT-2 nanosatellite for low-cost technology demonstration. In Proceedings of the 2nd International Conference on Recent Advances in Space Technologies, Istanbul, Turkey, 9–11 June 2005; pp. 173–179.
69. Kurtulus, C.; Baltaci, T.; Ulusoy, M.; Aydin, B.T.; Tutkun, B.; Inalhan, G.; Cetiner-Yildirim, N.O.; Karyot, T.B.; Yarim, C.; Edis, F.O.; et al. iTU-pSAT I: Istanbul Technical University Student Pico-Satellite Program. In Proceedings of the 2007 3rd International Conference on Recent Advances in Space Technologies, Istanbul, Turkey, 14–16 June 2007; pp. 725–732.
70. Ricco, A.; Parra, M.; Niesel, D.; McGinnis, M.; Ly, D.; Kudlicki, A.; Hines, J.; Piccini, M.; Timucin, L.; Beasley, C.; et al. PharmaSat: Drug dose dependence results from an autonomous microsystem-based small satellite in low Earth orbit. In *2010 Solid-State Sensors, Actuators, and Microsystems Workshop*; Transducer Research Foundation: Hilton Head Island, SC, USA, 2010; pp. 110–113.
71. Starr, G.J.; Wersinger, J.M.; Chapman, R.; Riggs, L.; Nelson, V.P.; Klingelhoeffer, J.; Stroud, C.E. Application of Embedded Systems in Low Earth Orbit for Measurement of Ionospheric Anomalies. Presented at the International Conference on Embedded Systems and Applications, Las Vegas, NV, USA, 13–16 July 2009.
72. Kuester, D.G.; Radhakrishna, P.N.P. *A 2.4 GHz High Speed Communications System for Cubesat Applications*; Colorado Space Grant Consortium; University of Colorado at Boulder: Boulder, CO, USA, 2007.
73. Dontchev, K.A.; Ghorakavi, K.; Haag, C.E.; Liu, T.M.; Ramos, R. M-cubed: University of michigan multipurpose minisatellite with optical imager payload. In Proceedings of the AIAA Space 2010 Conference & Exhibition, Anaheim, CA, USA, 30 August –2 September 2010.
74. Gregorio, A.; Bernardi, T.; Carrato, S.; Kostadinov, I.; Messerotti, M.; Stalio, R. AtmoCube: Observation of the Earth atmosphere from the space to study" space weather" effects. In Proceedings of the Recent Advances in Space Technologies, Istanbul, Turkey, 20–22 November 2003; pp. 188–193.
75. Aherne, M.; Barrett, T.; Hoag, L.; Teegarden, E.; Ramadas, R. Aeneas–Colony I Meets Three-Axis Pointing. In Proceedings of the 25th Annual AIAA/USU Conference on Small Satellite, Logan, UT, USA, 8–11 August 2011.
76. Balan, M.; Piso, M.-I.; Stoica, A.M.; Dragasanu, C.; Trusculescu, M.; Dumitru, C. Goliat space mission: Earth observation and near Earth environment monitoring using nanosatellites. In Proceedings of the 59th International Astronautical Congress, Glasgow, UK, 29 September–3 October 2008.
77. CHOMPTT (CubeSat Handling of Multisystem Precision Time Transfer). Available online: https://directory.eoportal.org/web/eoportal/satellite-missions/c-missions/chomptt (accessed on 20 January 2022).
78. Stolarski, M.; Dobrowolski, M.; Graczyk, R.; Kurek, K. Space platform for student cubesat pico-satellite. In *Photonics Applications in Astronomy, Communications, Industry, and High-Energy Physics Experiments*; SPIE: Wilga, Poland, 2009; Volume 7502.

79. EncinasPlaza, J.M.; VilanVilan, J.A.; AquadoAgelet, F.; BrandiaranMancheno, J.; LopezEstevez, M.; MartinezFernandez, C.; SarmientoAres, F. Xatcobeo: Small Mechanisms for CubeSat Satellites–Antenna and Solar Array Deployment. In Proceedings of the 40th Aerospace Mechanisms Symposium, NASA Kennedy Space Centre, Cocoa Beach, FL, USA, 12–14 May 2010; pp. 12–14.
80. E1P-2. eoPortal Directory. Available online: https://directory.eoportal.org/web/eoportal/satellite-missions/e/e1p-2 (accessed on 28 January 2022).
81. Francis, C.L. *ISM S-Band CubeSat Radio designed for the PolySat System Board*; California Polytechnic State University: San Luis Obispo, CA, USA, 2016.
82. De Jong, S.; Maddox, E.; Vollmuller, G.J.; Schuurbiers, C.A.; Van Swaaij, R.A.; Ubbels, W.J.; Hamann, R.J. The Delfi-n3Xt nanosatellite: Space weather research and qualification of microtechnology. In Proceedings of the 59th International Astronautical Congress, Scotland, UK, 29 September–3 October 2008.
83. Gao, S.; Clark, K.; Unwin, M.; Zackrisson, J.; Shiroma, W.A.; Akagi, J.M.; Maynard, K.; Garner, P.; Boccia, L.; Amendola, G.; et al. Antennas for modern small satellites. *IEEE Antennas Propag. Mag.* **2009**, *51*, 40–56. [CrossRef]
84. Lehmensiek, R.; Zyl, R.R.V.; Visser, D.F. The design of an HF antenna on a 1U CubeSat. In Proceedings of the 2013 Africon, Pointe aux Piments, Mauritius, 9–12 September 2013; pp. 1–5.
85. Move. Available online: https://www.move2space.de/missions/first-move/ (accessed on 29 January 2022).
86. Busch, S.; Bangert, P.; Dombrovski, S.; Schilling, K. UWE-3, in-orbit performance and lessons learned of a modular and flexible satellite bus for future pico-satellite formations. *Acta Astronaut.* **2015**, *117*, 73–89. [CrossRef]
87. Niederstrasser, C.; Hassan, A.; Hermle, J.; Kemp, A.; McGlothlin, A.; Samant, D.; Stein, J. TJ3Sat–The First Satellite Developed and Operated by High School Students. In Proceedings of the 23rd Annual AIAA/USU Conference on Small Satellite, Logan, UT, USA, 10–13 August 2009.
88. Sandau, R. Status and trends of small satellite missions for Earth observation. *Acta Astronaut.* **2010**, *66*, 1–12. [CrossRef]
89. Orr, N.; Eyer, J.; Larouche, B.; Zee, R. Precision Formation Flight: The CanX-4 and CanX-5 Dual Nanosatellite Mission. In Proceedings of the 21st Annual AIAA/USU Conference on Small Satellite, Logan, UT, USA, 13–16 August 2007.
90. Blackwell, W.; Allen, G.; Galbraith, C.; Hancock, T.; Leslie, R.; Osaretin, I.; Retherford, L.; Scarito, M.; Semisch, C.; Shields, M.; et al. Nanosatellites for earth environmental monitoring: The MicroMAS project. In Proceedings of the Microwave Radiometry and Remote Sensing of the Environment (MicroRad), Rome, Italy, 5–9 March 2012; pp. 1–4.
91. Veljovic, M.J.; Skrivervik, A.K. Aperture-Coupled Low-Profile Wideband Patch Antennas for CubeSat. *IEEE Trans. Antennas Propag.* **2019**, *67*, 3439–3444. [CrossRef]
92. Janson, S.; Welle, R. The NASA Optical Communication and Sensor Demonstration Program. In Proceedings of the 27th Annual AIAA/USU Conference on Small Satellite, Logan, UT, USA, 10–15 August 2013.
93. Rahmat-Samii, Y.; Manohar, V.; Kovitz, J.M. For Satellites, Think Small, Dream Big: A review of recent antenna developments for CubeSats. *IEEE Antennas Propag. Mag.* **2017**, *59*, 22–30. [CrossRef]
94. Tatomirescu, A.; Pedersen, G.F.; Christiansen, J.; Gerhardt, D. Antenna system for nano-satelite mission GOMX-3. In Proceedings of the 2016 IEEE-APS Topical Conference on Antennas and Propagation in Wireless Communications (APWC), Cairns, QLD, Australia, 19–23 September 2016; pp. 282–285.
95. Graves, J.; Perez, J.; Reed, H.; McLelland, A.; Kanipe, D.; Provence, R.; Runkle, T. AggieSat2 Student Satellite Mission. In Proceedings of the 50th AIAA Aerospace Sciences Meeting, AIAA, Nashville, TN, USA, 9–12 January 2012.
96. Kestilä, A.; Tikka, T.; Peitso, P.; Rantanen, J.; Näsilä, A.; Nordling, K.; Saari, H.; Vainio, R.; Janhunen, P.; Praks, J.; et al. Aalto-1 nanosatellite technical description and mission objectives. *Geosci. Instrum. Method. Data Syst.* **2013**, *2*, 121–130. [CrossRef]
97. Praks, J.; Kestilä, A.; Hallikainen, M.; Saari, H.; Antila, J.; Janhunen, P.; Vainio, R. Aalto-1—An experimental nanosatellite for hyperspectral remote sensing. In Proceedings of the 2011 IEEE International Geoscience and Remote Sensing Symposium, Vancouver, BC, Canada, 24–29 July 2011; pp. 4367–4370.
98. The UNSW-EC0 Cubesat. Australian Centre for Space Engineering Research (ACSER). Available online: https://www.acser.unsw.edu.au/system-basics (accessed on 2 February 2022).
99. Landsman, H.; Ruckman, L.; Varner, G.S. AURA—A radio frequency extension to IceCube. In *Nuclear Instruments and Methods in Physics Research Section A: Accelerators, Spectrometers, Detectors and Associated Equipment*; Elsevier: Amsterdam, The Netherlands, 2009; Volume 604, pp. S70–S75.
100. Fraile, J.J.F.; Laverón-Simavilla, A.; Calvo, D.; Benavides, E.M. The QBito CubeSat: Applications in Space Engineering Education at Technical University of Madrid. In Proceedings of the 40th COSPAR Scientific Assembly, Moscow, Russia, 2–10 August 2014; Volume 40, p. PE-2.
101. Hodges, R.E.; Radway, M.J.; Toorian, A.; Hoppe, D.J.; Shah, B.; Kalman, A.E. ISARA—Integrated Solar Array and Reflectarray CubeSat deployable Ka-band antenna. In Proceedings of the 2015 IEEE International Symposium on Antennas and Propagation & USNC/URSI National Radio Science Meeting, Vancouver, BC, Canada, 19–24 July 2015; pp. 2141–2142.
102. Park, J.P.; Park, S.Y.; Song, Y.B.; Kim, G.N.; Lee, K.; Oh, H.J.; Yim, J.C.; Lee, E.; Hwang, S.H.; Kim, S.; et al. Cubesat development for CANYVAL-X mission. In Proceedings of the 14th International Conference on Space Operations, Deajeon, South Korea, 16–20 May 2016; American Institute of Aeronautics and Astronautics Inc., AIAA: Reston, VA, USA, 2016.
103. Bangert, P.; Kramer, A.; Schilling, K. UWE-4: Integration State of the First Electrically Propelled 1U CubeSat. In Proceedings of the 31st Annual AIAA/USU Conference on Small Satellite, Logan, UT, USA, 5–10 August 2017.

104. CUTE-1. eoPortal Directory. Available online: https://directory.eoportal.org/web/eoportal/satellite-missions/c-missions/cute-i (accessed on 5 February 2022).
105. Chalermwisutkul, S.; Jirawattanaphol, A.; Jantarachote, V.; Arpanutud, K. Communication system development of the pioneer Thai CubeSat project: KNACKSAT. In Proceedings of the 2017 International Symposium on Antennas and Propagation (ISAP), Phuket, Thailand, 30 October–2 November 2017; pp. 1–2.
106. Peral, E.; Tanelli, S.; Haddad, Z.; Sy, O.; Stephens, G.; Im, E. Raincube: A proposed constellation of precipitation profiling radars in CubeSat. In Proceedings of the 2015 IEEE International Geoscience and Remote Sensing Symposium (IGARSS), Milan, Italy, 26–31 July 2015; pp. 1261–1264.
107. ION CubeSat Information Sheet. Available online: http://cubesat.ece.illinois.edu/Files/ION_Info_Sheet.pdf (accessed on 10 February 2022).
108. Fragnier, R.; Contreres, R.; Palacin, B.; Elis, K.; Bellion, A.; Romier, M.; Fur, G.L.; Maleszka, T. Collocated Compact UHF and L-Band Antenna for Nanosatellite Applications. In Proceedings of the 32nd Annual AIAA/USU Conference on Small Satellite, Logan, UT, USA, 4–9 August 2018.
109. Bellion, A.; Elis, K.; de Gaetano, S. New compact S-band antenna for Nanosatellite TeleMetry and TeleCommand applications-EyeSat program. In Proceedings of the 2016 10th European Conference on Antennas and Propagation (EuCAP), Davos, Switzerland, 10–15 April 2016; pp. 1–5.
110. Garcia-Talavera, M.R.; Sodnik, Z.; Lopez, P.; Alonso, A.; Viera, T.; Oppenhauser, G. Preliminary results of the in-orbit test of ARTEMIS with the Optical Ground Station. In *High-Power Lasers and Applications*; SPIE: Wilga, Poland, 2002; Volume 4635, p. 12.
111. SPATIUM (Space Precision Atomic-clock TIming Utility Mission). Available online: https://directory.eoportal.org/web/eoportal/satellite-missions/s/spatium (accessed on 28 March 2022).
112. Corvus-BC. Available online: https://spaceflight101.com/soyuz-kanopus-v-ik/corvus-bc/ (accessed on 28 March 2022).
113. Lemur-2. Nanosats Database. Available online: https://www.nanosats.eu/sat/lemur-2 (accessed on 28 March 2022).
114. Flock 1 Imaging Constellation. EoPortal Dicrectory. Available online: https://earth.esa.int/web/eoportal/satellite-missions/f/flock-1 (accessed on 29 March 2022).
115. Kepler Awarded Contributions for Small Satellite Phased Array Technology Development. Available online: https://kepler.space/2020/08/05/kepler-awarded-contributions-for-small-satellite-phased-array-technology-development/ (accessed on 29 March 2022).
116. TTU. The International Amateur Radio Union. Available online: http://www.amsatuk.me.uk/iaru/formal_detail.php?serialnum=565 (accessed on 30 March 2022).
117. NetSat. Nanosats Database. Available online: https://www.nanosats.eu/sat/netsat (accessed on 31 March 2022).
118. TRISAT (NANOsky, Misija Trisat). Nanosats Database. Available online: https://www.nanosats.eu/sat/trisat (accessed on 31 March 2022).
119. CICERO 1, . . . , 12/OSM 1 CICERO. Gunter's Space Page. Available online: https://space.skyrocket.de/doc_sdat/cicero.htm (accessed on 31 March 2022).
120. PICASSO (Pico-Satellite for Atmospheric and Space Science Observations). EoPortal Dicrectory. Available online: https://directory.eoportal.org/web/eoportal/satellite-missions/p/picasso (accessed on 1 April 2022).
121. AMICal-Sat. Gunter's Space Page. Available online: https://space.skyrocket.de/doc_sdat/amical-sat.htm (accessed on 2 April 2022).
122. Astrocast Commercial IoT Network Service. EoPortal Dicrectory. Available online: https://directory.eoportal.org/web/eoportal/satellite-missions/a/astrocast (accessed on 2 April 2022).
123. BeeSat 5, 6, 7, 8, 10, 11, 12, 13. Gunter's Space Page. Available online: https://space.skyrocket.de/doc_sdat/beesat-5.htm (accessed on 4 April 2022).
124. RADCUBE. Nanosats Database. Available online: https://www.nanosats.eu/sat/radcube (accessed on 4 April 2022).
125. ExoCube 1, 2 (CP 10, 12). Gunter's Space Page. Available online: https://space.skyrocket.de/doc_sdat/exocube.htm (accessed on 5 April 2022).
126. CM1 (Cesium Mission 1). eoPortal Dicrectory. Available online: https://directory.eoportal.org/web/eoportal/satellite-missions/c-missions/cesium-mission-1 (accessed on 5 April 2022).
127. CAS-9. The International Amateur Radio Union. Available online: http://www.amsatuk.me.uk/iaru/formal_detail.php?serialnum=804 (accessed on 5 April 2022).
128. PIXL-1 CubeSat Mission. eoPortal Dicrectory. Available online: https://eoportal.org/web/eoportal/satellite-missions/content/-/article/pixl-1 (accessed on 5 April 2022).
129. W-Cube Transmits the First 75 GHz Signal from Space. KUVA SPACE. Available online: https://kuvaspace.com/2021/09/01/w-cube-transmits-the-first-75-ghz-signal-from-space/ (accessed on 5 April 2022).
130. Fleet Space's Centauri 5 Satellite Successfully Launched from the SpaceX Transporter-5 Rideshare Mission. Stanews. Available online: https://news.satnews.com/2022/05/25/fleet-spaces-centauri-5-satellite-successfully-launched-from-the-spacex-transporter-5-rideshare-mission/ (accessed on 6 April 2022).
131. ELO Alpha. Gunter's Space Page. Available online: https://space.skyrocket.de/doc_sdat/elo-alpha.htm (accessed on 6 April 2022).

132. Loren, C.; Chi-Kuang, C.; Cheng-Ling, K.; Jann-Yenq, L.; Duann, Y.; Chandran, A.; Tzu-Wei, F.; Priyadarshan, H.; Kandi, K.A.; Evonosky, W.; et al. IDEASSat: The Ionosphere Dynamics Explorer and Attitude Subsystem Satellite. In Proceedings of the 31nd Annual AIAA/USU Conference on Small Satellite, Logan, UT, USA, 5–10 August 2017.
133. Kleos Space to Launch Development Technology into Orbi—3rd July 2020. sUAS News. Available online: https://www.suasnews.com/2020/06/kleos-space-to-launch-development-technology-into-orbit-3rd-july-2020/ (accessed on 7 April 2022).
134. Two NanoAvionics-Built Smallsats Were Passengers on the SpaceX Transporter-2 Mission Launch to Successful Orbit. Satnews. Available online: https://news.satnews.com/2021/07/06/two-nanoavionics-built-smallsats-were-passengers-on-the-spacex-transporter-2-mission-launch-to-successful-orbit/ (accessed on 8 April 2022).
135. Palmroth, M.; Praks, J.; Vainio, R.; Janhunen, P.; Kilpua, E.K.J.; Afanasiev, A.; Ala-Lahti, M.; Alho, A.; Asikainen, T.; Asvestari, E.; et al. FORESAIL-1 CubeSat Mission to Measure Radiation Belt Losses and Demonstrate Deorbiting. *J. Geophys. Res. Space Phys.* **2019**, *124*, 5783–5799. [CrossRef]
136. Into the Future Trend Beyond 5G: NCKU-Developed Satellite IRIS-A Successfully Launched into Space. Available online: https://web.ncku.edu.tw/p/406-1000-235668,r3344.php?Lang=en (accessed on 8 April 2022).
137. HYPSO (HYPerspectral Smallsat for Ocean Observation. EoPortal Dicrectory. Available online: https://directory.eoportal.org/web/eoportal/satellite-missions/h/hypso (accessed on 8 April 2022).
138. Thales Alenia Space to Build Two Prototype Satellites for Constellation Venture. Spacenews. Available online: https://spacenews.com/thales-alenia-space-to-build-two-prototype-satellites-for-constellation-venture/ (accessed on 9 April 2022).
139. SatNOGS DB. Available online: https://db.satnogs.org/satellite/BZWR-6172-2785-3507-5423 (accessed on 10 April 2022).
140. Planetum-1. Spacemanic. Available online: https://www.spacemanic.com/missions/planetum-1/ (accessed on 10 April 2022).
141. SpaceBEE 1, 2, 3, 4. Gunter's Space Page. Available online: https://space.skyrocket.de/doc_sdat/spacebee.htm (accessed on 10 April 2022).
142. Tsuda, Y.; Sako, N.; Eishima, T.; Ito, T.; Arikawa, Y.; Miyamura, N.; Tanaka, A.; Nakasuka, S. University of Tokyo's CubeSat Project-Its Educational and Technological Significance. In Proceedings of the 15th Annual AIAA/USU Conference on Small Satellites, Logan, UT, USA, 13–16 August 2001.
143. Costantine, J.; Tawk, Y.; Christodoulou, C.G.; Banik, J.; Lane, S. CubeSat Deployable Antenna Using Bistable Composite Tape-Springs. *IEEE Antennas Wirel. Propag. Lett.* **2012**, *11*, 285–288. [CrossRef]
144. Murphey, T.; Jeon, S.; Biskner, A.; Sanford, G. Deployable Booms and Antennas Using Bi-Stable Tape-Springs. In Proceedings of the 24th Annual AIAA/USU Conference on Small Satellite, Logan, UT, USA, 9–12 August 2010.
145. Murphey, T.W.; Sanford, G.E.; Jeon, S. Deployable Space Boom Using Bi-Stable Tape Spring Mechanism. U.S. Patent No. 8,770,522 B1, 8 July 2014.
146. Leao, T.F.C.; Mooney-Chopin, V.; Trueman, C.W.; Gleason, S. Design and Implementation of a Diplexer and a Dual-Band VHF/UHF Antenna for Nanosatellites. *IEEE Antennas Wirel. Propag. Lett.* **2013**, *12*, 1098–1101. [CrossRef]
147. Yousuf, H.J.; Haider, M.M.; Siddique, M.K.; Amin, M. Analysis of G-shape antennas mounted on a CUBESAT. In Proceedings of the 2008 2nd International Conference on Advances in Space Technologies, Islamabad, Pakistan, 29–30 November 2008; pp. 28–32.
148. Schraml, K.; Narbudowicz, A.; Chalermwisutkul, S.; Heberling, D.; Ammann, M.J. Easy-to-deploy LC-loaded dipole and monopole antennas for cubesat. In Proceedings of the 2017 11th European Conference on Antennas and Propagation (EUCAP), Paris, France, 19–24 March 2017; pp. 2303–2306.
149. Wilke, R.; Reiffenrath, M.; Parow-Souchon, K.; Heberling, D. S-Band, UHF and VHF Communication System for Cubesats including Ground Station Software. In Proceedings of the 8th Pico- and Nanosatellite Workshop, Würzburg, Germany, 15–16 September 2015.
150. Liu, S.; Raad, R.; Theoharis, P.I.; Tubbal, F. Dual-Band Folded-End Dipole Antenna for Plastic CubeSats. *IEEE J. Miniat. Air Space Syst.* **2020**, *1*, 172–178. [CrossRef]
151. Balanis, C.A. *Antenna Theory: Analysis and Design*; John Wiley & Sons: Hoboken, NJ, USA, 2016.
152. Babuscia, A.; Corbin, B.; Knapp, M.; Jensen-Clem, R.; van de Loo, M.; Seager, S. Inflatable antenna for cubesats: Motivation for development and antenna design. *Acta Astronaut.* **2013**, *91*, 322–332. [CrossRef]
153. Tubbal, F.E.; Raad, R.; Chin, K.W. A wideband F-shaped patch antenna for S-band CubeSats communications. In Proceedings of the 2016 10th International Conference on Signal Processing and Communication Systems (ICSPCS), Surfers Paradise, QLD, Australia, 19–21 December 2016; pp. 1–4.
154. Yao, Y.; Liao, S.; Wang, J.; Xue, K.; Balfour, E.A.; Luo, Y. A New Patch Antenna Designed for CubeSat: Dual feed, L/S dual-band stacked, and circularly polarized. *IEEE Antennas Propag. Mag.* **2016**, *58*, 16–21. [CrossRef]
155. Palacios, O.F.G.; Vargas, R.E.D.; Perez, J.A.H.; Erazo, S.B.C. S-band koch snowflake fractal antenna for cubesats. In Proceedings of the 2016 IEEE ANDESCON, Arequipa, Peru, 19–21 October 2016; pp. 1–4.
156. Tubbal, F.; Raad, R.; Chin, K.W.; Butters, B. S-band Planar Antennas for a CubeSat. *Int. J. Electr. Eng. Inform.* **2015**, *7*, 559–568.
157. Abulgasem, S.; Tubbal, F.; Raad, R.; Theoharis, P.I.; Liu, S.; Khan, M.U.A. A wideband metal-only patch antenna for CubeSat. *Electronics* **2020**, *10*, 50. [CrossRef]
158. Gunaseelan, S.; Murugan, M. High gain patch antenna for CubeSat. In Proceedings of the 2016 International Conference on Wireless Communications, Signal Processing and Networking (WiSPNET), Chennai, India, 23–25 March 2016; pp. 52–54.

159. Islam, M.T.; Cho, M.; Samsuzzaman, M.; Kibria, S. Compact Antenna for Small Satellite Applications [Antenna Applications Corner]. *IEEE Antennas Propag. Mag.* **2015**, *57*, 30–36. [CrossRef]
160. Kakoyiannis, C.G.; Constantinou, P. A compact microstrip antenna with tapered peripheral slits for CubeSat RF Payloads at 436MHz: Miniaturization techniques, design & numerical results. In Proceedings of the 2008 IEEE International Workshop on Satellite and Space Communications, Toulouse, France, 1–3 October 2008; pp. 255–259.
161. Yekan, T.; Baktur, R. Conformal Integrated Solar Panel Antennas: Two effective integration methods of antennas with solar cells. *IEEE Antennas Propag. Mag.* **2017**, *59*, 69–78. [CrossRef]
162. Mahmoud, M.N. *Integrated Solar Panel Antennas for Cube Satellites*; Utah State University: Logan, UT, USA, 2010.
163. Lim, E.H.; Leung, K.W. Transparent Dielectric Resonator Antennas for Optical Applications. *IEEE Trans. Antennas Propag.* **2010**, *58*, 1054–1059.
164. Montgomery, B.K.; Podilchak, S.K.; Antar, Y.M.M. Circularly polarized meshed patch antenna for cubesats and other small sattelites. In Proceedings of the 2016 IEEE International Symposium on Antennas and Propagation (APSURSI), Fajardo, PR, USA, 26 June–1 July 2016; pp. 1547–1548.
165. Tariq, S.; Baktur, R. Circularly polarized UHF up- and downlink antennas integrated with CubeSat solar panels. In Proceedings of the 2015 IEEE International Symposium on Antennas and Propagation & USNC/URSI National Radio Science Meeting, Vancouver, BC, Canada, 19–24 July 2015; pp. 1424–1425.
166. Yekan, T.; Baktur, R. Polarization reconfigurable antenna for small satellite application. In Proceedings of the 2016 United States National Committee of URSI National Radio Science Meeting (USNC-URSI NRSM), Boulder, CO, USA, 6–9 January 2016; pp. 1–2.
167. Neveu, N.; Garcia, M.; Casana, J.; Dettloff, R.; Jackson, D.R.; Chen, J. Transparent microstrip antennas for CubeSat applications. In Proceedings of the IEEE International Conference on Wireless for Space and Extreme Environments, Baltimore, MD, USA, 7–9 November 2013; pp. 1–4.
168. Liu, X.; Liu, J.; Jackson, D.R.; Chen, J.; Fink, P.W.; Lin, G.Y. Broadband transparent circularly-polarized microstrip antennas for CubeSats. In Proceedings of the 2016 IEEE International Symposium on Antennas and Propagation (APSURSI), Fajardo, PR, USA, 26 June–1 July 2016; pp. 1545–1546.
169. Montano, R.; Neveu, N.; Palacio, S.; Martinez, E.; Jackson, D.R.; Chen, J. Development of Low-profile Antennas for CubeSats. In Proceedings of the 28th Annual AIAA/USU Conference on Small Satellites, Logan, UT, USA, 2–7 August 2014.
170. Ochoa, D.; Hummer, K.; Ciffone, M. Deployable Helical Antenna for nano-Satellites. In Proceedings of the 28th Annual AIAA/USU Conference on Small Satellites, Logan, UT, USA, 2–7 August 2014.
171. Muri, P.; Challa, O.; McNair, J. Enhancing small satellite communication through effective antenna system design. In Proceedings of the Milcom 2010 Militar Communications Conference, San Jose, CA, USA, 31 October–3 November 2010; pp. 347–352.
172. Ernest, A.J.; Tawk, Y.; Costantine, J.; Christodoulou, C.G. A Bottom Fed Deployable Conical Log Spiral Antenna Design for CubeSat. *IEEE Trans. Antennas Propag.* **2015**, *63*, 41–47. [CrossRef]
173. Costantine, J.; Tawk, Y.; Christodoulou, C.G.; Maqueda, I.; Sakovsky, M.; Pellegrino, S. A new UHF deployable antenna for cubeSats. In Proceedings of the 2015 IEEE International Symposium on Antennas and Propagation & USNC/URSI National Radio Science Meeting, Vancouver, BC, Canada, 19–24 July 2015; pp. 1426–1427.
174. Costantine, J.; Tran, D.; Shiva, M.; Tawk, Y.; Christodoulou, C.G.; Barbin, S.E. A deployable quadrifilar helix antenna for CubeSat. In Proceedings of the 2012 IEEE International Symposium on Antennas and Propagation, Chicago, IL, USA, 8–14 July 2012; pp. 1–2.
175. Vourch, C.J.; Drysdale, T.D. V-Band Bull's Eye Antenna for CubeSat Applications. *IEEE Antennas Wirel. Propag. Lett.* **2014**, *13*, 1092–1095. [CrossRef]
176. Borthakur, M.; Khan, T.; Dash, S.K.K. Circularly polarized dual-band cylindrical dielectric resonator antenna for Cubesat applications. In Proceedings of the 2017 XXXIInd General Assembly and Scientific Symposium of the International Union of Radio Science (URSI GASS), Montreal, QC, Canada, 19–26 August 2017; pp. 1–4.
177. Dicandia, F.A.; Genovesi, S. A compact CubeSat antenna with beamsteering capability and polarization agility: Characteristic modes theory for breakthrough antenna design. *IEEE Antennas Propag. Mag.* **2020**, *62*, 82–93. [CrossRef]
178. González-Ovejero, D.; Chahat, N.; Sauleau, R.; Chattopadhyay, G.; Maci, S.; Ettorre, M. Additive Manufactured Metal-Only Modulated Metasurface Antennas. *IEEE Trans. Antennas Propag.* **2018**, *66*, 6106–6114. [CrossRef]
179. Warren, P.A.; Steinbeck, J.W.; Minelli, R.J.; Muller, C. Large, deployable S-band antenna for 6U CubeSat. In Proceedings of the 29th Annual AIAA/USU Conference on Small Satellites, Logan, UT, USA, 8–13 August 2015.
180. Balanis, C.A. *Antenna Theory: Analysis and Design*; Wiley-Interscience: Hopoken, NJ, USA, 2005.
181. Romanofsky, R. Array Phase Shifters: Theory and Technology. In *Antenna Engineering Handbook*; NASA, McGraw Hil: New York, NY, USA, 2018.
182. Mailloux, R.J. *Phased Array Antenna Array*, 2nd ed.; Artech House Boston: Norwood, MA, USA, 2005.
183. Budianu, A.; Meijerink, A.; Bentum, M.J. Swarm-to-Earth communication in OLFAR. *Acta Astronaut.* **2015**, *107*, 14–19. [CrossRef]
184. Alomar, W.; Degnan, J.; Mancewicz, S.; Sidley, M.; Cutler, J.; Gilchrist, B. An extendable solar array integrated Yagi-Uda UHF antenna for CubeSat platforms. In Proceedings of the 2011 IEEE International Symposium on Antennas and Propagation (APSURSI), Spokane, WA, USA, 3–8 July 2011; pp. 3022–3024.

185. Maciulis, L.; Buzas, V. LituanicaSAT-2: Design of the 3U in-Orbit Technology Demonstration CubeSat. *IEEE Aerosp. Electron. Syst. Mag.* **2017**, *32*, 34–45. [CrossRef]
186. Mishra, G.; Sharma, S.K.; Chieh, J.C.S.; Rowland, J. W-band circular polarized series fed single plane beamsteering array antenna with 4-bit phase shifter for cubesat applications. In Proceedings of the 2017 IEEE International Symposium on Antennas and Propagation & USNC/URSI National Radio Science Meeting, San Diego, CA, USA, 9–14 July 2017; pp. 2555–2556.
187. Liu, S.; Raad, R.; Theoharis, P.I.; Tubbal, F.E. A Printed Yagi Antenna for CubeSat with Multi-Frequency Tilt Operation. *Electronics* **2020**, *9*, 986. [CrossRef]
188. Klein, J.; Hawkins, J.; Thorsen, D. Improving cubesat downlink capacity with active phased array antennas. In Proceedings of the 2014 IEEE Aerospace Conference, Big Sky, MT, USA, 1–8 March 2014; pp. 1–8.
189. Pittella, E.; Pisa, S.; Pontani, M.; Nascetti, A.; D'Atanasio, P.; Zambotti, A.; Hadi, H. Feature article: Reconfigurable S-band patch antenna system for cubesat satellites. *IEEE Aerosp. Electron. Syst. Mag.* **2016**, *31*, 6–13. [CrossRef]
190. Nascetti, A.; Pittella, E.; Teofilatto, P.; Pisa, S. High-Gain S-band Patch Antenna System for Earth-Observation CubeSat Satellites. *IEEE Antennas Wirel. Propag. Lett.* **2015**, *14*, 434–437. [CrossRef]
191. Lehmensiek, R. Design of a wideband circularly polarized 2 x 2 array with shorted annular patches at X-band on a CubeSat. In Proceedings of the 2017 International Symposium on Antennas and Propagation (ISAP), Phuket, Thailand, 30 October–2 November 2017; pp. 1–2.
192. Shrestha, R.; Anagnostou, D.E.; Horst, S.J.; Hoffman, J.P. Two antenna arrays for remote sensing applications. In Proceedings of the 2017 IEEE Aerospace Conference, Big Sky, MT, USA, 4–11 March 2017; pp. 1–9.
193. Theoharis, P.I.; Raad, R.; Tubbal, F.; Abulgasem, S. High-Gain Circular Polarized Microstrip Patch Array for X-band CubeSat Applications. In Proceedings of the 2021 15th International Conference on Signal Processing and Communication Systems (ICSPCS), Sydney, Australia, 13–15 December 2021; pp. 1–5.
194. Hodges, R.E.; Hoppe, D.J.; Radway, M.J.; Chahat, N.E. Novel deployable reflectarray antennas for CubeSat communications. In Proceedings of the 2015 IEEE MTT-S International Microwave Symposium, Phoenix, AZ, USA, 17–22 May 2015; pp. 1–4.
195. Chahat, N.; Hodges, R.E.; Sauder, J.; Thomson, M.; Peral, E.; Rahmat-Samii, Y. CubeSat Deployable Ka-Band Mesh Reflector Antenna Development for Earth Science Missions. *IEEE Trans. Antennas Propag.* **2016**, *64*, 2083–2093. [CrossRef]
196. Hodges, R.E.; Chahat, N.; Hoppe, D.J.; Vacchione, J.D. A Deployable High-Gain Antenna Bound for Mars: Developing a new folded-panel reflectarray for the first CubeSat mission to Mars. *IEEE Antennas Propag. Mag.* **2017**, *59*, 39–49. [CrossRef]
197. Pozar, D.M. Wideband reflectarrays using artificial impedance surfaces. *Electron. Lett.* **2007**, *43*, 148–149. [CrossRef]
198. Arya, M.; Sauder, J.F.; Hodges, R.; Pellegrino, S. Large-Area Deployable Reflectarray Antenna for CubeSats. In Proceedings of the AIAA Scitech 2019 Forum AIAA SciTech Forum: American Institute of Aeronautics and Astronautics, San Diego, CA, USA, 7–11 January 2019.
199. Veljovic, M.J.; Skrivervik, A.K. Circularly Polarized Transmitarray Antenna for CubeSat Intersatellite Links in *K*-Band. *IEEE Antennas Wirel. Propag. Lett.* **2020**, *19*, 1749–1753. [CrossRef]
200. Chahat, N.; Sauder, J.; Hodges, R.; Thomson, M.; Samii, Y.R.; Peral, E. Ka-band high-gain mesh deployable reflector antenna enabling the first radar in a CubeSat: RainCube. In Proceedings of the 2016 10th European Conference on Antennas and Propagation (EuCAP), Davos, Switzerland, 10–15 April 2016; pp. 1–4.
201. Costantine, J.; Tawk, Y.; Ernest, A.; Christodoulou, C.G. Deployable antennas for CubeSat and space communications. In Proceedings of the 2012 6th European Conference on Antennas and Propagation (EUCAP), Prague, Czech Republic, 26–30 March 2012; pp. 837–840.
202. Akbar, P.R.; Saito, H.; Zhang, M.; Hirokawa, J.; Ando, M. Parallel-Plate Slot Array Antenna for Deployable SAR Antenna Onboard Small Satellite. *IEEE Trans. Antennas Propag.* **2016**, *64*, 1661–1671. [CrossRef]
203. Babuscia, A.; Choi, T.; Lee, C.; Cheung, K.M. Inflatable antennas and arrays for interplanetary communication using CubeSats and smallsats. In Proceedings of the 2015 IEEE Aerospace Conference, Big Sky, MT, USA, 7–14 March 2015; pp. 1–9.
204. Iwami, R.T.; Chun, T.F.; Tonaki, W.G.; Shiroma, W.A. A power-detecting, null-scanning, retrodirective array for a CubeSat platform. In Proceedings of the 2017 IEEE MTT-S International Microwave Symposium (IMS), Honolulu, HI, USA, 4–9 June 2017; pp. 662–665.
205. Murakami, B.; Ohta, A.; Tamamoto, M.; Shiroma, G.; Miyamoto, R.; Shiroma, W. Self-Steering Antenna Arrays for Distributed Picosatellite Networks. In Proceedings of the 17th Annual AIAA/USU Conference on Small Satellite, Logan, UT, USA, 11–14 August 2003.
206. Murakami, B.T.; Roque, J.D.; Sung, S.S.; Shiroma, G.S.; Miyamoto, R.Y.; Shiroma, W.A. A quadruple subharmonic phase-conjugating array for secure picosatellite crosslinks. In Proceedings of the 2004 IEEE MTT-S International Microwave Symposium Digest (IEEE Cat. No.04CH37535), Fort Worth, TX, USA, 6–11 June 2004; Volume 3, pp. 1687–1690.
207. Mizuno, T.J.; Roque, J.D.; Murakami, B.T.; Yoneshige, L.K.; Shiroma, G.S.; Miyamoto, R.Y.; Shiroma, W.A. Antennas for distributed nanosatellite networks. In Proceedings of the IEEE/ACES International Conference on Wireless Communications and Applied Computational Electromagnetics, Honolulu, HI, USA, 3–7 April 2005; pp. 606–609.
208. Bagri, D.S.; Statman, J.I.; Gatti, M.S. Proposed array-based deep space network for NASA. *Proc. IEEE* **2007**, *95*, 1916–1922.
209. Napier, P.J.; Thompson, A.R.; Ekers, R.D. The very large array: Design and performance of a modern synthesis radio telescope. *Proc. IEEE* **1983**, *71*, 1295–1320. [CrossRef]

210. Rogstad, D.H.; Pham, T.T.; Mileant, A. *Antenna Arraying Techniques in the Deep Space Network*; Deep-Space Communications and Navigation Series; No. xiii; John Wiley: Hoboken, NJ, USA, 2003; 166p.
211. Cheung, K.M. Eigen Theory for Optimal Signal Combining: A Unified Approach. *JPL Telecommunicayions Data Acquis. Prog. Rep.* **1996**, *2*, 1–9.
212. Rogstad, D.H. *The SUMPLE Algorithm for Aligning Arrays of Receiving Radio Antennas: Coherence Achieved with Less Hardware and Lower Combining Loss*; The Interplanetary Network Progress Report; Jet Propulsion Laboratory: Pasadena, CA, USA, 2005.
213. Lee, C.H.; Kar-Ming, C.; Vilnrotter, V.A. Fast eigen-based signal combining algorithms for large antenna arrays. In Proceedings of the IEEE Aerospace Conference (Cat. No.03TH8652), Helena, MY, USA, 8–15 March 2003; Volume 2, pp. 1123–1129.
214. Bai, Y.F.; Wang, X.H.; Gao, C.J.; Huang, Q.L.; Shi, X.W. Adaptive correlation algorithm for aligning antenna arrays in deep space communication. *Electron. Lett.* **2013**, *49*, 733–734. [CrossRef]

electronics

Review

A Review on 5G Sub-6 GHz Base Station Antenna Design Challenges

Madiha Farasat [1], Dushmantha N. Thalakotuna [1,*], Zhonghao Hu [2] and Yang Yang [1]

1 School of Electrical and Data Engineering, University of Technology, Sydney 2007, Australia; Madiha.Farasat@student.uts.edu.au (M.F.); yang.yang-1@uts.edu.au (Y.Y.)
2 Wireless Business Unit, Rosenberg Technology Australia, Northmead 2152, Australia; nathan.hu@rosenbergerap.com
* Correspondence: dushmantha.thalakotuna@uts.edu.au

Abstract: Modern wireless networks such as 5G require multiband MIMO-supported Base Station Antennas. As a result, antennas have multiple ports to support a range of frequency bands leading to multiple arrays within one compact antenna enclosure. The close proximity of the arrays results in significant scattering degrading pattern performance of each band while coupling between arrays leads to degradation in return loss and port-to-port isolations. Different design techniques are adopted in the literature to overcome such challenges. This paper provides a classification of challenges in BSA design and a cohesive list of design techniques adopted in the literature to overcome such challenges.

Keywords: base station antenna challenges; multiband antennas; multibeam antennas; antenna arrays

Citation: Farasat, M.; Thalakotuna, D.N.; Hu, Z.; Yang, Y. A Review on 5G Sub-6 GHz Base Station Antenna Design Challenges. *Electronics* **2021**, *10*, 2000. https://doi.org/10.3390/electronics10162000

Academic Editor: Massimo Donelli

Received: 19 July 2021
Accepted: 12 August 2021
Published: 19 August 2021

1. Introduction

Base station Antenna (BSA) is the edge element in the air interface towards the mobile terminal in all communication systems, from the first-generation (1G) AMTS (advanced mobile telephone systems) to the fifth-generation (5G) networks. A significant amount of research and development has been done on BSAs; however, it appears largely disseminated across the literature. Thus, this communication aims to collate, categorize, and discuss the latest development and challenges associated with the BSAs.

There are two basic types of BSAs used in cellular communication systems: omnidirectional and directional (sector) antennas. These variants are used in almost all wireless technologies, from 1G to 5G. Omni-directional antennas are preferred for low-capacity and extended coverage scenarios such as in rural areas. In contrast, directional antennas are used to serve a targeted coverage area while providing a higher capacity. This targeted coverage, commonly known as a sector, is determined during mobile coverage planning. The serving sector antenna specifications are derived considering sector parameters.

The scope of this communication is to provide a comprehensive summary of recent BSA antenna designs and challenges, with particular interest placed on lower microwave bands in sub-6 GHz range. Whilst there is a significant amount of published work on BSAs, a cohesive discussion of BSA evolution with mobile technologies is not available. Hence, the antenna community will benefit from a brief discussion on how BSA technologies have evolved through mobile generations. This is discussed in Section 2 prior to discussing BSA challenges in detail in Section 3. Concluding remarks are drawn in Section 5.

2. Evolution of BSA Technologies

First-generation (1G) networks had omnidirectional cells, as the main focus was coverage, not the capacity. Hence, the BSAs, otherwise known as base transceiver station (BTS) antennas, were omnidirectional. As the number of users increased in the second-generation (2G) networks, operators started to consider ways to increase the capacity. One of the techniques used for capacity improvement is sectorization. A common sectorization technique used in 2G is to divide the previous omnidirectional cell into three sectors of 120° each. As a result, three antennas, each having a 10dB beamwidth of 120°, were used in the BTS.

Another technique used in 2G networks to enhance capacity is the use of polarization diversity. The aim was to provide two orthogonal polarizations in the antenna array. The horizontal (H) and vertical (V) polarization were initially used, but ±45°, otherwise known as slant polarization, has been widely adopted in many BTS antennas since 2G. Figure 1a,b shows vertical and slant polarized dipole array configurations used in BSAs in 2G and following generations. The number of users increased rapidly, moving from 2G to 3G due to the introduction of mobile data services in 3G. As a result, operators had to explore further techniques to increase capacity. One solution was to further subdivide the sector into narrower sectors. This was done using narrow beam antennas with half-power beamwidth (HPBW) such as 65° or even 33°. One of the disadvantages was increased antenna loading on the tower.

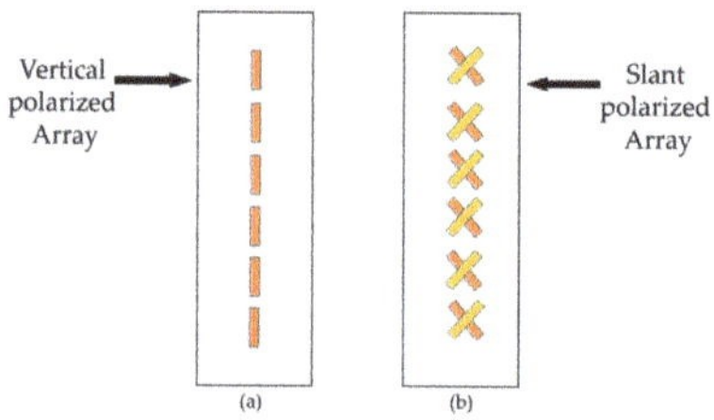

Figure 1. (**a**) Vertical polarized dipole array. (**b**) Slant polarized dipole array.

A solution to antenna loading was achieved by introducing multibeam panel antennas. Such antennas are similar in appearance to conventional sector antennas but have multiple narrow beams. As a result, increased capacity is achieved without the need for additional antennas, as shown in Figure 2. The twin beams or multibeams are achieved by introducing hybrid couplers into the feed network. These multibeam antennas have been a prevalent choice among operators for mobile networks since then.

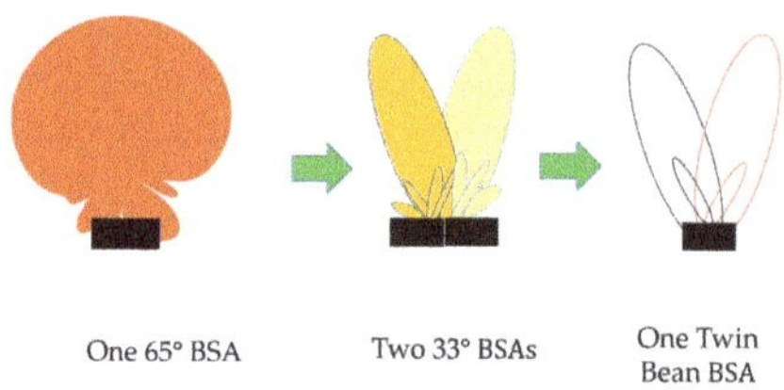

Figure 2. Patterns of a single-beam BSA, two narrow-beam BSAs, and a twin-beam BSA.

An increased number of cells/sectors requires comprehensive network planning to reduce cell edge interference and the need for operators to adjust the cell/sector coverage. This is achieved in a BSA antenna by providing beam tilting in the elevation plane. Most of the BSA antennas from 3G networks and later were equipped with a Remote Electrical Tilt (RET) feature, allowing the operator to remotely configure the elevation beam tilt in a 0–10° range to optimize the network. The elevation beam tilt in the antenna is achieved using a phase shifter that provides a phase gradient to the elements in the antenna array. Therefore, most of the BSA antennas from 3G networks and beyond had the capability of beam tilting in the elevation plane. However, the elevation beam steering feature achieved using such phase shifters is only suitable for network optimization, not for beamforming in the elevation plane. With 2G and 3G technologies, an additional spectrum was introduced beyond previously used 800 MHz and 900 MHz bands. Primarily occupied higher frequency bands worldwide were in 1710–2100 MHz bands. With 4G LTE, additional spectra up to 2.6 GHz were used worldwide. In 5G, sub-6 GHz bands have an additional spectrum in 700 MHz bands and 3.4–3.6 GHz bands. Therefore, with every generation, the existing spectrum is reused, while additional bands are introduced.

The radiating antenna elements used in BSAs, however, have limited bandwidth. Hence, a single element cannot operate in multiple bands. The most common approach in BSAs is to have two to three distinct radiating elements to cover these bands, i.e., one radiating element to cover lower bands 650–960 MHz [1] and another for 1695–2760 MHz [1–3] and a third element type to operate in 3.4–3.6 GHz [4]. As a result, each of these bands is provided by a separate antenna array. Although early 2G networks used separate antennas for each band, later generations used antennas with multiple bands in one housing. This technology is called multiband antennas. A multiband antenna has multiple arrays, each serving a different band within one enclosure. In the initial generations of multiband antennas, the arrays were physically separated as shown in Figure 3a. These multiband antennas provided space and weight savings compared to two separate antennas.

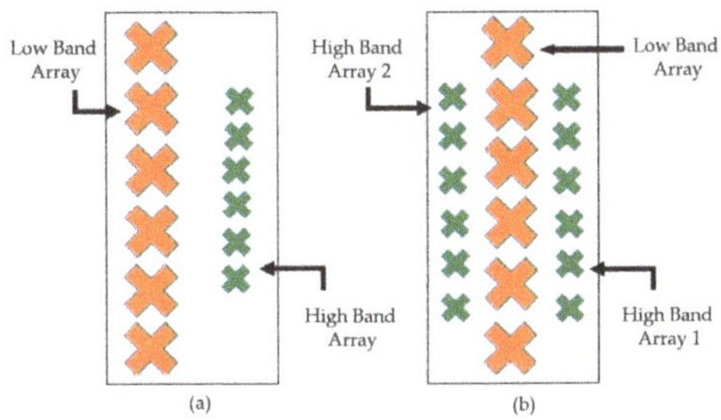

Figure 3. (**a**) Multiband antenna with physically separated arrays; (**b**) multiband antenna with interspersed arrays.

Later generations resulted in multiple interspersed arrays, as shown in Figure 3b, which further improved the space and weight savings. However, interspersed arrays present significant challenges during antenna designs, such as inter-band coupling and pattern distortions, discussed in detail in Section 3.2 of this paper.

The antenna arrays in higher frequency bands occupy less area since the vertical element spacing is much smaller in higher frequencies compared to lower frequency bands. Therefore, multiple higher-frequency band arrays can be accommodated with one low band array, as shown in Figure 4. The most common configurations were 1:2 or 1:4 array ratio between low band to high band. The antenna shown in Figure 4, has two ports for two polarizations used in low band array and two ports each for high band array, resulting in a total of 10 ports. This antenna architecture also allows multiple operators to share antenna arrays. Alternatively to such multiband arrays, antenna co-sharing was still practiced among operators using a diplexer to combine multiple bands into one BSA, even with single-band antennas. Some advanced BSAs in the 3G era comprised dual polar multiband and multibeam antennas.

Notable changes compared to previous mobile network generations in the air interface are first introduced with LTE-A. A significant increase in data rates is achieved in LTE-A due to MIMO capability. Up to eight layers of MIMO are first introduced in 3GPP Release 10. MIMO transmit data in parallel both in time and frequency in segregated streams. The BSAs are required to have spatially separated antenna arrays or polarization diversity to achieve the decorrelation in RF paths. The spatial separation has to be at least 0.7λ or more [5]. Some of the multiband antennas used in 3G networks at the time already had 2–4 high band arrays already, and the operators could use them for MIMO operation without upgrading the BSA design. High band arrays are commonly used for MIMO operation, but some BSA designs can support up to 4X MIMO for low band arrays as well [6].

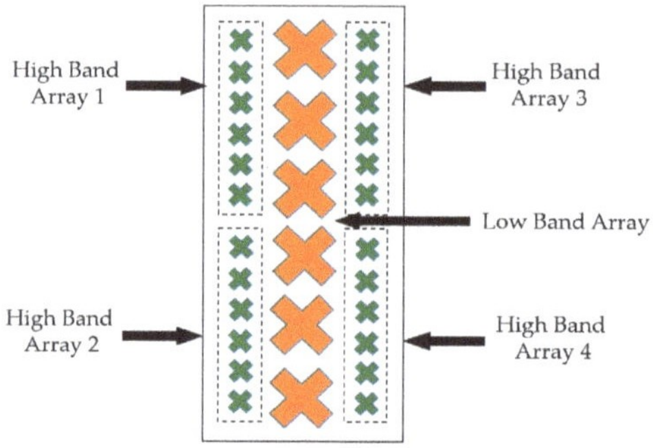

Figure 4. Low band to high band 1:4 antenna array configuration.

The 3GPP release 12 and 13 introduced active antenna systems (AAS) and massive MIMO (mMIMO) operation, which allowed real-time beamforming to provide increased capacity and reduced interference. These capabilities are adopted in 5G air interface and are commonly referred to as 5G antennas among the antenna community. The mMIMO with AASs are extensively used in mm-wave bands due to advancements in the Antenna in Package (AiP) and millimeter-wave integrated circuit (MMIC) technologies. The smaller wavelengths at mm-wave require only small antenna footprints, allowing them to be directly integrated with the transceivers. However, the peak power limitations, efficiencies at mm-wave transceivers and propagation and penetration losses at mm-wave frequencies limit the use of mm-wave BSAs to indoor use and small cells. The mMIMO in sub-6 GHz is also used by the operators for outdoor coverage due to low propagation and penetration losses at these frequencies. The challenges at mm-wave antennas have significant differences to the sub-6 GHz BSAs. Within the scope of this discussion, we present the challenges associated with the BSAs in sub-6 GHz and below, as these BSAs are the most widely used antennas by the operators to date.

The mMIMO aims to control the Signal-to-Noise Ratio (SNR) to each user by forming beams to each user unlike single-user MIMO and Multi-User MIMO in LTE and LTE-A. This requires a two-dimensional antenna array with control on the amplitude and phase to steer the beam in azimuth and elevation. The conventional beamforming antenna arrays require the element spacing to be 0.5λ to reduce the grating lobes. In contrast, the MIMO demands more spacing between the arrays to increase spatial diversity at least 0.7λ or more. In practice, a compromise is made, and a 0.65λ column spacing is used in most of the mMIMO antennas. The need to form the beam requires a phase and amplitude control at each radiating element, which can be done using either via analog beamforming architecture, digital beamforming architecture, or a hybrid beamforming architecture. The analog and digital beamforming architectures that are the most commonly used are shown in Figure 5. Analog beamforming has a significantly lower power consumption than the digital beamforming technique due to a lack of active components. However, in practice, analog beamforming cannot provide true mMIMO capability because the beams formed by analog beamforming are either multiple fixed beams or steer at a much slower rate compared to the digitally formed beams. In practice, analog beamforming is achieved by either a beamforming network such as Butler Matrix [7] or using a lens [8]. Digital beamforming is the preferred architecture in Sub-6 GHz mMIMO antennas as most of the beamformers can provide IF outputs up to 6 GHz. Hybrid beamforming architecture is mostly used in mm-wave 5G realizations in order to achieve power savings and reduce complexity in the designs. More details on the current sub-6 GHz adapted beamforming technologies and the challenges are discussed in Section 3.3.

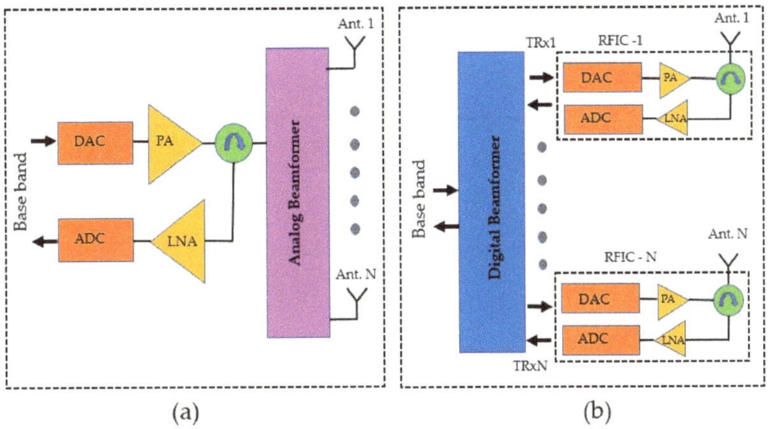

Figure 5. (**a**) Analog beamforming architecture; (**b**) digital beamforming architecture.

3. Challenges in BSA Design

The challenges associated with the BSAs are classified into three main categories, as shown in Figure 6. The "port measurements" include the design challenges in achieving wide impedance bandwidths and meeting the port-to-port isolation requirements. The challenges in radiation patterns are mostly due to the interspersed nature of the multiband arrays: the radiating elements of other bands causing scattering and undesirable effects on the radiation patterns. Therefore, the radiating elements of one band have to be made almost transparent for the other band and vice versa. The size constraints are another challenge faced by BSA designers. The real estate available on an antenna tower is very limited and costly; therefore, smaller compact antennas without any electrical performance degradation are always preferred.

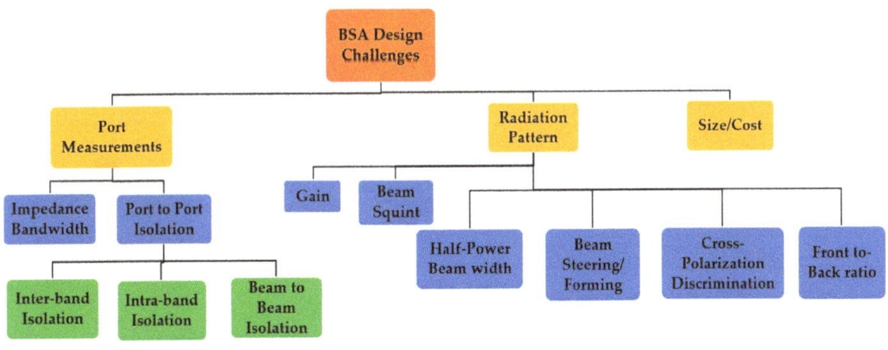

Figure 6. Classification of base station antenna challenges based on design.

Prior to proceeding with details in each challenge area, it is worth providing a general overview of the reported solutions in the literature for each of these challenges. Hence, a summary of available solutions/techniques reported in the literature for some of the key design challenges is provided in Table 1. These challenges highlighted in Figure 6 are discussed in detail in Sections 3.1–3.4.

Table 1. A summary of available solutions for design challenges in Base Station Antenna design.

Design Challenge	Solutions/Techniques	References
Achieving wide impedance bandwidth	Wideband balun design Modification to radiator shape Use of parasitic element/s to widen bandwidth	[9–12] [13–15] [16–20]
Achieving high isolation levels	Use of differential feed structure Use of decoupling network	[16,21,22] [15,23,24]
Stable HPBW in bandwidth of operation	Cavity shape reflector Convex shaped reflector	[16] [19]
Achieving high front-to-back ratio	Modifications to the radiator-Downward sloping dipoles	[25]
Minimizing Beam Squint	Enforce symmetric current distribution on the radiating element	[26–28]
Achieving high Cross polarization discrimination	Modifications to the radiator shape	[29,30]
Improving the gain	Modifications to the radiator-addition of notch metal wall	[30]
Achieving beam steering	Butler matrix Luneburg lens Digital beamforming-integrated RF transceiver	[7] [8] [31]
Achieving compact size designs/cost	Multiband compact radiating element design	[2,4,32,33]

3.1. Port Measurements: Impedance Bandwidth

One of the key challenges for the modern base station antennas is the bandwidth. While the spectrum is harmonized for certain mobile telecommunication bands, different countries tend to use different portions of the spectrum based on the license given to operators. It is desirable to design base station antennas to cover the entire allocated spectrum despite operators not having access to full bandwidth as it allows one antenna design to be used globally instead of variants for different geographic markets. However, it is difficult to design a single base radiating element that operates in all the frequency bands listed in Table 2. Hence, it is often the practice to use multiple radiating elements to cover multiple bands.

Table 2. Widely adapted frequency bands for mobile communication systems.

No.	Systems	Bands (MHz)
1	FDD 1700	1690–1710
2	DCS	1710–1880
3	PCS	1850–1990
4	UMTS	1920–2170
5	LTE	2300–2400
6	LTE+	2570–2700
7	GSM850	880–915
8	GSM900	925–960
9	Sub-6	3300–3600
10	mm-wave	24,000–28,000 (Non-contiguous)

Impedance bandwidth is an important measure in BSAs as it provides the frequency band in which a minimum return loss level is achieved. Typically this is about 14 dB for BSAs [34]. It can be challenging to achieve the return loss requirements over a wide band. This wideband matching needs to be achieved both for radiating elements as well as the feed network to meet the requirements. Generally, the techniques used to widen the impedance bandwidth can be categorized into three parts. (a) Use of a wideband Balun (b) modifications to radiator, and (c) use of parasitic elements.

3.1.1. Wideband Balun

A generic and the most adopted Balun design is shown in Figure 7. The 50 Ω transmission feed is transformed to a balanced feed through an impedance transformer as shown in Figure 7a. This is then fed to the dipole via a series LC resonator. This series LC resonator may be required depending on the impedance presented by the dipole. It is usually not required for a halfwave dipole but is used for longer dipoles. The implementation of the Balun on a microstrip feed or otherwise known as stalk is shown in Figure 7b. The 50 Ω feed is connected at the bottom of the stalk, which is underneath the reflector once mounted on the antenna (Figure 7c). The first transmission line (TL) and Open line (OL) at the input side of the transformer is implemented on the front side of the stalk. The backside balanced Short Line (SL), which is also connected to the reflector ground, acts as the ground for TL and OL microstrip lines printed on the front. The balanced Open Line (OL) then has a provision to include two inductors. The dipoles are soldered to one side of the capacitor on the top. The area of the parallel printed metals is adjusted to provide the required capacitance. It should be noted that Figure 7, only shows the feed for one polarization only. Two of such stalks can be combined together to feed a dual polar design.

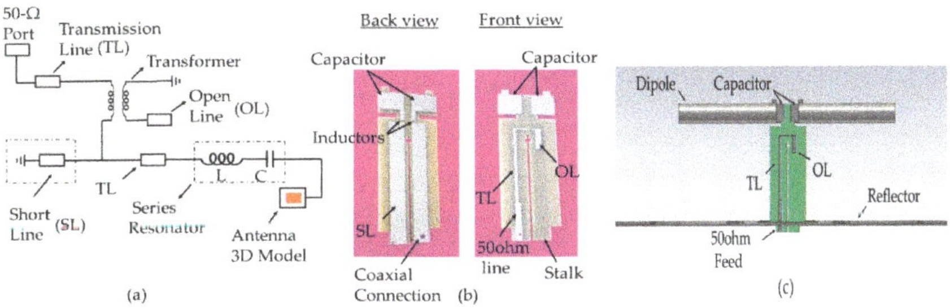

Figure 7. (**a**) Circuit representation of the matching circuit for the feed. (**b**) Microstrip implementation of the matching circuit (**c**) Side view of the radiating element.

The variations of this generic Balun are often used in many designs to achieve wider impedance bandwidths. Some of the reported works include L-probe feed with impedance bandwidth 54% [9], Y shaped feeding line with impedance bandwidth of 45% [10], T probe line [11] with impedance bandwidth of 71.17%, and shorted stub with impedance bandwidth 27.6% in [12]. In [35], microstrip to slot line balun is used to improve impedance matching with return loss better than 14 dB over the operating band 1710–2170 MHz. Slot line impedance is controlled through slot and ground width. The feed point height adjustment is a common parameter used during the Balun design to achieve a broader bandwidth [36].

3.1.2. Modifications to Radiator

Some of the commonly used techniques include various shapes of dipole [37], multi-dipole [13], and loop-shaped dipoles [10,14]; however, some of these techniques result in larger aperture size and limit the design freedom. In [15], a 55% wide impedance bandwidth is achieved from 1.65 to 2.9 GHz by using a fan-shaped etching slot and chamfering quadrants along the diagonal dipole arms. The resulting prototype increased the radiation area and extended the current path to help improve the bandwidth. A bowtie antenna modeled by a Bezier spline was proposed to have a bandwidth of 68% from 1.427–2.9 GHz [38]. The multi-dipole antenna proposed in [13] achieved wider impedance bandwidth of 60% from 1.55 to 2.87 GHz.

A multimode antenna with an embedded double loop configuration proposed in [14] has an impedance bandwidth of 51% from 1.68 to 2.83 GHz. In this design, a small loop inside an outer loop is added to generate a new resonant mode and hence widen the impedance bandwidth. In [39], a comparison of different patches with and without slots, including shorting strip, is performed to identify the structure that provides the widest impedance match. This comparison shows that the position of the coaxial feed cables, chamfer dimensions, and shape of the slotted patch contributed to 21.7% (0.82–1.02 GHz) and 49.5% (1.64–2.72 GHz) impedance bandwidth.

3.1.3. Use of Parasitic Elements

Another method is to use parasitic elements to improve the impedance bandwidth. An octagonal loop dipole in [10] achieved an impedance bandwidth of 45%. These loop dipoles show wider impedance bandwidth compared to fundamental quarter wave dipoles. When one loop dipole is excited, the other behaves like a parasitic element to improve the bandwidth. The length of the parasitic loop element is optimized in [16] to achieve an impedance bandwidth of 52% from 1.7 to 2.9 GHz.

In [17], an antenna array with a bandwidth of 70% from 1.32–2.74 GHz is designed. The antenna configuration is a U-shaped slot etched on each polarization leaf of the element. A parasitic element with four layers of circular metal disks is introduced to improve the impedance bandwidth. It is found that the number of layers of the parasitic elements directly improves the impedance matching. The operating principle follows the rule that the radiation resistance exhibited in a dipole is proportional to the square of the electrical length of the current path [18]. The addition of a parasitic element makes the reactance of impedance be tuned either capacitive or inductive to achieve the best match. In [19], a parasitic patch was placed above the folded dipole to enhance the bandwidth to 64.7% from 1.4–2.77 GHz. In [20], the antenna showed 63% impedance bandwidth within the 1.68–3.23 GHz range due to the parasitic element. In [40], the resonator-loaded dipole antenna with a U-shaped strip feed widened the bandwidth by moving two resonating modes closer to each other. The length of the resonator and distance between resonator and dipole is optimized to achieve a 37.5% impedance bandwidth from 0.67 to 0.98 GHz. Although inserting parasitic elements is a unique approach to solve the matching issue, it can adversely impact the radiation pattern performance.

3.2. *Port Measurements: Port to Port Isolation*

The increased use of multiband, multibeam, and dual-polarized base station antennas for cost and space savings leads to challenges in achieving isolation requirements. The port-to-port isolation indicates how well any two RF signals on a multiport or MIMO antenna are decoupled from each other. In general, isolation can be categorized into three main types: intra-band isolation, inter-band isolation, and beam-to-beam isolation.

Intra-band isolation is the coupling between the polarizations of the same/multiple antenna arrays within the same frequency band. This is often referred to as cross-polar isolation when measured within the same array. Figure 8a shows a two-port BSA where each port represents two orthogonal polarizations of the same array, while Figure 8b shows a four-port BSA with two antenna arrays of same band. Ports 1 and 2 feed orthogonal polarizations of array 1, while ports 3 and 4 feed the two polarizations of array 2. The isolation between any of the ports 1 to 4 represents intra-band isolation. Generally, the intra-band isolations are required to be greater than 25 dB or 30 dB [34], depending on the frequency range of operation and operator requirements.

Inter-band isolation denotes the coupling between the arrays of different bands. In Figure 9, a dual-band antenna example is given, in which ports 1 and 2 represent feeds for Band 1, and ports 3 and 4 represent feeds for Band 2. The $|S13|$, $|S14|$, $|S23|$, and $|S24|$ represent the inter-band isolation between ports of high and low band elements. In multiband antennas, the isolation is measured in all frequency bands of operations, and typically these levels need to be above 30 dB or higher.

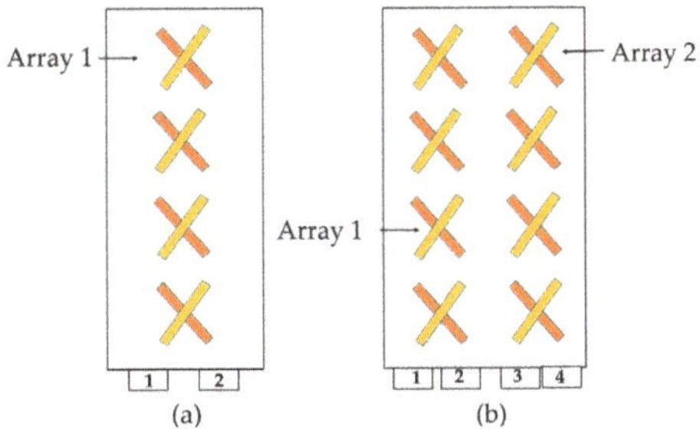

Figure 8. (**a**) Two-port single band antenna. (**b**) Four-port single band antenna.

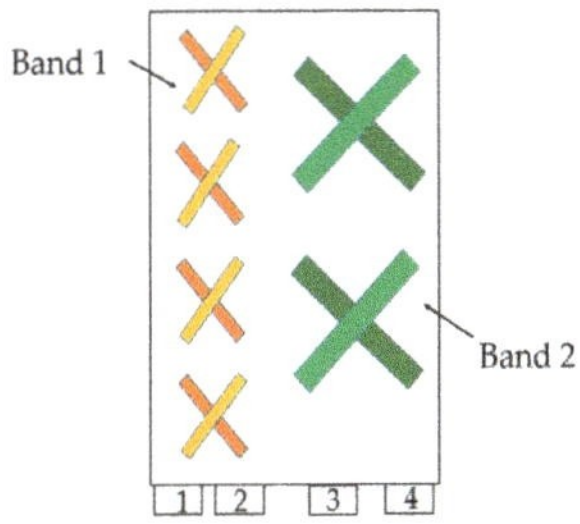

Figure 9. A dual-band antenna.

The higher the number of arrays in the antenna, the higher the port-to-port isolation combinations. Table 3 shows all possible port-to-port isolation combinations for a single-beam multiband antenna shown in Figure 4. The row entries in Table 3 are depicted as transmitting arrays, while the column entries are depicted as receiving arrays to make it simpler to distinguish the coupled band of interest. For example, $HB1_{tx}$-LB_{rx} indicate the HB array 1 coupling to LB array at High Band frequencies, while LB_{tx}-$HB1_{rx}$ indicate the LB array to HB array 1 coupling at Low Band frequencies. All diagonal entries in the table represent cross-polar isolations within the same band. All non-diagonal entries represent coupling between different arrays of the same band or a different band. It should also be noted that all non-diagonal entries can be further divided into four entries since the arrays are dual polarized. For example, LB_{tx}-$HB1_{rx}$ comprises LB_{+45}-HB_{+45}, LB_{+45}-HB_{-45}, LB_{-45}-HB_{+45}, and LB_{-45}-HB_{-45}, where subscripts +45 and −45 represent two slant polarizations. Usually, the isolation between orthogonal polarizations is always better compared to the same polarization.

Beam-to-beam isolation is a special case applied for multibeam antennas representing the coupling between each beam in the array. Typically, the beam-to-beam isolation needs to be 20 dB or higher.

The techniques used to improve isolation performance include differential feeding and decoupling network/structures. Differential feed structure, in theory, can provide infinite isolation in a dual-polarized symmetrical feed since an excitation of one port does not induce common mode or different voltage currents in the other port. In practice, intra-band isolation greater than 36 dB [21] and 26 dB [16] were achieved through a differential fed scheme. In [22], folded feeding lines based on differential feeding techniques are used to achieve intra-band isolation larger than 43 dB.

Decoupling networks are also another technique used to improve port-to-port isolations. Coupled resonator-based decoupling network is deployed in [41] to improve inter-band isolation from 8 to 10 dB. The currents induced by the coupled resonator helps to cancel the strong coupling between antennas operated in two frequency bands. Decoupling and matching network techniques implemented in [42] achieve inter-band isolation level over 10 dB in the 1.71 GHz t–1.76 GHz and 2.27–2.32 GHz bands.

Table 3. Port-to-port isolation combinations for the multiband antenna depicted in Figure 4.

Transmitter/Receiver	LB Array$_{rx}$	HB Array 1$_{rx}$	HB Array 2$_{rx}$	HB Array 3$_{rx}$	HB Array 4$_{rx}$
LB Array$_{tx}$	Cross polar isolation (LB)	LB$_{tx}$-HB1$_{rx}$	LB$_{tx}$-HB2$_{rx}$	LB$_{tx}$-HB3$_{rx}$	LB$_{tx}$-HB4$_{rx}$
HB Array 1$_{tx}$	HB1$_{tx}$-LB$_{rx}$	Cross polar Isolation (HB)	HB1$_{tx}$-HB2$_{rx}$	HB1$_{tx}$-HB3$_{rx}$	HB1$_{tx}$-HB4$_{rx}$
HB Array 2$_{tx}$	HB2$_{tx}$-LB$_{rx}$	HB2$_{tx}$-HB1$_{rx}$	Cross polar isolation (HB)	HB2$_{tx}$- HB3$_{rx}$	HB2$_{tx}$-HB4$_{rx}$
HB Array 3$_{tx}$	HB3$_{tx}$-LB$_{rx}$	HB3$_{tx}$-HB1$_{rx}$	HB3$_{tx}$-HB2$_{rx}$	Cross polar isolation (HB)	HB3$_{tx}$-HB4$_{rx}$
HB Array 4$_{tx}$	HB4$_{tx}$-LB$_{rx}$	HB4$_{tx}$-HB1$_{rx}$	HB4$_{tx}$-HB2$_{rx}$	HB4$_{tx}$-HB3$_{rx}$	Cross polar isolation (HB)

As discussed in Section 3.4, the compactness of BSA design embedded scheme can cause low isolation. Different techniques were tried to improve the isolation in such schemes. In [1], a ring-shaped baffle is placed between the lower band and high band elements to decouple the two bands and achieve port-to-port isolation of 23 dB in the low band (0.77 to 0.98 GHz) and 17.5 dB in the high band (1.65 to 2.9 GHz). In [43], a similar configuration, in which a high band element nested inside a lower band, is used. Four arc-shaped baffle plates are used in this work, which results in port isolation greater than 27.3 dB for the lower band 704–960 MHz and 28.3 dB for the high band 1710–2690 MHz. The $\pm45^{\circ}$ dual-polarized antenna with the dielectric cavity achieves crosspolar isolation of 40 dB in [44] using carefully positioned symmetrical two shorting pins in the coax feedline. An orthogonal coupled sectorial loop-antennas with a cavity is used to achieve >30 dB intra-band isolation over a 1710–2170 MHz band in [35]. Decoupling networks such as bandgap structure [45], band stop decoupling unit [23], and filtering antenna elements [24] are also some commonly used techniques.

A filtering technique with different configurations is employed to get better isolation. C-shaped filtering stubs as shown in Figure 10a are introduced in [15] for achieving port-to-port isolation >25. The purpose of the filtering stub is to control the current flow across the feeding line of the relevant port to act as a band stop for the specific band to achieve high isolation. Filtering response through parasitic elements as shown in Figure 10b is realized in [20], resulting in improved isolation of greater than 32 dB. The basic filtering structures include metasurface structure [46], slot [47], shorting vias [48], and parasitic elements [49], and defected ground structure [50] integrated with the radiator to achieve the filtering response for base station application. Although an extra filtering structure increases insertion loss, the filtering antenna as array elements without extra decoupling structure has been proposed in [24]. The radiating element realized the filtering response by adding the shorting pins and E-slot to achieve inter-band isolation of 35 dB. Balun design is modified to provide required filtering removing the need to have additional filters in [51]. Intra-band isolation >31 dB is achieved in [52] by carefully designing the dipoleto mutual coupling to complement the cross-polar isolation.

A configuration named lower-band–ground–upper-band (L–G–U), where the high-band antennas are located above the lower-band antenna separated by a low pass surface, is presented in [53]. It demonstrates inter-band isolation better than 30 dB in both working bands. In [54], a frequency selective surface is introduced between the high band and low band elements in the L–G–U configuration to achieve inter-band isolation >25 dB. The frequency-selective surface is optimized to serve as top capacitive loading for low-band 0.69–0.96 GHz and act as a reflector for high-band 3.5–4.9 GHz. In [55], beam-to-beam isolation > 32 dB was achieved through Luneburg lens. This Luneburg lens antenna operates from 1710–2690 MHz, made up of a special periodic structure to become suitable for base station application. In the literature, a Luneburg lens with different materials was designed such as in [56], which is configured with a metamaterial layer to make it compact at lower frequencies (0.8–6 GHz).

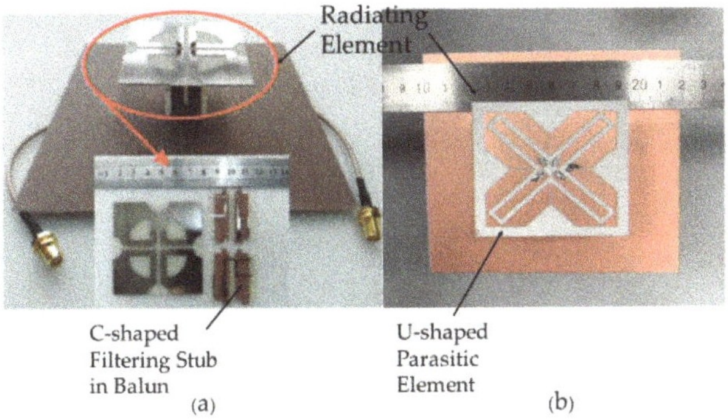

Figure 10. (**a**) Proposed antenna configuration [15]. (**b**) Parasitic element configuration in proposed antenna [20].

3.3. Radiation Patterns

The far-field radiation patterns are a very important, if not the most important, factor for an operator. Some critical parameters in patterns include 3 dB beamwidth, 10 dB beamwidth, beam squint, front-to-back (F/B) ratio, sidelobe levels, and cross-polarization discrimination (XPD). Some of these parameters are marked in Figure 11. The specifications provided by the operators to antenna designers can slightly change but mostly follow the industry-standard requirements listed in [34].

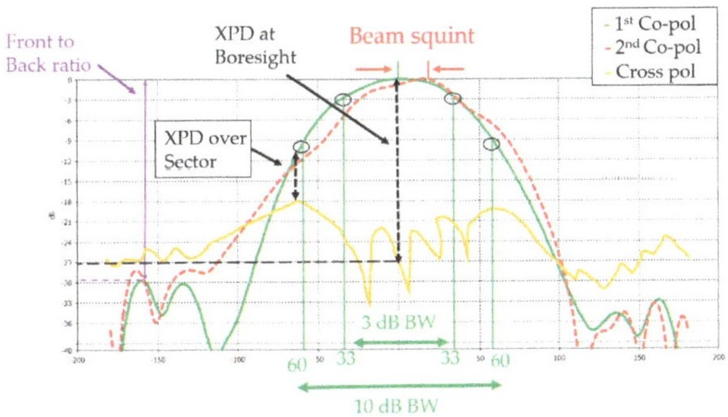

Figure 11. Illustration of some critical parameters for antenna radiation pattern.

The beamwidths 3 dB and 10 dB are important in network planning to mark sector footprints. Usually, single beam/sector antennas must have an HPBW requirement of 65° for three sector cell sites and 45° or 33° for six sector sites. The 10 dB beamwidth refers to the angular beamwidth at 10 dB below the peak values. It is desirable to have a 10 dB beamwidth of 120° in a three-sector site which implies that signal strength at the sector edge is 10 dB below the peak in boresight.

The radiation element most often needs modifications in its design to achieve the desired azimuth beamwidth. The important factors that affect the azimuth beamwidth are the length of the radiator and the distance between the radiator and reflector. These parameters are optimized to attain 3 dB beamwidth within the range of 65.7° ± 3.2° [19]. The use of dipole-type radiators is common in radiating element design. The umbrella-shaped dipoles are configured in [57] to achieve an HPBW (half-power beamwidth) of 63° ± 5° in the H-plane. Another technique to modify the azimuth beamwidth is the use of cavity-shaped reflector over a planer reflector [58]. Further, rectangular cavity-shaped reflectors in [16] and convex-shaped reflectors in [19] are used to achieve 3 dB beamwidth around 65° ± 5°. The length and height of the reflector primarily affect the 3 dB beamwidth in the H-plane. This effect is realized in [10] to achieve a 3 dB beamwidth 68° ± 2° at H-plane and V-plane. It is found that increasing the length of the box-shaped reflector narrows the beamwidth in lower frequencies and widens the beamwidth in high frequencies in the band 1.7 to 2.7 GHz.

Another critical parameter in radiation patterns is the front-to-back (F/B) ratio. A higher F/B ratio is desirable to minimize backward radiation that can contribute to co-channel interference. Typical values need to be 25 dB or more in most of the base station antenna designs. Having a large reflector helps improve the F/B ratio. However, most of the BSAs have size constraints in terms of antenna width. Different topologies have been adopted in the literature to achieve the desired F/B ratio. Radiating elements proposed in [25] have downward-sloping dipoles, which improves the F/B ratio. An F/B ratio better than 30 dB and sidelobe levels better than 25 dB were achieved. Placing a large reflector behind BSA can improve F/B; however, it can introduce other intricacies such as increased wind resistance, large antenna size, and antenna loading. An electromagnetic scattering structure applied on the Radome is used in [59] to reduce back lobe radiation.

Beam Squint is another critical performance parameter for antenna radiation patterns. The beam squint refers to the deviation of the main beam direction from its boresight in the azimuth plane. The beam squint is measured in ±degrees from the boresight direction. The squint can be measured as a 3 dB beam squint or a 10 dB beam squint, and the aim is to keep the squint as low as possible throughout the entire operating band for all elevation tilts. Generally, the squint gets worse with higher elevation tilts. In [26], beam squint up to 12° is reported at 10° elevation tilt. Different techniques are followed to minimize the beam squint. In [27], a combination of microstrip and a stripline PCB's are introduced in the radiation element to minimize the beam squint below 5° for downtilt measured at 7° and 0°. Another technique is to enforce symmetric current distribution on radiating elements to minimize squint. This is achieved in [38] by using the Pawsey stub balun to feed radiating elements and to reduce leaky current distortions to have HPBW within 54–76°. In [28], the beam squint < 4° with a maximum 10° down tilt is achieved through octagon-shape-folded dipoles as shown in Figure 12a.

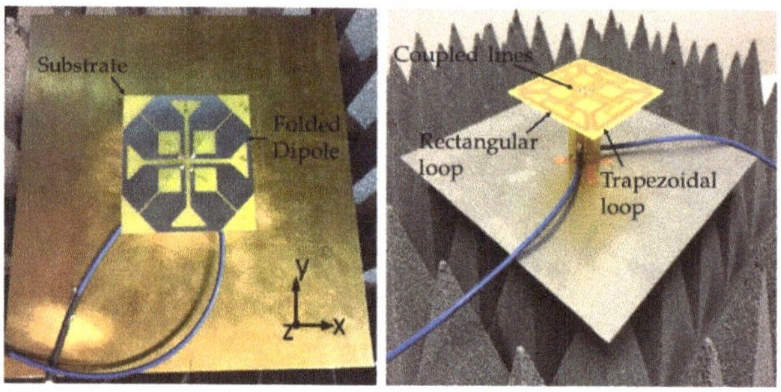

(a) (b)

Figure 12. (**a**) Fabricated prototype of folded dipole [28]. (**b**) Fabricated Prototype of magnetoelectric loop dipole [29].

Cross-polar discrimination (XPD) is another important parameter when looking at radiation pattern performance. In general, the XPD above 10 dB in the sector is recommended for base station application. The magnetoelectric layered loop dipole configuration shown in Figure 12b, is used in [29] to achieve XPD over 20 dB in the boresight direction. Antenna gain is also a vital characteristic in BSA design. Operators prefer a positive gain slope across the band as the higher gain at higher frequencies compensates for the additional free space losses. In the azimuth, the gain in sector edge directions is lower compared to boresight due to gain roll-off. A lower gain roll-off in azimuth patterns is achieved in [30] by introducing a notch metal wall to radiating element, which enhances the gain by 2 dB at sector edge $\pm$ 60° angle.

The aforementioned radiation pattern-related parameters are applicable for single-band and multiband antennas. However, in multiband antennas, radiation patterns can be severely impacted due to cross-band scattering when interleaved or embedded arrays. In a dual-band interspaced array, the high-frequency element patterns are impacted due to scattering from the currents induced in low-frequency elements and vice versa. The impact can be on multiple parameters such as beamwidth, squint, and XPD. Overcoming such impacts can be very difficult. It is often attempted to ensure that each radiating element is transparent to the other in their operating frequency bands, which is a challenging task. In [60], the high-frequency band (HB) pattern distortions caused by the lower frequency band (LB) radiating elements are minimized by introducing chokes into the LB element. These chokes are quarter-wavelength open circuit segments at high band frequencies, and this minimizes scattering. In [61], the printed dipole is segmented into smaller segments that are not resonant in the higher frequency band region, and each segment is connected to the other by inductive thin lines. This makes the lower band element transparent to the higher band radiating element. In [62], a cloaked antenna system is realized to minimize the scattering of closely located antennas. A dual-polarized mantle cover to cloak the dipole antenna is used in this work, and radiation performance is almost unaffected.

The aforementioned challenges for patterns are fundamental and equally applicable for single beam, multibeam, or steerable beams in BSAs. When it comes to 5G, there are some additional challenges and complexities associated with the mMIMO and beamforming. As mentioned earlier in Section 2, there are two main beamforming technologies, namely analog and digital, used in BSAs for sub-6 GHz. Although the analog beamforming is not true mMIMO, it is still used in some of the 5G base station antennas to form multiple beams. The true mMIMO in sub-6 GHz is achieved through digital beamforming in 5G BSAs.

The most popular techniques used for analog beamforming are based on either Butler-matrix circuits [63] or Luneburg lenses [64]. Compared to lenses, the Butler-matrix circuit implementations are compact, low-cost, and planar. The designers can incorporate the Butler matrix implementations with the feed distribution network. Therefore, it does not necessarily increase the antenna height. However, there are several challenges associated with the Butler matrix implementation such as dual-band operation, isolation between beams, side-lobe suppression, and wide operating bandwidth [65]. The branch line couplers used in the Butler matrix have inherent bandwidth limitations, and as a result, they cannot be designed to have multi-band operation. The approach is to have distinct Butler matrices for each band [66] to overcome this limitation. However, this comes with inherent crossband coupling challenges, which were discussed in Section 3.2. The narrowband challenges were addressed with wideband quadrature couplers and fixed-phase shifters in a Butler matrix [67]. Although there has been some reported literature on high-beam-to-beam isolation [68], it is still an ongoing challenge. One solution to minimize the sidelobes as well as grating lobes is achieved by changing the antenna element arrangement in [63].

In contrast to butler matrices, a careful design of a lens-based beamformer can provide stable radiation patterns, with low sidelobe levels in a wide band [8]. However, the size of the lens in front of the antenna increases the antenna height as well as weight. Therefore, low profile, low cost, and lightweight lenses remain a potential research topic for analog beamforming in BSAs. The other analog beamforming techniques reported in the literature include the use of metasurfaces [69], parabolic cylindrical reflectors [70], and reconfigurable parasitic radiators [71].

The digital beamforming architecture, which provides true mMIMO capability, has certain challenges as well. One of the key challenges is the design complexity. An example 4×4 Tx/Rx dual polar mMIMO configuration for a BSA is shown in Figure 13. Each radiating element requires an RFIC to provide amplification and filtering, which are then connected to the digital beamformer. This dual polar Tx/Rx array requires in total 32 RFICs and 8 baseband beamformers. All these components need to be placed closer to radiating element, making the integration of the antenna element and RF circuitry very complex. Unlike mm-wave designs, the sub-6 GHz designs cannot be realized as an integrated AiP in MMIC technology due to the large size of the antenna element. In addition, the large number of RFICs and digital processing results in high power consumption. Even with the state-of-the-art efficient power amplifiers available in the sub-6 GHz bands, the thermal dissipation from processors and RFICs remains a significant challenge that needs to be addressed in the design. Thermal vias and thermal pads are necessary at the back of the RF and digital electronics to dissipate the heat in these antennas [31]. The increased power consumption results in less value for money for the operators despite the capacity improvement. Some antenna designs have addressed this by limiting the real-time beamforming on the horizontal axis only [64]. The elevation beam tilting in this case is not done electronically and initially set by the remote electrical tilting mechanism using the legacy phase shifters. Then, the Azimuth beamforming is done using the phase and amplitude control among array columns.

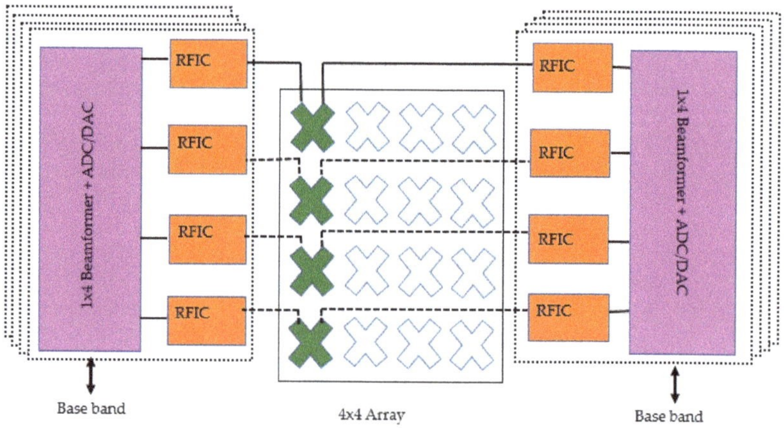

Figure 13. Components of a 4 × 4 Tx/Rx dual polar mMIMO configuration.

Another challenge with the mMIMO antenna design is the calibration. The phase needs to be calibrated with high accuracy along the entire RF transceiver chain in order to ensure the expected beamforming gains. The amplifiers can have varying phases based on their operating conditions such as bias points, temperature, and frequency of operation, and these variations can lead to deviations from the expected phase distribution at the elements. As a result, the patterns may not form nulls, where it is expected that leads to increased interference with adjacent users. The antenna testing is another challenge for mMIMO as the access to RF inputs of the antenna is difficult with the other components in the RF transceiver chain. Therefore, instead of doing antenna pattern testing with an RF input, Over-the-Air (OTA) testing is required with the baseband IQ data as the input/output.

3.4. Size and Cost

Operators are keen to maximize the performance per unit area in the tower space. Therefore, they require compact antennas with better overall performance at a low cost. This requires the designs to be compact and low in cost. The miniaturizations are often achieved by having interspersed multiband arrays [4]. A comparison of different arrangement schemes for high-frequency elements and low-frequency elements is performed [2]. As shown in Figure 14, side-by-side schemes, up-and-down coaxial schemes, and embedded schemes were considered. Although the up-and-down scheme is simpler, cable losses increase due to the increased length of the main feed line. The embedded scheme has the advantage of compactness as two antennas are located within a single band antenna volume. However, embedded schemes exhibit challenges of isolation and pattern distortions. A single radiating element for both the low and high bands is used in[72,73] to reduce the number of required radiating elements and footprint. However, the required vertical element spacing for optimum elevation patterns is hard to achieve in this approach. Hence, the grating lobe levels are high. The dual broadband planer BSA configuration is followed in [32], where high band elements are nested inside lower band elements to achieve compactness.

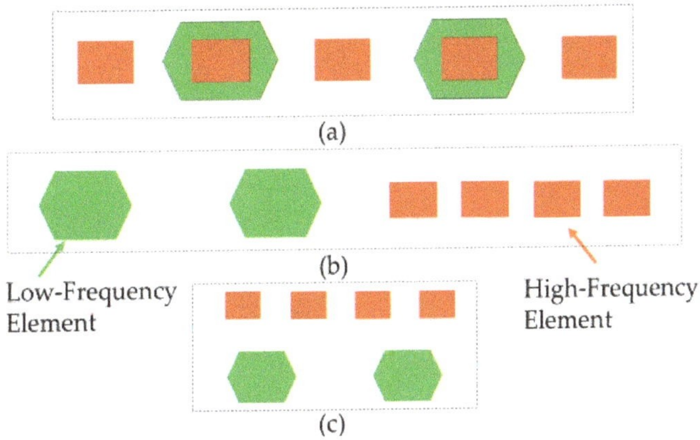

Figure 14. (**a**) Multiband embedded scheme; (**b**) multi band side-by-side scheme; (**c**) multiband -up-and-down coaxial scheme [2].

The lower cost in the antenna production is achieved by using low-cost, simple, printed antennas. The printed circuit board technologies reduces antenna assembly times. Another low-cost choice for enabling 5G BSA is 3D printing, which facilitates complex designs. Three-dimensional printing is an effective manufacturing method for designing MIMO antenna prototypes to reduce the cost. In [33], ±45 dual-polarized antenna is fabricated using 3D printing technology while achieving a wide bandwidth. Although printed antennas have a low-cost advantage, they can exhibit higher dielectric losses compared to die-cast radiating elements. Die-casting can be cost-effective if the entire element is made out from a cast reducing the assembly time. In addition to the cost incurred on radiating elements, the other RF components, such as phase shifters, may incur higher costs. A wiper phase shifter [74] can be a cost-effective solution as it is very compact but may have other disadvantages, such as limited control for null-filling in down tilts and the use of a large number of cables. Other types of phase shifters such as [75,76] can overcome some of these limitations while reducing fabrication costs.

4. Future Trends in 5G Base Station Antennas

The foundation of any BSA antenna is the radiating elements in the arrays. Therefore, it is important to select wideband radiating elements with stable radiation patterns. Table 4 contains some of the recent state-of-the-art radiating elements that show wideband performance. Some of the design concepts used in the design of these elements can be used as the basis to develop more improved radiating elements for current and future 5G antennas.

Table 4. Comparison of a selected set of wideband radiating elements used in BSAs.

Reference	Antenna Element Type	Frequency Band (GHz)	Size *	Impedance Bandwidth	Isolation (dB)	HPBW	XPD (dB)	Gain (dBi)
[20]	Cross-dipole antenna with U-shaped parasitic element	1.68–3.23	$0.38\lambda \times 0.38\ \lambda$	63%	>32	$65 \pm 5°$	<16	8.5
[28]	Octagon shape folded dipoles	1.69–2.71	$0.50\lambda \times 0.50\lambda$	46.4%	>28	$66.5° \pm 5.5°$	25	9.8
[29]	Magneto-electric loop antenna	1.7–2.7	$0.43\lambda \times 0.43\lambda$	45.5%	NG #	$66.5° \pm 3.5°$	>20	NG #
[3]	Folded Dipole with coplanar stripline	1.7–2.25	$0.52\lambda \times 0.52\lambda$	27.8%	>25	$66.3° \pm 2.9°$	>16	>8
[16]	Loop radiator, cross shaped feeding with loop parasitic element	1.7–2.9	$0.56\lambda \times 0.56\lambda$	52%	>26	$66.2° \pm 3.7°$	NG #	8.5
[10]	Octagonal shaped loop radiator with Y-shaped feeding line	1.7–2.7	$0.39\lambda \times 0.39\lambda$	45%	>25	$68° \pm 2°$	NG #	8.2
[39]	Slotted patch with shorting strips	0.82–0.99	$0.42\lambda \times 0.42\lambda$	18.7%	>30	$65° \pm 10°$	NG #	9.9
[38]	Spline Edged bowtie radiator	1.42–2.9	$0.50\lambda \times 0.50\lambda$	68%	>20	$65° \pm 11°$	20	8
[37]	crossed stepped-width loop dipoles	1.68–2.94	$0.41\lambda \times 0.41\lambda$	54.5%	>28.5	$66.2° \pm 3.7°$	NG #	8.5
[52]	Square-loop shape dipole	1.63–2.95	$0.36\lambda \times 0.36\lambda$	58%	>31	$58.1° \pm 12°$	<27	8.8
[1]	Folded dipole	0.79–1	$0.44\lambda \times 0.44\lambda$	23.5%	>30	69°	NG #	7.7
[15]	Orthogonal dipoles with fan-shaped slots	2.27–2.53	$0.49\lambda \times 0.49\lambda$	52.6%	>25.4	60°	NG #	7.6
[17]	Leaf clover antenna with round metal disks and U-shape slot	1.39–2.8	$0.42\lambda \times 0.42\lambda$	67%	30	$65° \pm 5°$	NG #	9

* size in terms of wavelength(λ) at midband frequency, # Not Given.

The massive MIMO is one of the most popular topics among the 5G BSA community. Currently, the massive MIMO used in the 5G network is typically standalone, i.e., contains only one frequency band commonly referred to as mid-band (2.6 GHz–4.2 GHz) with 32 or 64 ports (eight dual-polarized columns with two or four rows). No other bands are integrated into those active antennas. The massive MIMO antennas require the radios to be used for each of the ports. Despite the high throughput advantage of these massive MIMO antennas, the extreme power consumption puts a significant loading on the electrical network. The operators cannot remove the existing 4G BSAs to reduce the power consumption since the current 5G massive MIMO antennas cannot still serve all the frequency bands. Therefore, the industry is leaning towards integrating legacy multiband 4G antennas with 5G massive MIMO antennas. One potential approach is to reduce the number of ports in massive MIMO BSAs from 32 to 16 and provide some space to integrate a legacy antenna. The aim is to strike a balance with operational cost and performance. The combination of legacy 4G and 5G massive MIMO still poses the challenges highlighted in the paper. For example, the low band (under 1 GHz) needs to be decoupled in an even wider band from 1.4 GHzto 4.2 GHz, which is very difficult. Moreover, the antenna profile is required to be slimmer for less wind loading, making the technical design even harder. Possibilities remain open to design novel wideband frequency selective surfaces to overcome some of these challenges to reduce coupling and improve patterns since more bands are now required to be integrated under one radome.

5. Conclusions

Base Station Antennas have evolved from simpler Omni antennas to multiband multi-beam sectorized antennas over the last three decades. The antenna complexity has increased with each wireless generation, leading to more design challenges to antenna engineers. From LTE-A leading to 5G, sub-6 GHz BSA designs have multiple antenna arrays. This is due to several reasons, including the need to support MIMO capability and the need to maximize value for money by covering multiple bands. To achieve size savings, the multiple bands are interspersed in the design. The presence of multiple radiators introduces scattering, degrading the pattern performance of each band. Coupling between the close proximity elements leads to challenges in meeting impedance and isolation specifications. As a result, multiple techniques are applied by antenna designers such as cloaked dipoles, parasitic elements, FSSs, and filters to mitigate these challenges.

Author Contributions: All authors conceived and proposed the idea; all authors contributed to the structure of the paper. Wrote the paper under the supervision of D.N.T. All authors have read and agreed to the published version of the manuscript.

Funding: This research received no external funding.

Data Availability Statement: The data presented in this article can be obtained from the relevant references provided in this article.

Conflicts of Interest: The authors declare no conflict of interest.

References

1. Huang, H.; Liu, Y.; Gong, S. A Novel Dual-Broadband and Dual-Polarized Antenna for 2G/3G/LTE Base Stations. *IEEE Trans. Antennas Propag.* **2016**, *64*, 4113–4118. [CrossRef]
2. He, Y.; Pan, Z.; Cheng, X.; He, Y.; Qiao, J.; Tentzeris, M.M. A Novel Dual-Band, Dual-Polarized, Miniaturized and Low-Profile Base Station Antenna. *IEEE Trans. Antennas Propag.* **2015**, *63*, 5399–5408. [CrossRef]
3. Wen, D.; Zheng, D.; Chu, Q. A Dual-polarized Planar Antenna Using Four Folded Dipoles and Its Array for Base Stations. *IEEE Trans. Antennas Propag.* **2016**, *64*, 5536–5542. [CrossRef]
4. Liu, Y.; Wang, S.; Li, N.; Wang, J.; Zhao, J. A Compact Dual-Band Dual-Polarized Antenna With Filtering Structures for Sub-6 GHz Base Station Applications. *IEEE Antennas Wirel. Propag. Lett.* **2018**, *17*, 1764–1768. [CrossRef]
5. Mandhyan, A. *4G and 5G Capacity Solutions Comparative Study*; Commscope: Hickory, NC, USA, 2019.
6. Kelly, I.; Zimmerman, M.; Butler, R.; Zheng, Y. *Base Station Antenna Selection for LTE Networks*; Commscope: Hickory, NC, USA, 2017.

7. Zhu, H.; Sun, H.; Ding, C.; Guo, Y.J. Butler Matrix Based Multi-Beam Base Station Antenna Array. In Proceedings of the 2019 13th European Conference on Antennas and Propagation (EuCAP), Krakow, Poland, 31 March–5 April 2019; pp. 1–4.
8. Ansari, M.; Jones, B.; Zhu, H.; Shariati, N.; Guo, Y.J. A Highly Efficient Spherical Luneburg Lens for Low Microwave Frequencies Realized With a Metal-Based Artificial Medium. *IEEE Trans. Antennas Propag.* **2021**, *69*, 3758–3770. [CrossRef]
9. Mak, K.M.; Lai, H.W.; Luk, K.M. A 5G Wideband Patch Antenna With Antisymmetric L-shaped Probe Feeds. *IEEE Trans. Antennas Propag.* **2018**, *66*, 957–961. [CrossRef]
10. Chu, Q.; Wen, D.; Luo, Y. A Broadband ±45 Dual-Polarized Antenna With Y-Shaped Feeding Lines. *IEEE Trans. Antennas Propag.* **2015**, *63*, 483–490. [CrossRef]
11. Wang, W.; Liu, X.; Wu, Y.; Liu, Y. A Broadband Filtering Patch Antenna Using T-Probe, Transverse Stubs, and U-Slots. *IEEE Access* **2019**, *7*, 7502–7509. [CrossRef]
12. Wu, W.; Fan, R.; Zhang, Z.; Zhang, W.; Zhang, Q. A shorted dual-polarized cross bowtie dipole antenna for mobile communication Systems. In Proceedings of the 2014 XXXIth URSI General Assembly and Scientific Symposium (URSI GASS), Beijing, China, 16–23 August 2014; pp. 1–4.
13. Chu, Q.; Luo, Y. A Broadband Unidirectional Multi-Dipole Antenna with Very Stable Beamwidth. *IEEE Trans. Antennas Propag.* **2013**, *61*, 2847–2852. [CrossRef]
14. Zheng, D.; Chu, Q. A Multimode Wideband ± 45° Dual-Polarized Antenna with Embedded Loops. *IEEE Antennas Wirel. Propag. Lett.* **2017**, *16*, 633–636. [CrossRef]
15. Huang, H.; Liu, Y.; Gong, S. A Broadband Dual-Polarized Base Station Antenna with Anti-Interference Capability. *IEEE Antennas Wirel. Propag. Lett.* **2017**, *16*, 613–616. [CrossRef]
16. Wen, D.; Zheng, D.; Chu, Q. A Wideband Differentially Fed Dual-Polarized Antenna with Stable Radiation Pattern for Base Stations. *IEEE Trans. Antennas Propag.* **2017**, *65*, 2248–2255. [CrossRef]
17. Cui, Y.; Wu, L.; Li, R. Bandwidth Enhancement of a Broadband Dual-Polarized Antenna for 2G/3G/4G and IMT Base Stations. *IEEE Trans. Antennas Propag.* **2018**, *66*, 7368–7373. [CrossRef]
18. Kraus, J.D.; Marhefkas, R.J. *Antennas: For All Applications*, 3rd ed.; Tempe, A.Z., Ed.; McGrw-Hill: New York, NY, USA, 2002; pp. 165–196.
19. Ye, L.H.; Zhang, X.Y.; Gao, Y.; Xue, Q. Wideband Dual-Polarized Four-Folded-Dipole Antenna Array With Stable Radiation Pattern for Base-Station Applications. *IEEE Trans. Antennas Propag.* **2020**, *68*, 4428–4436. [CrossRef]
20. Ding, C.F.; Zhang, X.Y.; Yu, M. Simple Dual-Polarized Filtering Antenna With Enhanced Bandwidth for Base Station Applications. *IEEE Trans. Antennas Propag.* **2020**, *68*, 4354–4361. [CrossRef]
21. Xue, Q.; Liao, S.W.; Xu, J.H. A Differentially-Driven Dual-Polarized Magneto-Electric Dipole Antenna. *IEEE Trans. Antennas Propag.* **2013**, *61*, 425–430. [CrossRef]
22. Liu, Y.; Wang, S.; Wang, X.; Jia, Y. A Differentially Fed Dual-Polarized Slot Antenna with High Isolation and Low Profile for Base Station Application. *IEEE Antennas Wirel. Propag. Lett.* **2019**, *18*, 303–307. [CrossRef]
23. Chiu, C.; Cheng, C.; Murch, R.D.; Rowell, C.R. Reduction of Mutual Coupling Between Closely-Packed Antenna Elements. *IEEE Trans. Antennas Propag.* **2007**, *55*, 1732–1738. [CrossRef]
24. Zhang, Y.; Zhang, X.Y.; Ye, L.; Pan, Y. Dual-Band Base Station Array Using Filtering Antenna Elements for Mutual Coupling Suppression. *IEEE Trans. Antennas Propag.* **2016**, *64*, 3423–3430. [CrossRef]
25. Fu, W.; Thalakotuna, D.; Liversidge, P. Stadium Antenna. WO Patent 201 605 4672 A1, 1 March 2016.
26. Timofeev, I.; Ai, X.; Teillet, A. Dual Polarized Three-Sector Base Station Antenna with Variable Beam Tilt. U.S. Patent 7 196 674 B2, 27 March 2007.
27. Bisiules, P.; Shooshtari, A.S. Full Wave Dipole Array Having Improved Squint Performance. WO Patent 2016 137 526 A1, 1 August 2017.
28. Sun, H.; Ding, C.; Jones, B.; Guo, Y.J. A Wideband Base Station Antenna Element with Stable Radiation Pattern and Reduced Beam Squint. *IEEE Access* **2017**, *5*, 23022–23031. [CrossRef]
29. Ding, C.; Sun, H.; Ziolkowski, R.W.; Guo, Y.J. A Dual Layered Loop Array Antenna for Base Stations with Enhanced Cross-Polarization Discrimination. *IEEE Trans. Antennas Propag.* **2018**, *66*, 6975–6985. [CrossRef]
30. Yin, J.Y.; Zhang, L. Design of a Dual-Polarized Magneto-electric Dipole Antenna with Gain Improvement at Low Elevation Angle for a Base Station. *IEEE Antennas Wirel. Propag. Lett.* **2020**, *19*, 756–760. [CrossRef]
31. Walter, H. Sub-6 GHz mMIMO base stations meet 5G's size and weight challenges. *Microw. J.* **2019**, *62*, 40–52.
32. Cui, Y.; Li, R.; Wang, P. Novel Dual-Broadband Planar Antenna and Its Array for 2G/3G/LTE Base Stations. *IEEE Trans. Antennas Propag.* **2013**, *61*, 1132–1139. [CrossRef]
33. Li, Y.; Wang, C.; Yuan, H.; Liu, N.; Zhao, H.; Li, X. A 5G MIMO Antenna Manufactured by 3-D Printing Method. *IEEE Antennas Wirel. Propag. Lett.* **2017**, *16*, 657–660. [CrossRef]
34. Fournier, A.F.A. *Recommendation on Base Station Antenna Standards (V11.1)*; NGMN Alliance: Frankfurt am Main, Germany, 2019.
35. Elsherbini, A.; Wu, J.; Sarabandi, K. Dual Polarized Wideband Directional Coupled Sectorial Loop Antennas for Radar and Mobile Base-Station Applications. *IEEE Trans. Antennas Propag.* **2015**, *63*, 1505–1513. [CrossRef]
36. Li, R.; Wu, T.; Pan, B.; Lim, K.; Laskar, J.; Tentzeris, M.M. Equivalent-Circuit Analysis of a Broadband Printed Dipole With Adjusted Integrated Balun and an Array for Base Station Applications. *IEEE Trans. Antennas Propag.* **2009**, *57*, 2180–2184.

37. Zheng, D.; Chu, Q. A Wideband Dual-Polarized Antenna with Two Independently Controllable Resonant Modes and Its Array for Base-Station Applications. *IEEE Antennas Wirel. Propag. Lett.* **2017**, *16*, 2014–2017. [CrossRef]
38. Zhang, Q.; Gao, Y. A Compact Broadband Dual-Polarized Antenna Array for Base Stations. *IEEE Antennas Wirel. Propag. Lett.* **2018**, *17*, 1073–1076. [CrossRef]
39. Nie, L.Y.; Lin, X.Q.; Chen, Y.J.; Zhang, J.; Wang, B.; Yang, Z.Q.; Fan, Y. A Low-Profile Coplanar Dual-Polarized and Dual-Band Base Station Antenna Array. *IEEE Trans. Antennas Propag.* **2018**, *66*, 6921–6929. [CrossRef]
40. Wu, R.; Chu, Q. Resonator-Loaded Broadband Antenna for LTE700/GSM850/GSM900 Base Stations. *IEEE Antennas Wirel. Propag. Lett.* **2017**, *16*, 501–504. [CrossRef]
41. Zhao, L.; Wu, K. A Dual-Band Coupled Resonator Decoupling Network for Two Coupled Antennas. *IEEE Trans. Antennas Propag.* **2015**, *63*, 2843–2850. [CrossRef]
42. Tang, X.; Mouthaan, K.; Coetzee, J.C. Dual-band decoupling and matching network design for very closely spaced antennas. In Proceedings of the 2012 42nd European Microwave Conference, Amsterdam, The Netherlands, 29 October–1 November 2012; pp. 49–52.
43. Huang, H.; Liu, Y.; Gong, S. A Dual-Broadband, Dual-Polarized Base Station Antenna for 2G/3G/4G Applications. *IEEE Antennas Wirel. Propag. Lett.* **2017**, *16*, 1111–1114. [CrossRef]
44. Li, M.; Chen, X.; Zhang, A.; Kishk, A.A. Dual-Polarized Broadband Base Station Antenna Backed With Dielectric Cavity for 5G Communications. *IEEE Antennas Wirel. Propag. Lett.* **2019**, *18*, 2051–2055. [CrossRef]
45. Mavridou, M.; Feresidis, A.P.; Gardner, P. Tunable Double-Layer EBG Structures and Application to Antenna Isolation. *IEEE Trans. Antennas Propag.* **2016**, *64*, 70–79. [CrossRef]
46. Pan, Y.M.; Hu, P.F.; Zhang, X.Y.; Zheng, S.Y. A Low-Profile High-Gain and Wideband Filtering Antenna with Metasurface. *IEEE Trans. Antennas Propag.* **2016**, *64*, 2010–2016. [CrossRef]
47. Jin, J.Y.; Liao, S.; Xue, Q. Design of Filtering-Radiating Patch Antennas With Tunable Radiation Nulls for High Selectivity. *IEEE Trans. Antennas Propag.* **2018**, *66*, 2125–2130. [CrossRef]
48. Wu, T.L.; Pan, Y.M.; Hu, P.F.; Zheng, S.Y. Design of a Low Profile and Compact Omnidirectional Filtering Patch Antenna. *IEEE Access* **2017**, *5*, 1083–1089. [CrossRef]
49. Sun, G.; Wong, S.; Zhu, L.; Chu, Q. A Compact Printed Filtering Antenna With Good Suppression of Upper Harmonic Band. *IEEE Antennas Wirel. Propag. Lett.* **2016**, *15*, 1349–1352. [CrossRef]
50. Yang, W.; Chen, S.; Xue, Q.; Che, W.; Shen, G.; Feng, W. Novel Filtering Method Based on Metasurface Antenna and Its Application for Wideband High-Gain Filtering Antenna With Low Profile. *IEEE Trans. Antennas Propag.* **2019**, *67*, 1535–1544. [CrossRef]
51. Yang, S.J.; Cao, Y.F.; Pan, Y.M.; Wu, Y.; Hu, H.; Zhang, X.Y. Balun-Fed Dual-Polarized Broadband Filtering Antenna without Extra Filtering Structure. *IEEE Antennas Wirel. Propag. Lett.* **2020**, *19*, 656–660. [CrossRef]
52. Bao, Z.; Nie, Z.; Zong, X. A Novel Broadband Dual-Polarization Antenna Utilizing Strong Mutual Coupling. *IEEE Trans. Antennas Propag.* **2014**, *62*, 450–454. [CrossRef]
53. Chen, Y.; Zhao, J.; Yang, S. A Novel Stacked Antenna Configuration and its Applications in Dual-Band Shared-Aperture Base Station Antenna Array Designs. *IEEE Trans. Antennas Propag.* **2019**, *67*, 7234–7241. [CrossRef]
54. Zhu, Y.; Chen, Y.; Yang, S. Decoupling and Low-Profile Design of Dual-Band Dual-Polarized Base Station Antennas Using Frequency-Selective Surface. *IEEE Trans. Antennas Propag.* **2019**, *67*, 5272–5281. [CrossRef]
55. Afanasyev, P.; Matitsine, S.; Sledkov, V.; Lagoiski, P.; Matytsine, L.; DeMarco, T.; Tse-Tong, C. Multi-beam Luneburg lens antenna for cellular communications. In Proceedings of the 2015 9th European Conference on Antennas and Propagation (EuCAP), Lisbon, Portugal, 13–17 April 2015; pp. 1–4.
56. Matytsine, L.; Lagoiski, P.; Matytsine, M.; Matitsine, S. Large size, lightweight, Luneburg Lenses for multi-beam antenna applications. In Proceedings of the 2012 6th European Conference on Antennas and Propagation (EUCAP), Prague, Czech Republic, 26–30 March 2012; pp. 2266–2270.
57. Su, D.; Fu, D.; Wang, T.N.C.; Yang, H. Broadband Polarization Diversity Base Station Antenna for 3G Communication System. In Proceedings of the 2007 International Symposium on Microwave, Antenna, Propagation and EMC Technologies for Wireless Communications, Honolulu, HI, USA, 16–17 August 2007; pp. 593–596.
58. Su, D.; Qian, J.J.; Hua, Y.; Fu, D. A novel broadband polarization diversity antenna using a cross-pair of folded dipoles. *IEEE Antennas Wirel. Propag. Lett.* **2005**, *4*, 433–435.
59. Wu, C.; Yang, C.F. Method and Apparatus for Improving Antenna Radiation Patterns. U.S. Patent 7 081 865 B2, 25 July 2006.
60. Sun, H.H.; Ding, C.; Zhu, H.; Jones, B.; Guo, Y.J. Suppression of Cross-Band Scattering in Multiband Antenna Arrays. *IEEE Trans. Antennas Propag.* **2019**, *67*, 2379–2389. [CrossRef]
61. Isik, O.; Gripo, P.; Thalakotuna, D.; Liversidge, P. Cloaked Low Band Elements for Multiband Radiating Arrays. U.S. Patent 10 439 285 B2, 8 October 2019.
62. Soric, J.C.; Monti, A.; Toscano, A.; Bilotti, F.; Alù, A. Dual-Polarized Reduction of Dipole Antenna Blockage Using Mantle Cloaks. *IEEE Trans. Antennas Propag.* **2015**, *63*, 4827–4834. [CrossRef]
63. Li, T.; Li, Q.; Zhang, M.; Zhang, A.X.; Zhang, J.; Chen, X. Four-beam Antenna Array with Low Side-lobe for Base Station. In Proceedings of the Application, 2019 Photonics & Electromagnetics Research Symposium-Fall (PIERS-Fall), Xiamen, China, 17–20 December 2019; pp. 2796–2805.
64. Hamdy, D.M.N. *An Introduction to LTE Smart Base Station Antennas*; Commscope: Hickory, NC, USA, 2017.

65. Wu, Z.; Wu, B.; Su, Z.; Zhang, X. Development challenges for 5G base station antennas. In Proceedings of the 2018 International Workshop on Antenna Technology (iWAT), Nanjing, China, 5–7 March 2018; pp. 1–3.
66. Zhang, X.; Xue, D.; Ye, L.; Pan, Y.; Zhang, Y. Compact Dual-Band Dual-Polarized Interleaved Two-Beam Array With Stable Radiation Pattern Based on Filtering Elements. *IEEE Trans. Antennas Propag.* **2017**, *65*, 4566–4575. [CrossRef]
67. Zhu, H.; Sun, H.; Jones, B.; Ding, C.; Guo, Y.J. Wideband Dual-Polarized Multiple Beam-Forming Antenna Arrays. *IEEE Trans. Antennas Propag.* **2019**, *67*, 1590–1604. [CrossRef]
68. Shen, L.; Wang, H.; Lotz, W.; Jamali, H. Dual Polarization 4x4 MIMO Sub-6GHz Multi-Beam Base Station Antennas. In Proceedings of the 2019 International Symposium on Antennas and Propagation (ISAP), Xi'an, China, 27–30 October 2019; pp. 1–3.
69. Gu, L.; Yang, W.; Che, W.; Chen, D.; Zhang, Y.; Feng, W. A Dual-Steerable-Beam Multi-Slot Coupled Metasurface Antenna. In Proceedings of the 2018 IEEE International Conference on Computational Electromagnetics (ICCEM), Chengdu, China, 26–28 March 2018; pp. 1–3.
70. Sanad, M.; Hassan, N. A Sub-6 GHz Multi-Beam Base Station Antenna for 5G with an Arbitrary Beam-Tilting for Each Beam. In Proceedings of the 2019 IEEE Radio and Wireless Symposium (RWS), Orlando, FL, USA, 20–23 January 2019; pp. 1–4.
71. Kayani, H.A.; Gueuning, Q.; Goreux, N.; Vanhoenacker-Janvier, D.; Oestges, C.; Craeye, C. Reconfigurable Cellular Base Station Antenna Consisting of Parasitic Radiators. *IEEE Trans. Ind. Electron.* **2020**, *67*, 7083–7093. [CrossRef]
72. Tefiku, F.; Grimes, C.A. Design of broad-band and dual-band antennas comprised of series-fed printed-strip dipole pairs. *IEEE Trans. Antennas Propag.* **2000**, *48*, 895–900. [CrossRef]
73. Qi, X.; Fushun, Z.; Baohua, S.; Yanlin, Z.; Qizhong, L. A novel dual-band Yagi-Uda antenna for wireless communications. In Proceedings of the 9th International Symposium on Antennas, Propagation and EM Theory, Guangzhou, China, 29 November–2 December 2010; pp. 289–292.
74. Ding, G.; Zimmerman, M.; Yu, J.; Qin, H. Base Station Antennas including Wiper Phase Shifters. EP Patent 3 588 670A1, 1 January 2020.
75. Xiao, W.; Xiao, Z.; Su, W. Base Station Antenna. U.S. Patent 2016/0248156 A1, 25 August 2016.
76. He, Y.; Li, J.; Wong, S.W.; Pan, X.; Zhang, L.; Chen, Z.N. A Miniaturized Base Station Antenna with Novel Phase Shifter for 3G/LTE Applications. *IEEE Access* **2018**, *6*, 52877–52888. [CrossRef]

electronics

MDPI

Article

A Sub-6 GHz MIMO Antenna Array for 5G Wireless Terminals

Jalal Khan [1], Sadiq Ullah [1,*], Farooq A. Tahir [2], Faisel Tubbal [3,4] and Raad Raad [3]

1 Telecommunication Engineering Department, University of Engineering and Technology, Mardan 23200, Pakistan; jalal@uetmardan.edu.pk
2 Research Institute for Microwave and Millimeter-Wave Studies (RIMMS), National University of Sciences and Technology (NUST), Islamabad 44000, Pakistan; farooq.tahir@rimms.nust.edu.pk
3 School of Electrical, Computer and Telecommunication Engineering, University of Wollongong, Wollongong, NSW 2522, Australia; faisel@uow.edu.au (F.T.); raad@uow.edu.au (R.R.)
4 Technological Projects Department, The Libyan Center for Remote Sensing and Space Science, Tripoli 21218, Libya
* Correspondence: sadiqullah@uetmardan.edu.pk

Abstract: This paper presents a novel antenna with its array and MIMO configuration for the 5G sub-6 GHz applications. The proposed antenna element operates at the central frequency of 5.57 GHz dedicated for Sub-6 GHz 5G communication applications. The antenna element holds a circular-shaped radiating portion with an inner-circular slot, plus a rectangular slot at its right edge to make the proposed design resonate at the desired frequency band. The RT5880 substrate is used with a thickness of 0.787 mm, and the low-loss tangent of 0.0009. To achieve a desired gain of 12 dB, a four-element array configuration is adopted, which improved a bore side gain to 12.4 dB from 6.66 dB. Then, the two-port configuration is adopted such that the isolation achieved between them is more than −30 dB. The total efficiency of the proposed antenna array is observed to be more than 80% within the operating bandwidth. Moreover, the Specific Absorption Rate (SAR) analysis is also presented for the proposed MIMO configuration, obeying the standard value (i.e., <2 W/kg for any 10 g of tissue). The measured results are in good agreement with the simulated results. All the simulations of the proposed design are performed in the CST MWS software.

Keywords: 5G; antenna array; CST; sub-6 GHz; MIMO; SAR

Citation: Khan, J.; Ullah, S.; Tahir, F.A.; Tubbal, F.; Raad, R. A Sub-6 GHz MIMO Antenna Array for 5G Wireless Terminals. *Electronics* **2021**, *10*, 3062. https://doi.org/10.3390/electronics10243062

Academic Editor: Christos J. Bouras

Received: 5 November 2021
Accepted: 5 December 2021
Published: 9 December 2021

1. Introduction

The day-by-day increase in the demand for higher data rates and bandwidth has created several issues for the current fourth-generation wireless communication infrastructure [1], as the current wireless communication infrastructure lacks the potential to deliver a higher data rate with a low latency [2]. Thus, the fifth generation has been introduced, which is promising to fulfil the requirement for higher data rates with the desired low latency level [3].

At the moment, most of the designs reported are single element, which possess quite low gain as compared to the desired level (>12 dB) required for 5G communication [4–8]. Likewise, several multiple-port antennas have been presented to improve data rates, but they could not achieve satisfactory gain [9–13]. In the future, these Sub-6 GHz 5G antennas will be integrated with their predecessors and some other higher-band application antennas on the same printed circuit board (PCB). Thus, the high-level of gain can be quite helpful to efficiently deliver a stronger signal at the user side, instead of congesting the lower frequencies of the spectrum [14]. Along with this, the multiple input multiple output (MIMO) configuration is quite helpful to provide a good channel capacity and data rate as compared to the single port antennas, which is one of the basic challenges in 5G transmission.

Recently, some MIMO antennas have been reported for operation in the Sub-6 GHz band [15–20]. Different configurations have been adopted—such as monopole, open-slot, or loop antenna design method to achieve operation in the desired frequency band. As in [15], the size of the proposed design noted is 130 mm × 100 mm with an operating band of 5.15–5.925 GHz. The gain obtained ranges from 2.5–4.2 dB within the operating band. Similarly, another antenna with an overall size of 150 mm × 75 mm and operational bandwidth of 5.15–5.85 GHz is presented in [16]. The peak gain noted is 4.62 dB. Moreover, an antenna with a size of 136 mm × 68 mm is presented in [17]. The proposed design adopted a monopole configuration in order to operate in the band of 5.15–5.925 GHz, while the gain achieved is not discussed. Apart from this, the main drawback observed is a low efficiency in [18–20], while in [21–23], the efficiency attained is even below 50%. In [24], the efficiency obtained ranges between 50–82%, while the antenna element is extended to the multi-port configuration over the same PCB with an overall size of 150 mm × 75 mm. However, the minimum isolation level observed is −12 dB, which can be a great cause of signal deterioration on the user side.

In this paper, we have presented a microstrip circular patch antenna array designed at the 5.5 GHz band with four elements in the array structure using microstrip methods, and then its MIMO configuration for 5G communication applications. The transmission line and quarter wavelength transformation techniques are used for the impedance and phase matching. Moreover, the SAR analysis is also presented for the proposed MIMO configuration. The single element, four-element array, and MIMO antennas are analyzed using CST Microwave Studio. The remainder of the paper is organized as follows: Section 2 presents the design methodology of the proposed antenna. The results and discussion have been presented in Section 3, and Section 4 discusses the SAR analysis. Section 5 concludes the paper and gives recommendations for the future.

2. Antenna Geometry

2.1. Single Element

The front view of the antenna element is shown in Figure 1. RT5880 with a thickness of 0.787 mm is used as a substrate. The length and width of the substrate is 50 × 40 mm^2. The radiating portion of the antenna exhibits a circular, patch-like geometry, with a circular-shaped slot in the center with an inner radius of 6.75 mm and an outer radius of 11.54 mm. A slot-termed gap is created at the edge of the patch to tune the resonant frequency to the desired frequency band. The feed contains an impedance transformer for the purpose of matching the transmission line with the radiating portion of the antenna. The back side of the proposed antenna is fully copper-coated. The parameters in Table 1 have been calculated via the known equations of the transmission line theory for the circular patch [25] and have then been optimized to achieve resonance at the desired frequency band.

$$\text{rad}_1 = \frac{\text{F}}{\left\{1 + \frac{2h}{\pi\varepsilon_r \text{F}}\left[\ln\left(\frac{\pi \text{F}}{2h}\right) + 1.7726\right]\right\}^{\frac{1}{2}}} \quad (1)$$

$$\text{F} = \frac{8.791 \times 10^9}{f_r\sqrt{\varepsilon_r}} \quad (2)$$

Table 1. Summary of the dimensions of the proposed design.

Parameter	Value (mm)	Parameter	Value (mm)
L	6.00	F	1.00
gap	1.00	R1	11.54
W1	2.38	L1	10.0
L2	9.7	R2	6.75

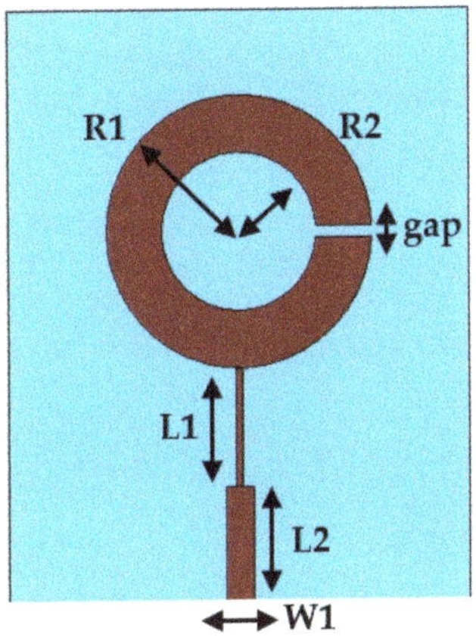

Figure 1. Front view of the antenna element.

The voltage standing wave ratio (VSWR) and reflection coefficient graphs are shown in Figure 2. It is seen that the magnitude of the reflection coefficient is below 15 dB, which is acceptable for optimum power transmission. The VSWR is also observed below 2 dB within the operating bandwidth, which depicts low loss and a good matching between the feedline and the radiating portion of the proposed antenna. The −10 dB bandwidth of almost 50 MHz is obtained with a range from 5.55 to 5.6 GHz, with the central frequency noted to be 5.57 GHz, which is dedicated for 5G Sub-6 GHz applications [18].

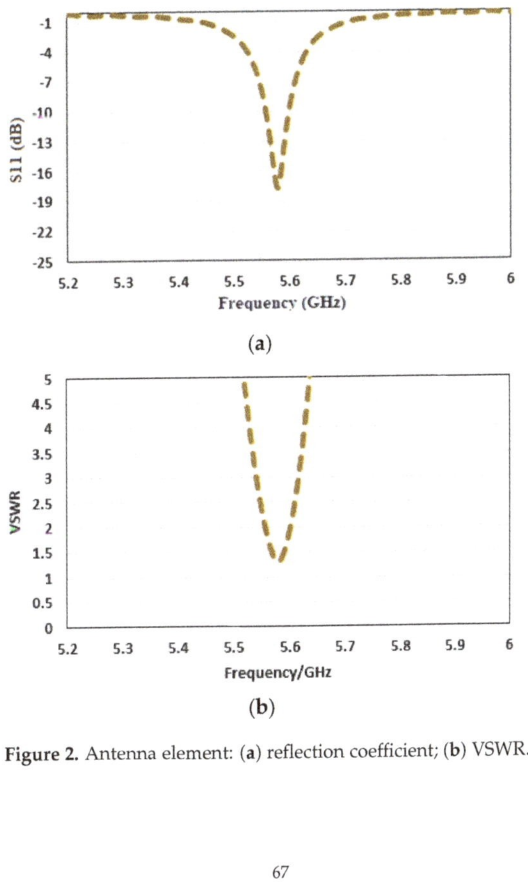

Figure 2. Antenna element: (**a**) reflection coefficient; (**b**) VSWR.

The parametric study for the proposed antenna element by varying the values of different key parameters is shown in Figure 3. It is observed that the value of the parameters' variations has a significant effect on the reflection coefficient of the proposed design. All three parameters, such as R1, R2 and gap are helpful to bring the antenna element to resonate at 5.57 GHz. In the case of R1, the resonance for the selected parameter is at 5.6 GHz, while simultaneously varying R2 and gap parameter with the R1 moves the resonance to 5.57 GHz. Alongside, the matching circuit helps to improve the magnitude of the reflection coefficient. Thus, after careful consideration and analysis, final values are taken, and the antenna element can proceed towards the array concept.

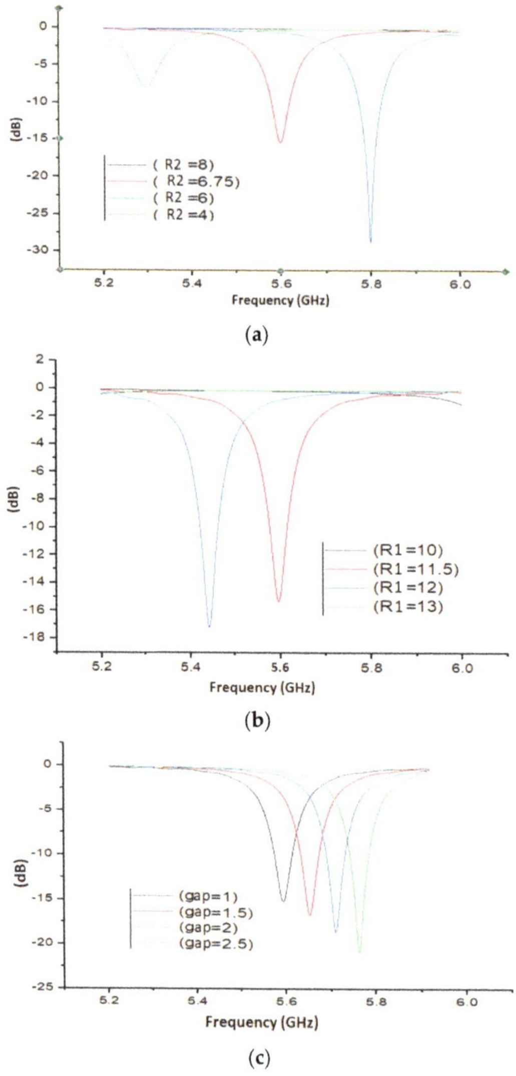

Figure 3. Antenna element reflection coefficient analysis by (**a**–**c**) varying different parameters.

The peak gain patterns for the proposed antenna element are presented in Figure 4. The polar and 3D gain patterns are observed in Figure 4a and Figure 4b, respectively. The gain pattern in the 90-degree plane is analyzed, and the side lobe level is observed to be −13.8 dB, while the magnitude of the main lobe is 6.65 dB. Further, the angular 3 dB beamwidth is seen to be 110.4 degrees. The 0-degree plane gain pattern is with the side lobe level of −24.1 dB, whereas a main lobe magnitude of 5.93 dB with a 3 dB beamwidth is 69.9 degrees. The gain achieved is 6.66 dB as shown in the 3D radiation pattern (Figure 4b) of the proposed antenna. The greater level of side lobe reduction makes the proposed design less lossy, whereas the back lobes level is also very low as shown in the E plane of Figure 4a.

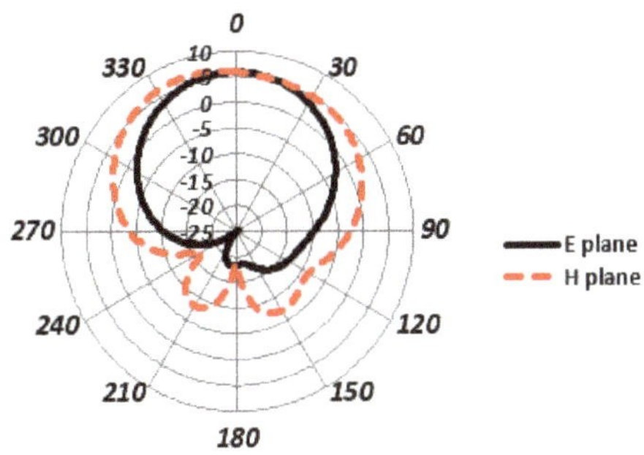

(a)

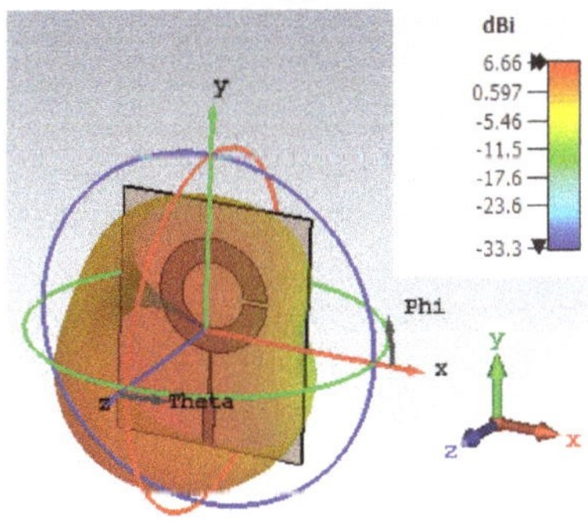

(b)

Figure 4. Antenna element: (**a**) gain pattern (polar); (**b**) gain pattern (3D).

2.2. Multiple-Element Antenna Array

The design of single element is extended to another type of the array technique, which is a most famous gain enhancement technique. In Figure 5, the corporate feed array is presented, where each element holding a symmetry is placed at a distance of 0.477λ0 to achieve a particular gain enhancement. The gap between the neighbor patches is optimized step by step to have minimal performance degradation effect on each other, and finally, 0.477λ0 is chosen as an optimized gap where reflection coefficient, gain, and efficiency could be quite satisfactory. The back of the substrate is fully coated with copper material. Some of the design parameters of the corporate feed array have been shown in Table 2.

Table 2. Design parameters of array antenna.

Parameter	Value (mm)	Parameter	Value (mm)
a	9.0	d	62.0
b	10.00	e	7.0
c	30.0	f	2.4
g	160	h	70.0

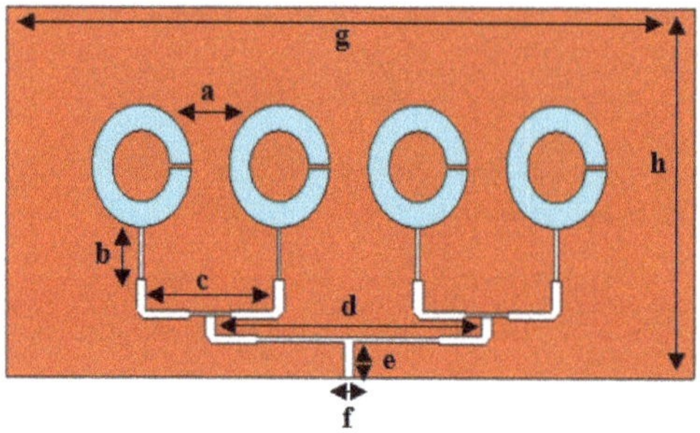

Figure 5. Corporate feed sub-6 GHz array.

The width of the transmission line is calculated by using the following relationship [26].

$$w_{Z_o} = \left(\frac{377}{Z_c \sqrt{\varepsilon_r}} - 2 \right) \times h \tag{3}$$

where ε_r is the relative permittivity and Z_c is the characteristic impedance of the transmission line. In the proposed antenna design a 50 Ω transmission line is used.

2.3. MIMO Configuration

The array design achieved utilizing the corporate feed technique is extended to the MIMO configuration of two ports as shown in Figure 6. The arrays are placed with a 90-degree shift with respect to each other in order to achieve pattern diversity and good isolation among antenna arrays, which can cause issues when extending arrays to MIMO. The dimensions of the MIMO antenna are outlined in Table 3.

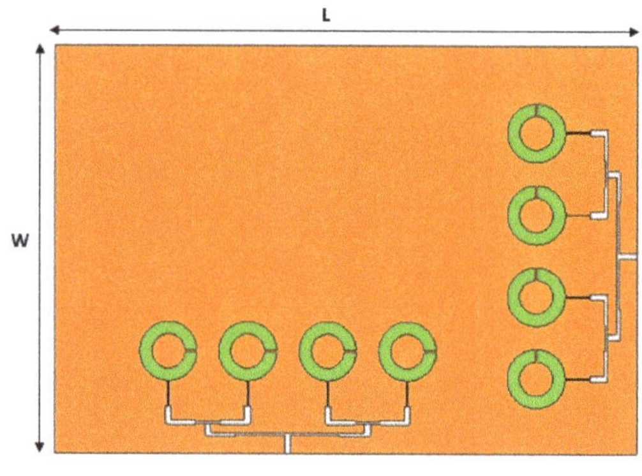

Figure 6. Corporate feed array MIMO configuration.

Table 3. Design parameters of MIMO array antenna.

Parameter	Value (mm)	Parameter	Value (mm)
W	165	L	235

3. Results and Discussion

The results obtained through simulation and measurements for the proposed corporate array depicted in Figure 5 are analyzed in this section. In Figure 7, the measurement setup is shown for the proposed corporate array results analysis. As shown in Figure 8, the reflection coefficient obtained for the corporate feed array is showing resonance at the desired band of 5.57 GHz with a magnitude of 25 dB. In case of corporate feed structure, the arrangement of elements is done such that—in terms of spacing between the elements—

the coupling among them should be negligible. The bandwidth covered by the proposed antenna array ranges from 5.54 to 5.59 GHz.

(**a**)

(**b**)

Figure 7. Proposed antenna array under testing: (**a**) measuring reflection coefficient and (**b**) radiation measurements inside the anechoic chamber.

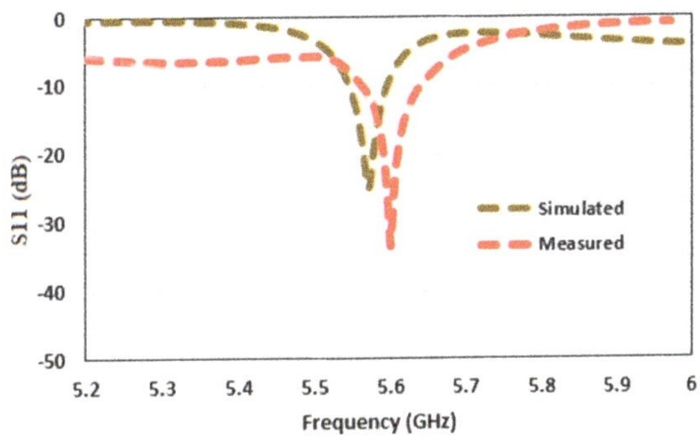

Figure 8. Corporate feed array reflection coefficient.

The measured reflection coefficient follows the simulated one quite efficiently, with an improvement in the magnitude level. Furthermore, a 3 dB offset is observed between the simulated and measured reflection coefficient due to the losses by connecters. The minor shift in the frequency observed is due to fabrication tolerances. The VSWR is also below 2 for the operating bandwidth while at the central frequency, the value of it is observed to be 1.12, which shows that there is minimum reflection of the transmitted power from the radiating structure to the feed, as shown in Figure 9.

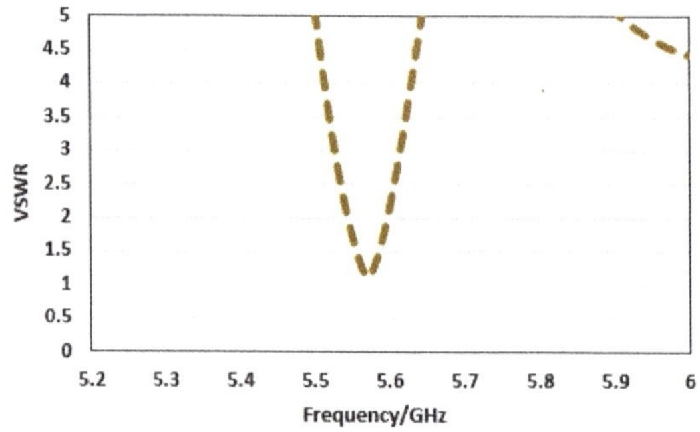

Figure 9. Corporate feed array VSWR.

The radiation pattern is highly directive, with a gain of 12.4 dB as shown in Figure 10. It can also be seen that the side lobe level is highly reduced in this type of array structure. According to the 0-degree plane of radiation, the side lobe level is −17.1 dB, which is quite satisfactory for 5G communication, and the main lobe direction is towards 0 degree, while the magnitude of the main lobe is found to be 12.6 dB. The angular width (3-dB beamwidth) is achieved at 22 degrees.

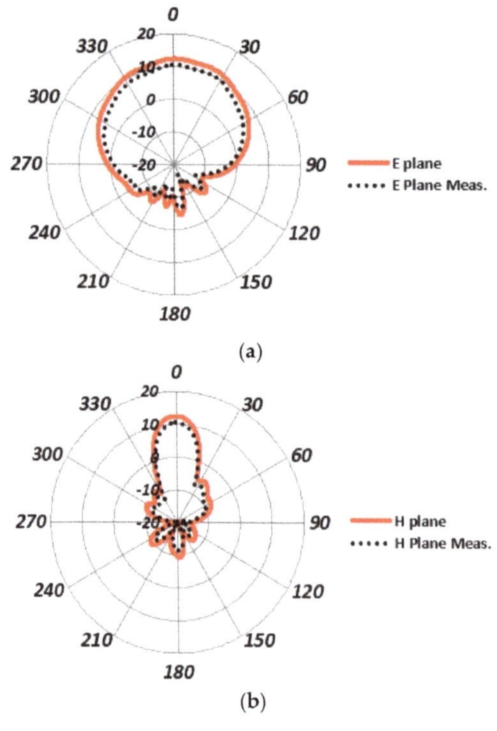

(a)

(b)

Figure 10. Corporate feed array gain pattern: (**a**) *E*-plane; (**b**) *H*-plane.

However, in the 90-degree plane of radiation, the main lobe direction is along a 2-degree angle with a side lobe level of −17.1 dB. The angular width (3-dB beamwidth) in this plane is 82.1 degrees, and the main lobe magnitude is 12.6 dB. Furthermore, the back lobe level in this plane is highly reduced, making the antenna perfect for the 5G communication environment. The measured results follow the simulated with a quite good coherence due to the better fabrication and measurement setup calibration.

The 3D gain pattern further helps to understand the above 2D gain pattern, as shown in Figure 11a using the Computer Simulation Technology (CST) software. The radiation and total efficiency within the operating bandwidth are 98.2% and 85.1%, respectively. The current distribution for the corporate feed structure is observed in Figure 11b, which uses the Computer Simulation Technology (CST) software to show that the current is mostly concentrated along the edges of the circular-shaped patch, and feedline while the current is negligible between the patches; this proves that the coupling among the elements is minimized by employing the corporate feeding technique. Thus, unwanted resonances are avoided in the return loss of the proposed antenna array, and a high gain is achieved with the high efficiency. Moreover, the current distribution in Figure 11b of the whole array ensures that the power distribution is equal to each antenna element. The distribution of the electric (E) and magnetic field (H) are shown in Figure 11c and Figure 11d, respectively.

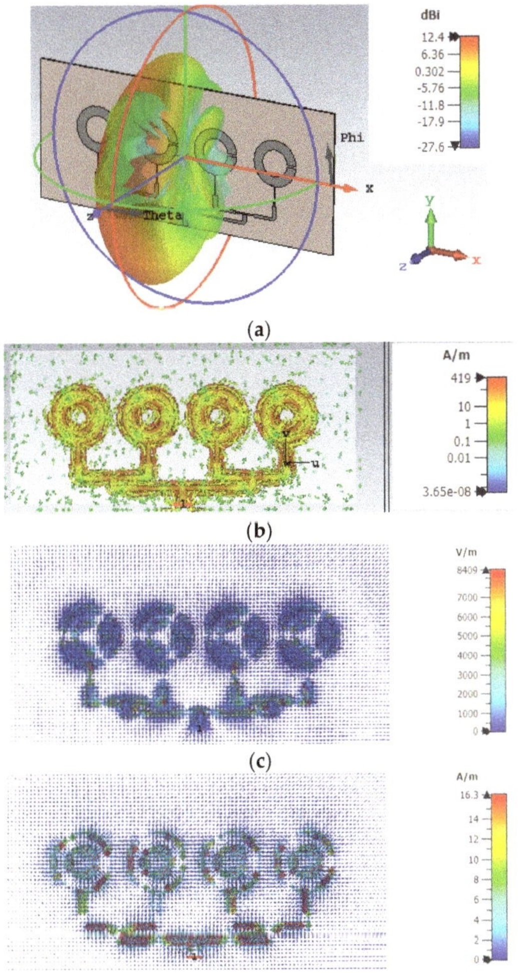

Figure 11. (**a**) The 3D gain pattern of the proposed antenna array, (**b**) current distribution, (**c**) E field and (**d**) H field of the proposed array structure.

The reflection coefficient for the proposed MIMO structure is presented in Figure 12a. For port-1, the antenna resonates with the central frequency of 5.57 GHz, gives a bandwidth from 5.59 to 5.54 GHz, and a magnitude of −25 dB. Similarly, for port-2, the antenna resonates with the central frequency of 5.56 GHz, with a reflection coefficient magnitude of −26 dB. The magnitude of the return loss remains below −10 dB, within the resonant bandwidth (5.58 to 5.53 GHz). The isolation for the proposed MIMO antenna is presented and analyzed in Figure 12b for both the ports with respect to each other. The isolation is observed to be greater than 40 dB within the operating bandwidth, which clearly indicates that the isolation among the arrays employed is quite minimum.

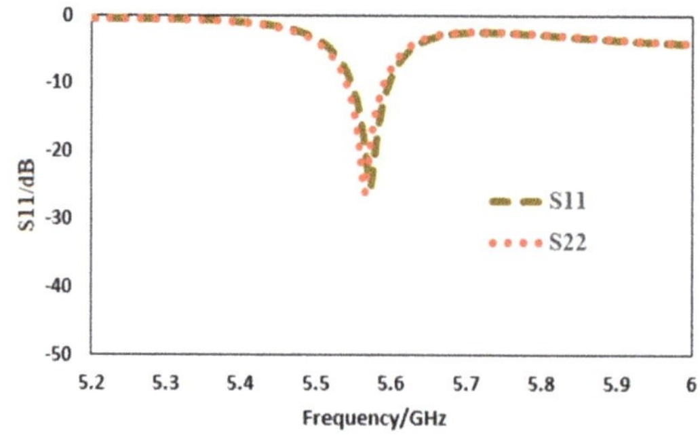

(**a**)

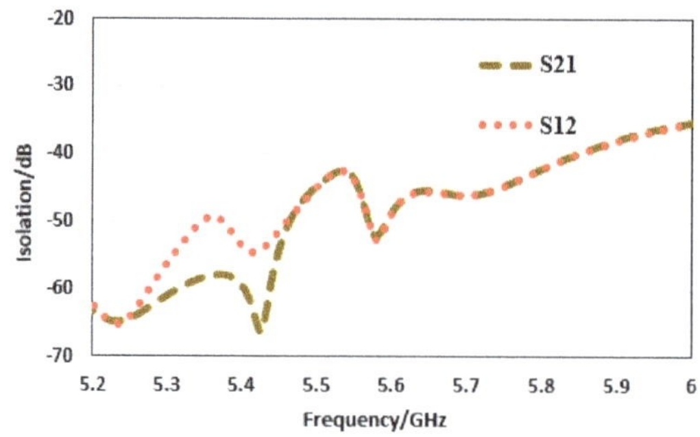

(**b**)

Figure 12. MIMO: (**a**) reflection coefficient; (**b**) isolation.

The gain patterns are analyzed and presented in Figure 13a in the 0- and 90-degree planes for the port-1. Initially, in the 90-degree plane, the antenna array main lobe direction is located at 2.0 degrees, while a very good side lobe level of −17.1 dB is achieved with a 3-dB beamwidth of 82.1 degrees. The radiation pattern is quite directive, with minimum back lobes and side lobes. The 0-degree plane analysis shows that the main lobe is directed towards the 0-degree angle and the side lobe level is −17.1 dB. Furthermore, a 3-dB beamwidth is 22 degrees. In this particular plane, the radiated beam is also seen to be highly directive with a low level of back lobes.

Similarly, the radiation pattern analysis in the case of port-2 is presented in Figure 13b for the two main planes: E and H. The main lobe direction for the 0-degree plane is along the 355-degree angle, and the side lobe level is −3.1 dB, while the angular width is 83 degrees in this particular plane. Likewise, the direction of the main lobe is along 1.0-degree angle in the 90-degree plane, with an angular width of 22.4 degree. The side lobe level is −15.1 degrees in this plane. Further in-depth analysis of the radiation feature of the proposed MIMO configuration (antenna array employing a corporate feeding technique) is presented in Figure 14. The port-1 or antenna-1 has a narrow and directive beam with a peak gain of 12.6 dB.

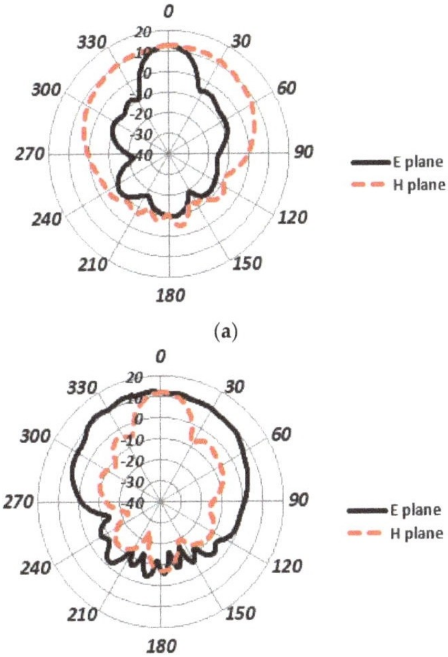

Figure 13. MIMO polar radiation pattern: (**a**) port-1; (**b**) port-2.

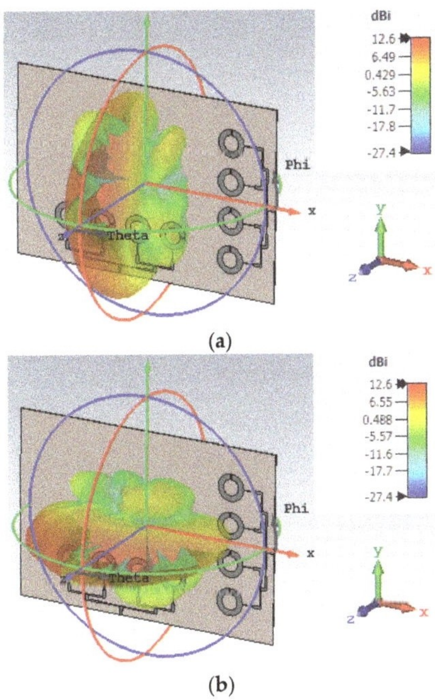

Figure 14. MIMO 3D radiation pattern: (**a**) port-1; (**b**) port-2.

The radiation efficiency and the total efficiency is 98.2% and 85.1%, respectively. Furthermore, for port-2 the radiation beam is shifted by a 90-degree angle, which shows that pattern diversity) is achieved with the proposed MIMO structure. The peak gain of 12.6 dB is achieved with radiation and total efficiency of 97.8 and 79.6%, respectively.

The Envelope Correlation Coefficient (ECC) is another parameter that describes the level of correlation among multiple antennas which come into close proximity of each other. For the proposed MIMO configuration of two port antenna array, the ECC value is below 0.005 for the entire operating bandwidth, which satisfies the standard criteria of a value of <0.5, and depicts that isolation among the antenna arrays is minimum as shown in Figure 15a. Another performance metric of MIMO is Diversity Gain (DG)—meeting or coming close to the standard value of 10 dB shows that reduction in the transmitted power will have no major effect on the quality of transmission or MIMO performance. As seen in Figure 15b, the DG value is close to 10 for the entire operating bandwidth, which satisfies the standard criteria. All the results discussed above show that MIMO-based corporate feed arrays are very suitable for future 5G Sub-6 GHz communication.

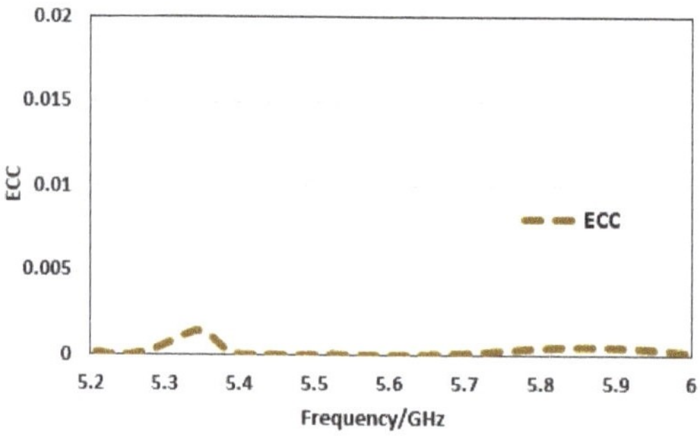

(a)

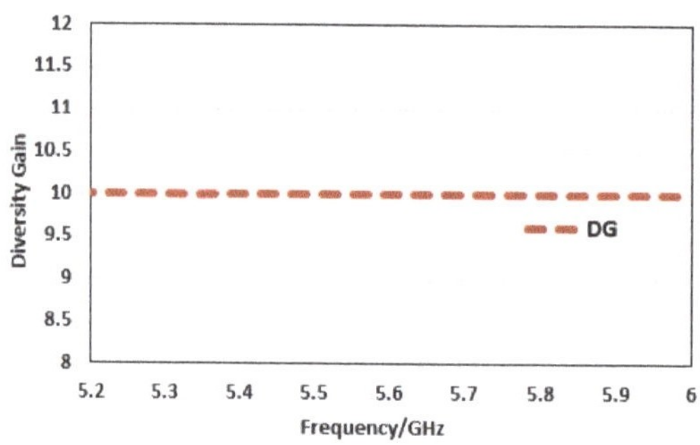

(b)

Figure 15. MIMO: (**a**) ECC; (**b**) DG.

4. SAR Analysis for the Proposed MIMO Antenna

The SAR analysis for the proposed MIMO array antenna is discussed in this section. The proposed model contains skin, fat and muscle with a thickness of 1 mm, 3 mm and 30 mm, respectively as shown in Figure 16. Skin is used as the topmost layer, while the fat and muscle are placed behind the skin in that order. The proposed MIMO configuration (i.e., corporate feed array) is placed at the center of the body model and the performance is analyzed. A conductivity of 3.5 S/m and relative permittivity of 35.3 have been used for the skin, while for fat and muscle, a conductivity and relative permittivity of 0.27 S/m, 4.69 S/m and 4.97, 48.7, respectively.

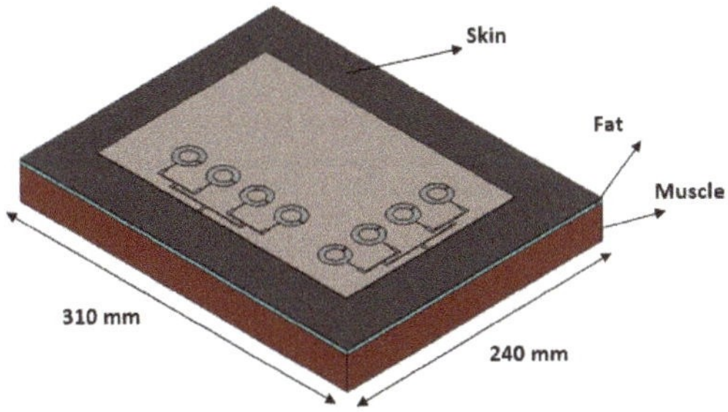

Figure 16. Corporate array MIMO placed on belly model.

Initially, for port-1, Figure 17a shows that the SAR is 0.0989 W/kg which is less than the value of 2 W/kg for the 10 g of tissue, as specified by IEC standards. Hence the SAR lies in the safe range, and no human tissue can be harmed. For the case of port-2 (Figure 17b), the SAR value obtained is 0.101 W/kg, which is also under the safe range, and again obeys the standard value set by the IEC (<2 W/kg). The reference power of 0.5 W has been used, and the lower value of SAR shows that the proposed antenna also qualifies for the human tissue's safety terms.

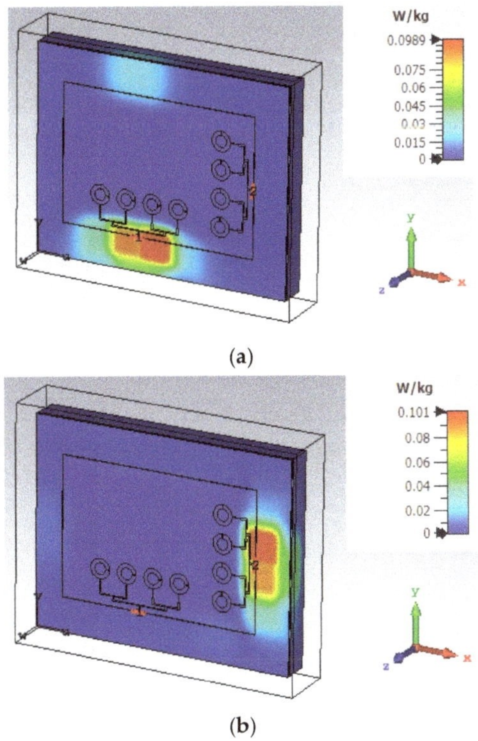

Figure 17. SAR of Corporate Array MIMO Placed on Belly Model: (**a**) Port-1; (**b**) Port-2.

Table 4 provides a performance comparison between the proposed MIMO design and the reported one in the literature working on the Sub-6 GHz band. It is observed that the reported work does not consider any gain improvement within the desired operating band, and apart from this, low isolation is achieved by the reported antennas. The proposed work, however, does provide a high gain antenna with a quite good total efficiency and high isolation level within the operating band, which makes the proposed antenna quite competent for 5G-based communication applications.

Table 4. Comparison with reported work based on sub-6 GHz.

Ref.	Frequency Band (GHz)	Size (mm^2)	Isolation (dB)	Efficiency (%)
[15]	5.15–5.925	130 × 100	>−15	70
[16]	5.15–5.85	150 × 75	>−14	60
[17]	5.15–5.925	136 × 68	<−10	41–69
[18]	4.8–5.1	150 × 75	<−12	40–85
[19]	5.147–5.95	150 × 80	>−10	49–75
[27]	3.3–5.8	150 × 75	<−15	55–87
Prop.	5.6–5.67	160 × 70	>−30	85.1

5. Conclusions

In this paper, a circular-shaped antenna is demonstrated with an inner circular slot, plus a rectangular slot at the right edge of it to make the proposed design resonate at the 5.57 GHz dedicated for Sub-6 GHz 5G communication applications. To achieve a desired gain, four-element array configuration is adopted which improves a bore side gain to 12.4 dB from 6.66 dB. Then, the two-port configuration is adopted such that the isolation achieved between them is more than −30 dB. The total efficiency of the proposed antenna array is observed to be more than 80% within the operating bandwidth, and the measured results of the proposed array are in good agreement with the simulated ones. Moreover, the SAR analysis is also presented for the proposed MIMO configuration; obeying the standard value of <2 W/kg for any 10 g of tissue. Thus, the proposed design becomes a potential candidate for 5G mobile phones and handheld devices. In the future, the MIMO antenna will be fabricated and tested to validate the results, and will be further extended utilizing the metamaterials concept.

Author Contributions: Conceptualization, S.U.; methodology, J.K.; software, J.K. and F.A.T.; validation, F.A.T. and J.K.; formal analysis, F.T.; investigation, J.K.; resources, S.U.; writing—original draft preparation, J.K. and F.A.T.; writing—review and editing, S.U., F.T. and R.R.; visualization, F.T.; supervision, S.U.; project administration, S.U.; funding acquisition, R.R. All authors have read and agreed to the published version of the manuscript.

Funding: This research received no external funding.

Institutional Review Board Statement: Not applicable.

Informed Consent Statement: Not applicable.

Data Availability Statement: The data presented in this study are available on request from the corresponding author.

Conflicts of Interest: The authors declare no conflict of interest.

References

1. Khan, J.; Sehrai, D.A.; Khan, M.A.; Khan, H.A.; Ahmad, S.; Ahmad, S.; Ali, A.; Arif, A.; Memon, A.A.; Khan, S. Design and performance comparison of rotated Y-shaped antenna using different metamaterial surfaces for 5G mobile devices. *Comput. Mater. Contin.* **2019**, *2*, 409–420. [CrossRef]
2. Sharaf, M.H.; Zaki, A.I.; Hamad, R.K.; Omar, M.M.M. A Novel Dual-Band (38/60 GHz) Patch Antenna for 5G Mobile Handsets. *Sensors* **2020**, *20*, 2541. [CrossRef] [PubMed]
3. Barreto, A.N.; Faria, B.; Almeida, E.; Rodriguez, I.; Lauridsen, M.; Amorim, R.; Vieira, R. 5G–Wireless Communications for 2020. *J. Commun. Inf. Syst.* **2016**, *31*, 146–163. [CrossRef]
4. Patnaik, P.; Sarkar, D.; Saha, C. A multi-band 5G antenna for Smart phones operating at Sub-6 GHz frequencies. In Proceedings of the IEEE International Symposium on Antennas & Propagation (APSYM), Cochin, India, 14–16 December 2020; pp. 32–35.
5. Nam, H.J.; Lim, S.; Yoon, Y.J.; Kim, H. Tunable triple-band antenna for Sub-6 GHz 5G mobile phone. In Proceedings of the IEEE International Symposium on Antennas and Propagation and North American Radio Science Meeting, Montreal, QC, Canada, 5–10 July 2020; pp. 1455–1456.

6. Jiang, J.; Li, Y. A wideband kanji patch antenna for 5G Sub-6-GHz applications. In Proceedings of the IEEE 13th UK-Europe-China Workshop on Millimetre-Waves and Terahertz Technologies (UCMMT), Tianjin, China, 29 August–1 September 2020; pp. 1–2.
7. Ghaffar, A.; Li, X.J.; Ahmad, T.; Hussain, N.; Alibakhshikenari, M.; Limiti, E. Circularly polarized pattern reconfigurable flexible antenna for 5G-sub-6-GHz applications. In Proceedings of the IEEE Asia-Pacific Microwave Conference (APMC), Hong Kong, China, 8–11 December 2020; pp. 625–627.
8. Zhao, X.; Yeo, S.P.; Ong, L.O. Decoupling of inverted-F antennas with high-order modes of ground plane for 5G mobile MIMO platform. *IEEE Trans. Antennas Propag.* **2017**, *66*, 4485–4495. [CrossRef]
9. Li, M.Y.; Xu, Z.Q.; Ban, Y.L.; Sim, C.Y.D.; Yu, Z.F. Eight-port orthogonally dual-polarized MIMO antenna using loop structures for 5G smartphone. *IET Microw. Antennas Propag.* **2017**, *11*, 1810–1816. [CrossRef]
10. Li, M.Y.; Ban, Y.L.; Xu, Z.Q.; Guo, J.; Yu, Z.F. Tri-polarized 12-Antenna MIMO array for future 5G smartphone applications. *IEEE Access* **2017**, *6*, 6160–6170. [CrossRef]
11. Jiang, W.; Cui, Y.; Liu, B.; Hu, W.; Xi, Y. A dual-band MIMO antenna with enhanced isolation for 5G smartphone applications. *IEEE Access* **2019**, *7*, 112554–112563. [CrossRef]
12. Li, H.; Tsiaras, A.; Lau, B.K. Analysis and estimation of MIMO-SAR for multi-antenna mobile handsets. *IEEE Trans. Antennas Propag.* **2017**, *65*, 1522–1527. [CrossRef]
13. Li, M.Y.; Li, C.; Ban, Y.L.; Kang, K. Multiple antennas for future 4G/5G smartphone applications. In Proceedings of the 2016 IEEE MTT-S International Microwave Workshop Series on Advanced Materials and Processes for RF and THz Applications (IMWS-AMP), Chengdu, China, 20–22 July 2016.
14. Ojaroudiparchin, N.; Shen, M.; Pedersen, G.F. Beam-steerable microstrip-fed bow-tie antenna array for fifth generation cellular communications. In Proceedings of the IEEE 10th European Conference on Antennas and Propagation (EuCAP), Davos, Switzerland, 10–15 April 2016; pp. 1–5.
15. Li, Y.; Zou, H.; Wang, M.; Peng, M.; Yang, G. Eight-element MIMO antenna array for 5G/Sub-6GHz indoor micro wireless access points. In Proceedings of the IEEE International Workshop on Antenna Technology (iWAT), Nanjing, China, 5–7 March 2018; pp. 1–4.
16. Ren, Z.; Wu, S.; Zhao, A. Triple band MIMO antenna system for 5G mobile terminals. In Proceedings of the IEEE International Workshop on Antenna Technology (iWAT), Miami, FL, USA, 3–6 March 2019; pp. 163–165.
17. Sim, C.; Liu, H.; Huang, C. Wideband MIMO antenna array design for future mobile devices operating in the 5G NR frequency bands n77/n78/n79 and LTE band 46. *IEEE Antennas Wirel. Propag. Lett.* **2020**, *19*, 74–78. [CrossRef]
18. Li, Y.; Sim, C.Y.D.; Luo, Y.; Yang, G. 12-Port 5G Massive MIMO antenna array in sub-6GHz mobile handset for LTE bands 42/43/46 applications. *IEEE Access* **2017**, *6*, 344–354. [CrossRef]
19. Guo, J., Cui, L.; Li, C.; Sun, B. Side-edge frame printed eight-port dual-band antenna array for 5G smartphone applications. *IEEE Trans. Antennas Propag.* **2018**, *66*, 7412–7417. [CrossRef]
20. Li, Y.; Yang, G. Dual-mode and triple-band 10 antenna handset array and its multiple-input multiple-output performance evaluation in 5G. *Int. J. RF Microw. Comput. Aided Eng.* **2019**, *29*, e21538 [CrossRef]
21. Zhang, X.; Li, Y.; Wang, W.; Shen, W. Ultra-wideband 8-port MIMO antenna array for 5G metal-frame smartphones. *IEEE Access* **2019**, *7*, 72273–72282. [CrossRef]
22. Cai, Q.; Li, Y.; Zhang, X.; Shen, W. Wideband MIMO antenna array covering 3.3–7.1 GHz for 5G metal-rimmed smartphone applications. *IEEE Access* **2019**, *7*, 142020–142084. [CrossRef]
23. Sun, L.; Li, Y.; Zhang, Z.; Feng, Z. Wideband 5G MIMO antenna with integrated orthogonal-mode dual-antenna pairs for metal-rimmed smartphones. *IEEE Trans. Antennas Propag.* **2020**, *68*, 2494–2503. [CrossRef]
24. Chen, H.; Tsai, Y.; Sim, C.; Kuo, C. Broadband eight-antenna array design for Sub-6 GHz 5G NR bands metal-frame smartphone applications. *IEEE Antennas Wirel. Propag. Lett.* **2020**, *19*, 1078–1082. [CrossRef]
25. Ballanis, C.A. *Antenna Theory Analysis and Design*; John Willey and Son's Inc.: New York, NY, USA, 2016.
26. Khan, J.; Sehrai, D.A.; Ali, U. Design of dual band 5G antenna array with SAR analysis for future mobile handsets. *J. Electr. Eng. Technol.* **2019**, *14*, 809–816. [CrossRef]
27. Zheng, Z.; Ntawangaheza, J.D.; Sun, L. Wideband MIMO antenna system for Sub-6 GHz cell phone. In Proceedings of the IEEE International Conference on Electronics, Circuits and Information Engineering (ECIE), Zhengzhou, China, 22–24 January 2021; pp. 1–5.

electronics

Article

A Compact mmWave MIMO Antenna for Future Wireless Networks

Muhammad Imran Khan [1,*], Sarmadullah Khan [2], Saad Hassan Kiani [3,4], Naser Ojaroudi Parchin [5,*], Khalid Mahmood [6], Umair Rafique [7] and Muhammad Mansoor Qadir [8]

1 Department of Electrical Engineering, CECOS University, Peshawar 25000, Pakistan
2 School of Computer Science and Informatics, De Montfort University, The Gateway, Leicester LE1 9BH, UK
3 Department of Electrical Engineering, IIC University of Technology, Phnom Penh 121206, Cambodia
4 Smart Systems Engineering Laboratory, College of Engineering, Prince Sultan University, Riyadh 11586, Saudi Arabia
5 School of Engineering and the Built Environment, Edinburgh Napier University, Edinburgh EH10 5DT, UK
6 Department of Electrical Engineering, Abasyn University, Peshawar 25000, Pakistan
7 Department of Information Engineering, Electronics and Telecommunications, Sapienza University of Rome, 00184 Rome, Italy
8 Department of Computer Science, Iqra National University, Peshawar 25000, Pakistan
* Correspondence: imran@cecos.edu.pk (M.I.K.); n.ojaroudiparchin@napier.ac.uk (N.O.P.)

Abstract: This article presents a four-element multiple-input multiple-output (MIMO) antenna design for next-generation millimeter-wave (mmWave) communication systems. The single antenna element of the MIMO systems consists of a T-shaped and plow-shaped patch radiator designed on an ultra-thin Rogers RT/Duroid 5880 substrate. The dimensions of the single antenna are 10×12 mm^2. The MIMO system is designed by placing four elements in a polarization diversity configuration whose overall dimensions are 24×24 mm^2. From the measured results, it is observed that the MIMO antenna provides 9.23 GHz impedance bandwidth ranging from 22.43 to 31.66 GHz. In addition, without the utilization of any decoupling network, a minimum isolation of 25 dB is achieved between adjacent MIMO elements. Furthermore, the proposed MIMO antenna system is fabricated, and it is noted that the simulated results are in good agreement with the measured results. Through the achieved results, it can be said that the proposed MIMO antenna system can be used in 5G mmWave radio frequency (RF) front-ends.

Keywords: MIMO; mmWave; plow-shape; polarization diversity

Citation: Khan, M.I.; Khan, S.; Kiani, S.H.; Ojaroudi Parchin, N.; Mahmood, K.; Rafique, U.; Qadir, M.M. A Compact mmWave MIMO Antenna for Future Wireless Networks. *Electronics* **2022**, *11*, 2450. https://doi.org/10.3390/electronics11152450

Academic Editor: Christos J. Bouras

Received: 27 June 2022
Accepted: 3 August 2022
Published: 6 August 2022

1. Introduction

The new communication era is based on fifth-generation (5G) technology, which has been deemed the fastest data-rate-providing technology when compared to third-generation (3G) and fourth-generation (4G) technologies. The International Telecommunication Union (ITU) divided the 5G communication spectrum into two parts. One of them is named the sub-6 GHz spectrum [1], whereas the second one is called the millimeter-wave (mmWave) spectrum [2–4]. The reason for choosing sub-6 GHz spectrum is to achieve high data rates with existing communication technologies [5], but the antenna size is quite large and also the gain of the antenna will be low [6]. In contrast to this, at mmWave frequency bands, the small wavelength provides an extra advantage for designing compact antennas for 5G communication systems. In addition, the higher frequency bands provide large bandwidth, which ultimately leads to achieving a high data rate [7,8]. At mmWave spectrum, the propagation loss is high due to small wavelengths [9,10], and this problem can be solved by designing high-gain antennas. In addition, high data rates can be achieved by utilizing multiple-input multiple-output (MIMO) technology. For this purpose, many researchers have reported different MIMO antenna designs in the literature for the mmWave spectrum.

In [11], a coplanar waveguide (CPW)-fed antenna array was presented for mmWave networks. The authors designed a linear MIMO array by utilizing a T-shaped radiator with a defected ground plane. The use of a defected ground plane enhanced the isolation between antenna elements, which tends to achieve high channel capacity, but it reduces the directivity of the MIMO antenna. In [12], a four-element MIMO antenna was designed for mmWave applications. A nature-inspired patch radiator and a partial ground plane with a semi-circular shell was used to achieve a wideband response. The presented results show that the single antenna provides a bandwidth of 4 GHz in the frequency range of 26–30 GHz. In the MIMO configuration, a minimum isolation of 24 dB was observed between the MIMO elements. In [13], a quad-port MIMO antenna design was presented for 5G mmWave applications. The MIMO's single element consists of a 1 × 2 array of patch radiators. To achieve wideband response and high isolation, a defected ground structure (DGS) configuration was utilized. It was observed that the use of DGS tends to achieve a wideband response from 25.5 to 29.6 GHz, whereas the isolation between antenna elements was noted to be less than 10 dB. The same kind of MIMO configuration was designed in [14]. In this design, the authors used a slotted zig-zag decoupling structure to achieve >25 dB of isolation in the band of interest. In [15], a tree-shaped patch-based MIMO antenna system was reported for future mmWave communication systems. The radiating structure was designed in such a way so that it can provide a wideband impedance bandwidth ranging from 23 to 40 GHz. It was also observed that the designed patch antenna offered a high gain of 11.45 dBi in the band of interest.

In [16], a MIMO antenna was designed by using the design technique presented in [13]. The presented MIMO antenna resonates well in the frequency range of 24.55–26.5 GHz. To achieve high isolation (>25 dB), an array of metasurface was placed above the MIMO antenna. The same kind of technique was utilized in [17]. In this design, the authors used a novel patch shape with a partial ground plane to achieve a wideband response ranging from 23.5 to 29.4 GHz. Furthermore, the isolation between the antenna elements was noted to be >20 dBi for the entire operating bandwidth. For high gain, a 2 × 2 array of metasurface was used and achieved a peak gain of 10.44 dBi. Although the MIMO structures provide high isolation and high gain, their bulky configuration restricts their use in compact communication devices. In [18], a novel MIMO antenna configuration was adopted to achieve high isolation. The MIMO design was able to operate at a very wideband frequency range of 36.83–40.0 GHz. Furthermore, an isolation of >18 dB was achieved with a peak gain of 6.5 dBi. However, the presented configuration can only be used in arbitrary-shaped communication devices. In [19], three-element and four-element MIMO antenna designs were presented for 5G mmWave applications. The authors used an inverted-L shaped patch element to achieve a wideband response in the frequency range of 26–40 GHz. In the three-element MIMO design, they employed slots in the ground plane to achieve an isolation of >15 dB between antenna elements, whereas in the four-element design, semi-arc was designed within the slots to obtain an isolation of >22 dB. Although the presented designs offer wideband response and high isolation, they suffer due to low antenna efficiency.

Based on the above-presented discussion, in this article, a T-shaped and plow-shaped patch-based wideband four-element MIMO antenna design is presented for mmWave communication systems. The proposed design is based on a polarization diversity technique and occupies an overall size of 24 × 24 mm^2, which enables its use in compact communication devices. The wideband response is achieved by utilizing a partial ground plane with a square notch. It is observed from the presented results that the MIMO antenna operates well from 22.43 to 31.66 GHz and provides measured impedance bandwidth of 9.23 GHz. Furthermore, without the use of any decoupling or isolation enhancement network, an isolation of 25 dB is achieved between adjacent antenna elements.

2. Antenna Design

2.1. Single Element

The proposed single antenna element design is shown in Figure 1. The front-side of the antenna consists of a T-shaped and plow-shaped patch radiator (see Figure 1a), whereas the back side consists of a partial ground plane with a square notch, as shown in Figure 1b. The plow-shape is comprised of a T-shaped resonator and two extended strips bent at 15°. The partial ground plane is utilized to achieve wideband repose in the band of interest, whereas the square notch is used to enhance the impedance matching. A low-loss ultra-thin Rogers RT/Duroid 5880 dielectric substrate is used for the design of an antenna. The thickness of the dielectric substrate is chosen to be 0.254 mm, and its relative permittivity is equal to 2.2. The overall dimensions of the proposed plow-shaped antenna are noted to be 10×12 mm^2. For the feeding of a plow-shaped patch, a 50 Ω microstrip feeding line is used, whose length and width are equal to 6.75 mm and 0.787 mm, respectively (see Figure 1a). The feed line length is extended to use the connector in an efficient manner. The rest of the design parameters with their respective values are shown in Figure 1.

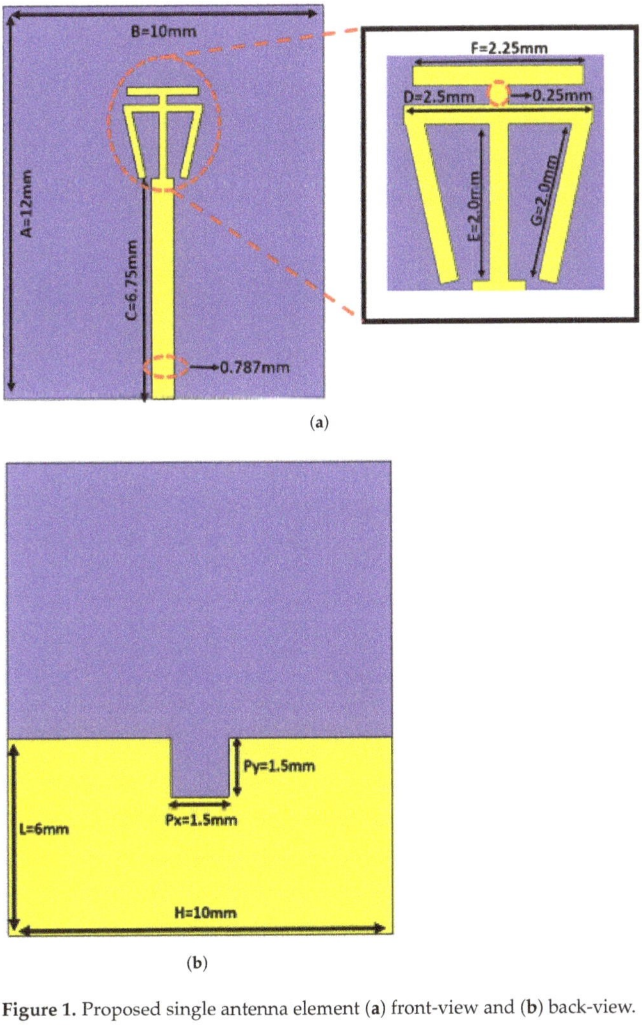

Figure 1. Proposed single antenna element (**a**) front-view and (**b**) back-view.

The simulated reflection coefficient (S_{11}) response of the proposed single antenna element is illustrated in Figure 2a. One can note from the figure that, according to −10 dB impedance bandwidth criteria, the antenna operates well from 22.24 to 31.76 GHz, providing an impedance bandwidth of 9.52 GHz and a fractional bandwidth of 35.25%. The simulated radiation and total efficiency responses of the proposed antenna are shown in Figure 2b. The radiation efficiency is noted to be >95% for the desired operating bandwidth. On the other hand, the total efficiency fluctuates in the range of 83–97%. In addition, the proposed antenna realized gain is shown in Figure 2c. It is noted that the gain of the antenna varies from 2 to 3.5 dBi in the operating bandwidth (see Figure 2c).

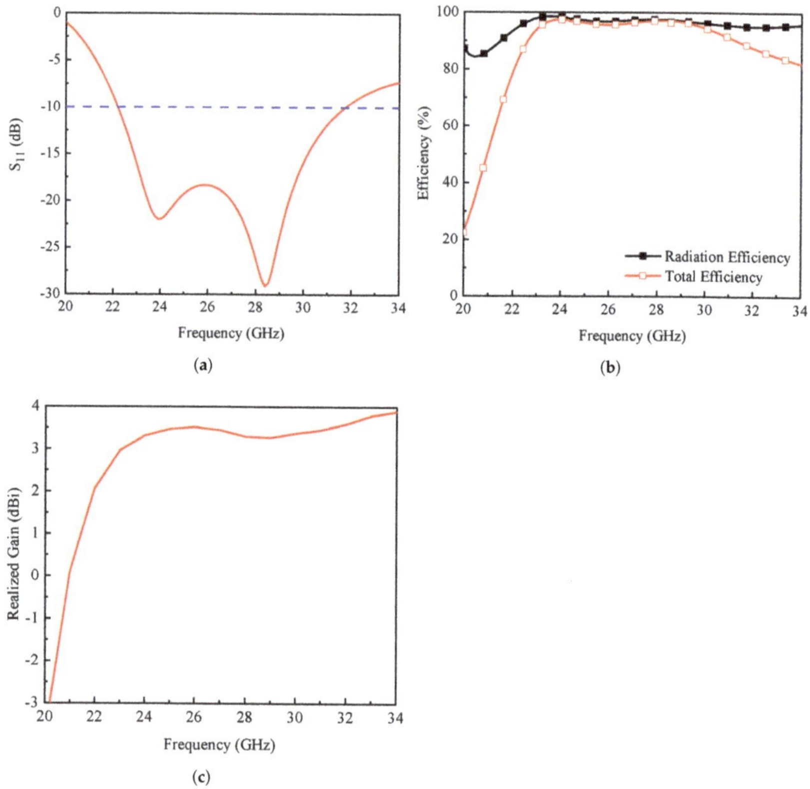

Figure 2. Simulated (**a**) S_{11}, (**b**) radiation and total efficiency, and (**c**) realized gain of the proposed antenna.

The single antenna element was developed through a series of parametric studies. The parameters optimized to obtain the desired response are: square notch modeling in the ground plane, the angle of the open strips attached at both ends of the plow-shaped resonator, the horizontal strip of the plow-shaped resonator, and the length of the T-shaped resonator. Figure 3a shows the effect of ground slot on an antenna's performance. The parameter "Px" value has been changed from 1.3 to 1.7 mm. As the value of "Px" increases, a shift in the lower frequency is observed, as shown in Figure 3a. The optimum response is achieved at Px = 1.5 mm, where the antenna provides acceptable bandwidth and impedance matching (see Figure 3a). For values greater than 1.5 mm, the bandwidth of the antenna is going to decrease, as depicted in Figure 3a.

Figure 3b shows the parametric modeling of the side suspended strips whose length is named as "G". The performance is assessed by changing the rotation angle of the strips from 0° to 20°. As seen from Figure 3b, this angle plays a vital role in producing a desired response. As the angle increases, an increase in impedance bandwidth is observed (see Figure 3b). From the presented study, the desired response is achieved at an angle of 15° angle, as shown in Figure 3b. The effect of parameter "D" on antenna's performance is depicted in Figure 3c. As the value of "D" changed from 2.3 to 2.7 mm, an improved impedance matching is observed on the expense of reduce antenna bandwidth. For D = 2.3 mm, the antenna offers wide impedance bandwidth but provides low impedance matching in the band of interest (see Figure 3c). The optimized response is achieved for D = 2.5 mm, as shown in Figure 3c.

Figure 3d depicts the effect of T-shaped resonator length "F" on the antenna's performance. As the length of the T-shaped resonator increases, a drastic change is observed in the impedance bandwidth, as shown in Figure 3d. For F = 1.75 mm, the antenna provides a wide impedance bandwidth, but the impedance matching is low. An acceptable performance in terms of bandwidth and impedance matching is achieved for F = 2.25 mm (see Figure 3d).

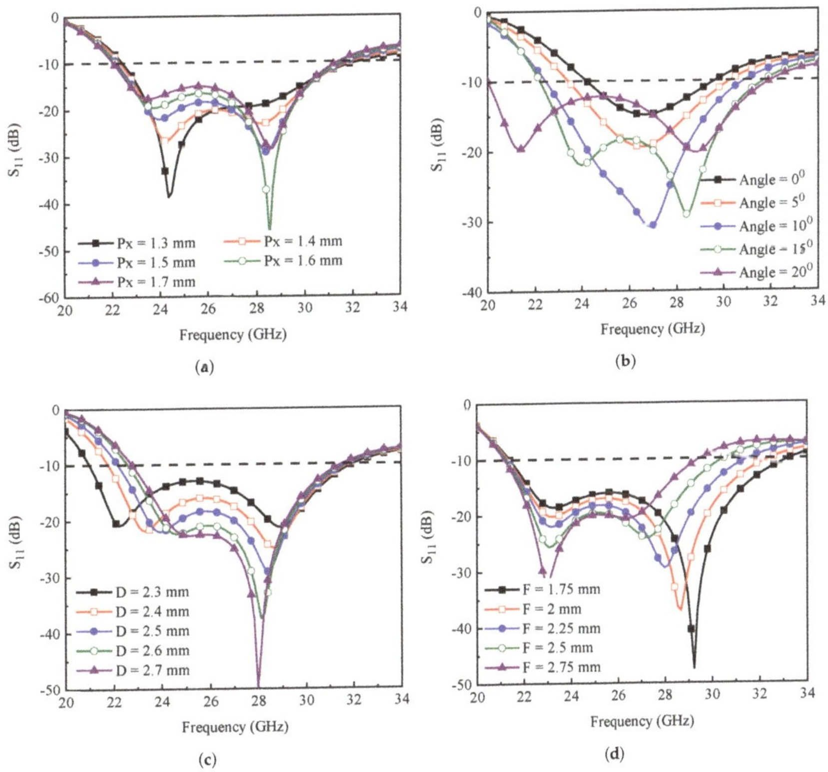

Figure 3. Effect of (**a**) square notch width "Px", (**b**) strip angle, (**c**) plow-shaped horizontal strip "D", and (**d**) T-shaped resonator length "F" on antenna's performance.

2.2. MIMO Antenna Configuration

The proposed single antenna element was transformed into a four-element MIMO configuration to achieve a high data rate and high channel capacity in the mmWave spectrum. Figure 4 shows the design of the proposed MIMO antenna. The MIMO antenna is designed by placing the above-presented antenna element in a polarization diversity configuration, as shown in Figure 4a, whereas the back side consists of independent partial ground planes loaded with a square notch (see Figure 4b). The dimensions of the proposed MIMO antenna are noted to be 24 × 24 mm^2. From the figure, one can observe that there is no decoupling network designed between the antenna elements to achieve high isolation.

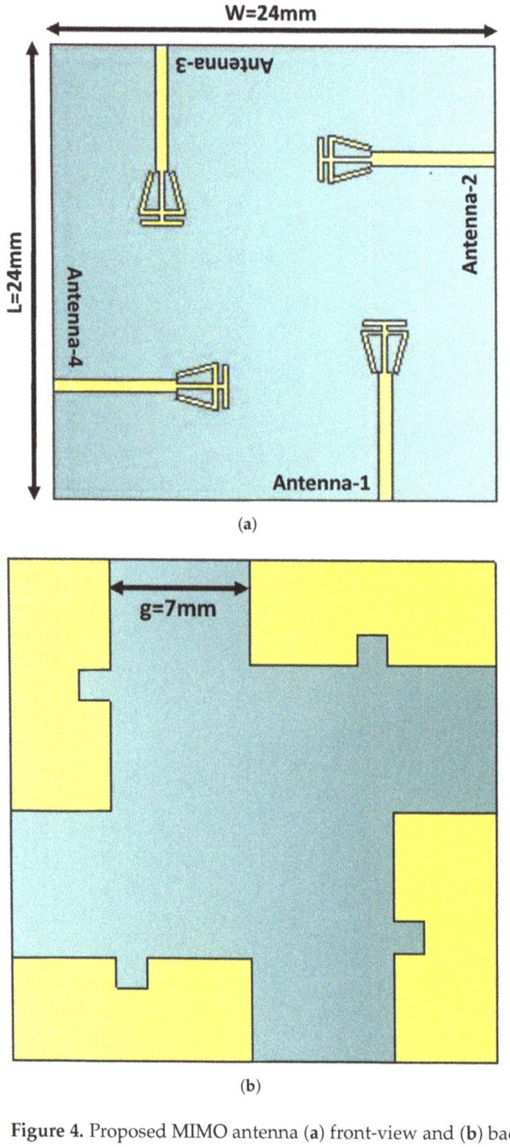

Figure 4. Proposed MIMO antenna (**a**) front-view and (**b**) back-view.

3. Experimental Results and Discussion

3.1. S-Parameters

To verify the simulated data, the proposed MIMO antenna was fabricated and tested using an in-house facility. Figure 5 shows the proposed antenna fabricated prototype. The S-parameters of the proposed MIMO antenna, both simulated and measured, are shown in Figure 6. For simplicity, the reflection coefficients of only two antenna elements are shown. Figure 6a shows the simulated and measured reflection coefficients of antenna-1 and antenna-2. From the simulated reflection coefficients, shown in Figure 6a, it is observed that antenna-1 resonates from 22.2 to 31.66 GHz, whereas antenna-2 operates in the frequency range of 22.4–32.57 GHz. On the other hand, the measured results show that antenna-1 and antenna-2 provide impedance bandwidths ranging from 22.95 to 31.53 GHz and 22.43 to 31.66 GHz, respectively (see Figure 6a). The simulated and measurement isolation results are depicted in Figure 6b. The isolation between adjacent antenna elements is noted to be greater than 25 dB for the entire operating bandwidth. The discrepancies between the results may arise due to SMA connector losses, fabrication tolerances, and indoor scattering environments.

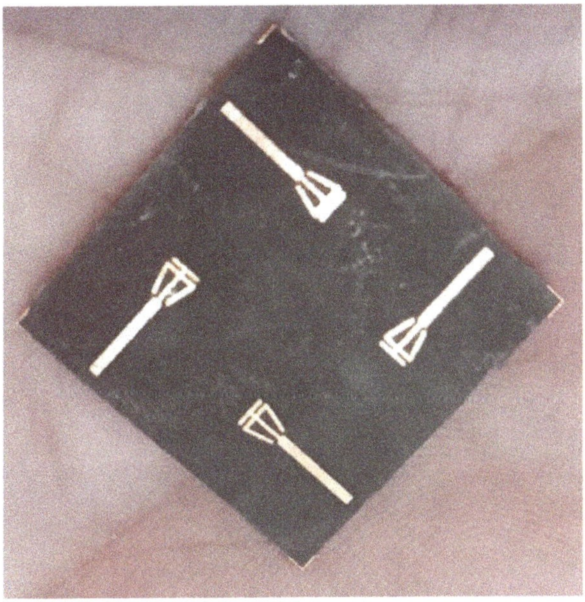

Figure 5. Fabricated prototype of proposed MIMO antenna.

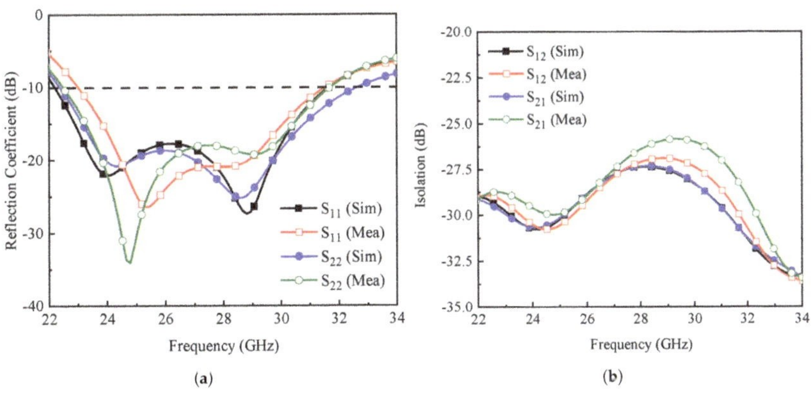

Figure 6. Simulated and measured (**a**) reflection coefficients and (**b**) port isolations of the proposed MIMO antenna.

3.2. Radiation Characteristics

The far-field radiation characteristics of the presented MIMO antenna are assessed using a traditional method. As a reference antenna, a dual-ridge horn antenna with a frequency range of 15–40 GHz is employed, and the designed MIMO antenna is mounted on a turntable that is positioned on the opposite side. RF absorbers are used to cover the anechoic chamber's walls in order to eliminate reflections.

The simulated and measured gain patterns for $\phi = 0^\circ$ and $\phi = 90^\circ$ are shown in Figure 7. The radiation patterns are extracted at 24.5 GHz. It is observed from Figures 7a,c that antenna-1 and antenna-3 offer omnidirectional radiation properties for $\phi = 0^\circ$ and bi-directional (monopole-like) radiation characteristics for $\phi = 90^\circ$. The same kinds of radiation patterns are observed for antenna-2 and antenna-4 (see Figures 7b,d). In this case, the proposed MIMO antenna shows polarization diversity behavior, as shown in Figures 7b,d. In addition to this, the proposed MIMO antenna system is also able to provide pattern diversity in both the planes at the same time, which is clear from the results of Figure 7. This effect can also be observed from Figure 8 where three-dimensional (3-D) radiation characteristics of the proposed antenna are plotted. From the results of Figure 8, it can also be observed that the gain of the MIMO antenna varies in the range of 5.23–5.46 dBi.

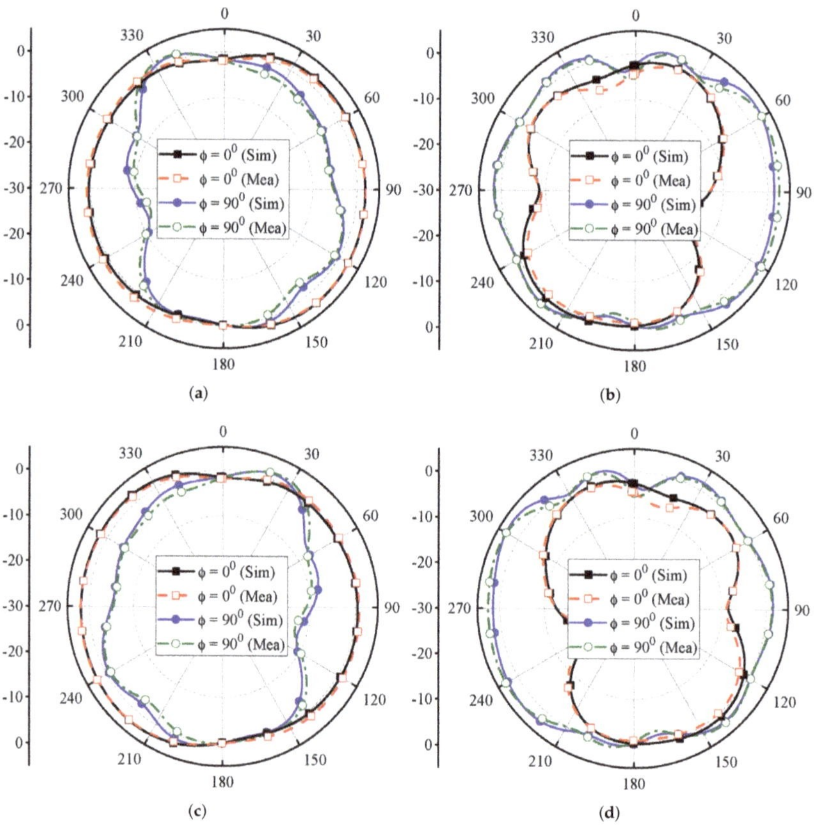

Figure 7. Far-field radiation characteristics of the proposed MIMO antenna at 24.5 GHz for (**a**) antenna-1, (**b**) antenna-2, (**c**) antenna-3, and (**d**) antenna-4.

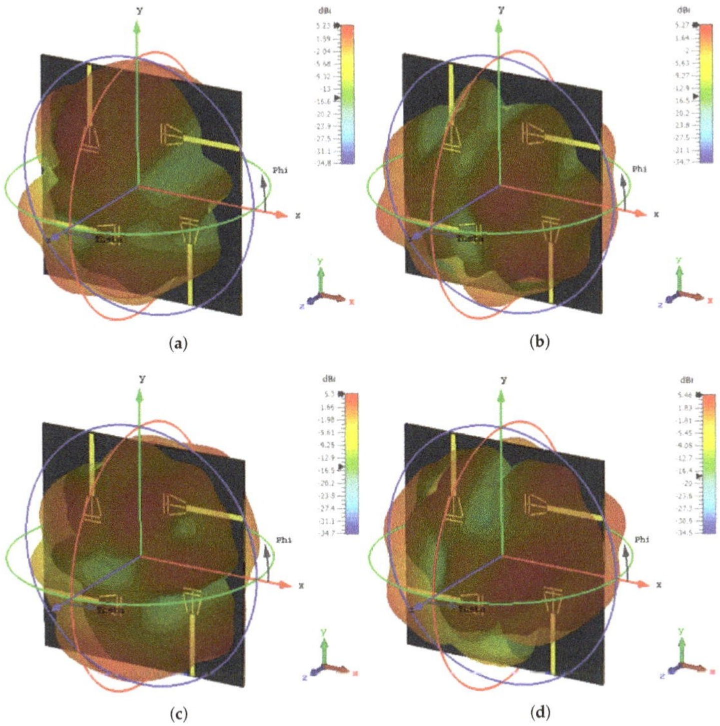

Figure 8. 3-D radiation patterns of the proposed MIMO antenna at 24.5 GHz for (**a**) antenna-1, (**b**) antenna-2, (**c**) antenna-3, and (**d**) antenna-4.

3.3. Total Efficiency and Realized Gain

In Figure 9a, the total efficiency of all the antenna elements is plotted, whereas the overall MIMO realized gain is depicted in Figure 9b. From Figure 9a, it can be observed that the total efficiency of the antenna elements is greater than 85% for the entire operating bandwidth. On the other hand, the simulated gain of the MIMO antenna fluctuates in the range of 5.23–5.46 dBi, whereas the measured gain varies from 4.98 to 5.66 dBi (see Figure 9b).

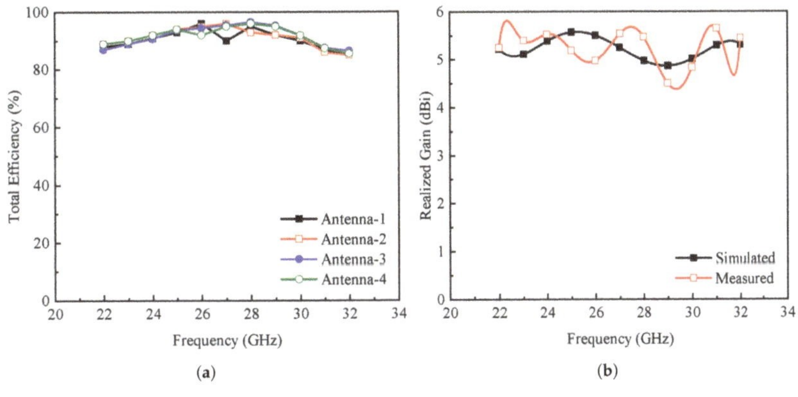

Figure 9. (**a**) Total efficiency and (**b**) realized gain of the proposed MIMO antenna.

3.4. Surface Current

The simulated surface current distribution of the proposed MIMO antenna for all the antenna elements is depicted in Figure 10. It can be noted that for all antenna elements, maximum current is distributed around the feedline, the radiating structure, and the square notch etched on the ground plane. So, it can be said that the square notch plays an important role in achieving a wideband response.

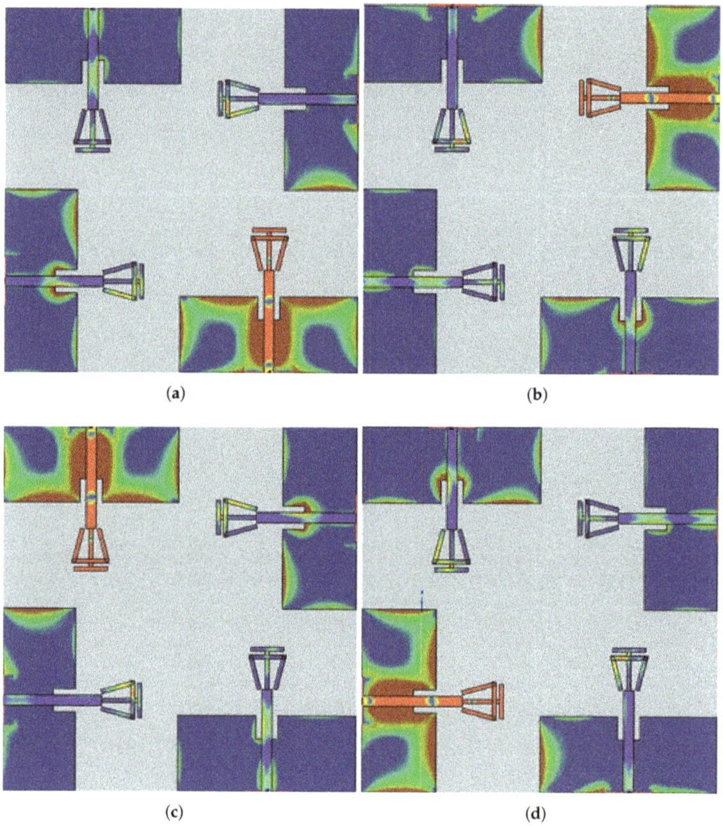

Figure 10. Surface current distribution for (**a**) antenna-1, (**b**) antenna-2, (**c**) antenna-3, and (**d**) antenna-4.

4. MIMO Parameters

MIMO diversity parameters are important to discuss while designing a MIMO antenna system. These MIMO parameters include envelope correlation coefficient (ECC), diversity gain (DG), and mean effective gain (MEG). ECC is the measure of how well the antennas are isolated and can be calculated using the far-field radiation characteristics of the MIMO antenna [1]. In Figure 11a, the ECC of the presented MIMO antenna system is presented. Within the desired operating band, the ECC value is observed to be less than 0.008, which follows the standard value of ≤ 0.5. The low value of ECC also ensures that the proposed MIMO system will have independent channel operation. Furthermore, the DG of the designed MIMO antenna is shown in Figure 11b. The DG can be calculated through ECC as mentioned in [1]. For the presented MIMO antenna system, the DG value is greater than 9.95 dB for the band of interest (see Figure 11b).

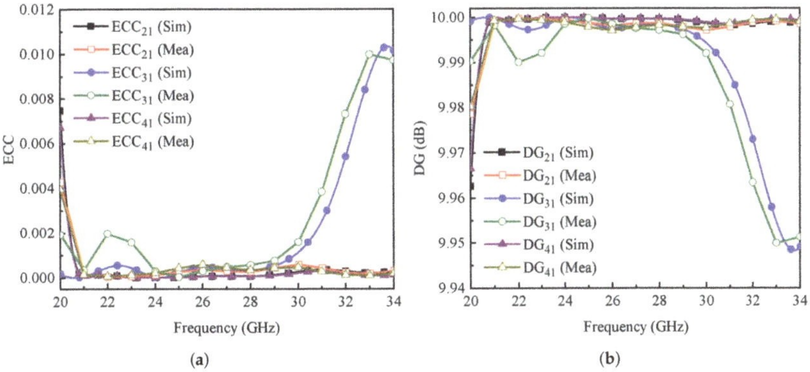

Figure 11. (**a**) Envelope correlation coefficient and (**b**) diversity gain of the proposed MIMO antenna.

MEG is another important MIMO parameter, which is calculated using the equation given in [3] at frequencies of 28 GHz, 28.5 GHz, and 29 GHz, respectively, for all the four antenna elements. MEG actually provides information about how much power is received by each antenna in the MIMO system. Through Table 1, the MEG of the proposed MIMO antenna system is less than −3 dB, which satisfies the convention.

Table 1. Mean effective gain of the proposed MIMO antenna.

Frequency (GHz)	MEG1	MEG2	MEG3	MEG4
28	−3.50	−3.21	−3.26	−3.10
28.5	−3.48	−3.17	−3.25	−3.33
29	−2.87	−3.1	3.47	−3.88

The comparison among previously presented and designed MIMO antenna systems is listed in Table 2. It is noted from the table that the size of the proposed MIMO antenna is small compared to the designs of [13,14,16,19]. Furthermore, the proposed MIMO antenna system offers high bandwidth and acceptable isolation performance compared to the designs reported in [12–14,16,17].

Table 2. Comparison among previously presented and proposed mmWave MIMO antennas.

Ref.	Size (mm^2)	Bandwidth (GHz)	Isolation (dB)	Efficiency (%)	Gain (dBi)	ECC
[12]	25 × 15	3	25	>90	8	<0.001
[13]	35 × 30	4	>20	>80	8.3	<0.04
[14]	35 × 30	1	>30	>80	12	<0.0008
[16]	43 × 30	1.95	>30	–	10.21	<0.0008
[17]	24 × 24	5.9	>25	>80	10.44	<0.01
[19]	30 × 30	14	>20	60	8	<0.018
This Work	24 × 24	9.23	>25	>85	5.66	<0.008

5. Conclusions

In this article, a wideband four-element MIMO antenna system is presented for mmWave communication systems. The MIMO antenna system is designed in such a way that it can provide both pattern and polarization diversity at the same time. From the results, it is noted that the bandwidth of the proposed MIMO antenna system is noted to be 9.23 GHz, ranging from 22.43 to 31.66 GHz. The isolation between adjacent elements and diagonal elements is observed to be >25 dB and >18.5 dB, respectively. The MIMO parameters are also evaluated and their values lie within an acceptable range. Therefore, the proposed MIMO antenna system could be used as a potential candidate for future mmWave wireless systems.

Author Contributions: Conceptualization, S.H.K. and N.O.P.; methodology, M.I.K. and K.M.; software, S.H.K. and K.M.; validation, N.O.P. and K.M.; formal analysis, N.O.P. and U.R.; investigation, S.K.; resources, S.K. and N.O.P.; writing—original draft preparation, S.H.K. and M.M.Q.; writing—review and editing, N.O.P., U.R., and M.M.Q.; visualization, U.R. and M.M.Q.; supervision, S.H.K. and N.O.P.; project administration, K.M. and U.R.; funding acquisition, N.O.P. and U.R. All authors have read and agreed to the published version of the manuscript.

Funding: This project received no external funding.

Conflicts of Interest: The authors declare no conflict of interest.

References

1. Rafique, U.; Khan, S.; Ahmed, M.M.; Kiani, S.H.; Abbas, S.M.; Saeed, S.I.; Alibakhshikenari, M.; Dalarsson, M. Uni-Planar MIMO Antenna for Sub-6 GHz 5G Mobile Phone Applications. *Appl. Sci.* **2022**, *12*, 3746. [CrossRef]
2. Munir, M.E.; Al Harbi, A.G.; Kiani, S.H.; Marey, M.; Parchin, N.O.; Khan, J.; Mostafa, H.; Iqbal, J.; Khan, M.A.; See, C.H.; et al. A New mm-Wave Antenna Array with Wideband Characteristics for Next Generation Communication Systems. *Electronics* **2022**, *11*, 1560. [CrossRef]
3. Kiani, S.H.; Alharbi, A.G.; Khan, S.; Marey, M.; Mostafa, H.; Khan, M.A. Wideband Three Loop Element Antenna Array for Future 5G mmwave Devices. *IEEE Access* **2022**, *10*, 22472–22479. [CrossRef]
4. Luo, Y.; Shen, Y.; Cai, X.; Qian, F.; Xu, S.; Cui, H.; Yang, G. Substrate integrated coaxial line design for mmWave antenna with multilayer configuration. *Int. J. RF Microw. Comput.-Aided Eng.* **2022**, *32*, e23090. [CrossRef]
5. Kiani, S.H.; Iqbal, A.; Wong, S.W.; Savci, H.S.; Alibakhshikenari, M.; Dalarsson, M. Multiple Elements MIMO Antenna System With Broadband Operation for 5th Generation Smart Phones. *IEEE Access* **2022**, *10*, 38446–38457. [CrossRef]
6. Farasat, M.; Thalakotuna, D.N.; Hu, Z.; Yang, Y. A review on 5G sub-6 GHz base station antenna design challenges. *Electronics* **2021**, *10*, 2000. [CrossRef]
7. Ishteyaq, I.; Muzaffar, K. Multiple input multiple output (MIMO) and fifth generation (5G): An indispensable technology for sub-6 GHz and millimeter wave future generation mobile terminal applications. *Int. J. Microw. Wirel. Technol.* **2021**, *14* 932–948. [CrossRef]
8. Wang, Y.; Li, J.; Huang, L.; Jing, Y.; Georgakopoulos, A.; Demestichas, P. 5G mobile: Spectrum broadening to higher-frequency bands to support high data rates. *IEEE Veh. Technol. Mag.* **2014**, *9*, 39–46. [CrossRef]
9. Osseiran, A.; Boccardi, F.; Braun, V.; Kusume, K.; Marsch, P.; Maternia, M.; Queseth, O.; Schellmann, M.; Schotten, H.; Taoka, H.; et al. Scenarios for 5G mobile and wireless communications: The vision of the METIS project. *IEEE Commun. Mag.* **2014**, *52*, 26–35. [CrossRef]
10. Sulyman, A.I.; Nassar, A.T.; Samimi, M.K.; MacCartney, G.R.; Rappaport, T.S.; Alsanie, A. Radio propagation path loss models for 5G cellular networks in the 28 GHz and 38 GHz millimeter-wave bands. *IEEE Commun. Mag.* **2014**, *52*, 78–86. [CrossRef]
11. Jilani, S.F.; Alomainy, A. Millimetre-wave T-shaped MIMO antenna with defected ground structures for 5G cellular networks. *IET Microwaves Antennas Propag.* **2018**, *12*, 672–677. [CrossRef]
12. Rahman, S.; Ren, X.c.; Altaf, A.; Irfan, M.; Abdullah, M.; Muhammad, F.; Anjum, M.R.; Mursal, S.N.F.; AlKahtani, F.S. Nature inspired MIMO antenna system for future mmWave technologies. *Micromachines* **2020**, *11*, 1083. [CrossRef] [PubMed]
13. Khalid, M.; Iffat Naqvi, S.; Hussain, N.; Rahman, M.; Mirjavadi, S.S.; Khan, M.J.; Amin, Y. 4-Port MIMO antenna with defected ground structure for 5G millimeter wave applications. *Electronics* **2020**, *9*, 71. [CrossRef]
14. Bilal, M.; Naqvi, S.I.; Hussain, N.; Amin, Y.; Kim, N. High-Isolation MIMO antenna for 5G millimeter-wave communication systems. *Electronics* **2022**, *11*, 962. [CrossRef]
15. Sehrai, D.A.; Abdullah, M.; Altaf, A.; Kiani, S.H.; Muhammad, F.; Tufail, M.; Irfan, M.; Glowacz, A.; Rahman, S. A novel high gain wideband MIMO antenna for 5G millimeter wave applications. *Electronics* **2020**, *9*, 1031. [CrossRef]
16. Tariq, S.; Naqvi, S.I.; Hussain, N.; Amin, Y. A metasurface-based MIMO antenna for 5G millimeter-wave applications. *IEEE Access* **2021**, *9*, 51805–51817. [CrossRef]

17. Sehrai, D.A.; Asif, M.; Shah, W.A.; Khan, J.; Ullah, I.; Ibrar, M.; Jan, S.; Alibakhshikenari, M.; Falcone, F.; Limiti, E. Metasurface-based wideband MIMO antenna for 5G millimeter-wave systems. *IEEE Access* **2021**, *9*, 125348–125357. [CrossRef]
18. Sehrai, D.A.; Asif, M.; Shoaib, N.; Ibrar, M.; Jan, S.; Alibakhshikenari, M.; Lalbakhsh, A.; Limiti, E. Compact quad-element high-isolation wideband MIMO antenna for mm-wave applications. *Electronics* **2021**, *10*, 1300. [CrossRef]
19. Patel, A.; Vala, A.; Desai, A.; Elfergani, I.; Mewada, H.; Mahant, K.; Zebiri, C.; Chauhan, D.; Rodriguez, J. Inverted-L Shaped Wideband MIMO Antenna for Millimeter-Wave 5G Applications. *Electronics* **2022**, *11*, 1387. [CrossRef]

electronics

Article

MTM-Inspired Graphene-Based THz MIMO Antenna Configurations Using Characteristic Mode Analysis for 6G/IoT Applications

Sherif A. Khaleel [1,2,*], Ehab K. I. Hamad [2,*], Naser Ojaroudi Parchin [3,*] and Mohamed B. Saleh [1]

1 College of Engineering and Technology, Arab Academy for Science, Technology and Maritime Transport, Aswan 81511, Egypt; besheersaleh@aast.edu

2 Electrical Engineering Department, Faculty of Engineering, Aswan University, Aswan 81542, Egypt

3 School of Engineering and the Built Environment, Edinburgh Napier University, Edinburgh EH10 5DT, UK

* Correspondence: sherif.abdalla@aast.edu (S.A.K.); e.hamad@aswu.edu.eg (E.K.I.H.); n.ojaroudiparchin@napier.ac.uk (N.O.P.)

Abstract: 6G wireless communications will be immersed in the future with different applications. It is expected to support all IoT services and satellite communications, and it is expected to support artificial intelligence (AI) and machine learning (ML). The THz frequency band has a vital role in 6G communication. In this study, a new graphene plasmonic two-port Terahertz (THz) MIMO antenna is analyzed by the characteristic mode theory (CMA), which gives a better insight into the physical behavior of the MIMO configurations. The proposed MIMO antenna is compact and designed on a Teflon substrate of 130×85 μm^2. The antenna provides a wide impedance bandwidth of 0.6 THz (3.2–3.8 THz). The CMA is applied to clarify the position at which the mutual coupling gives a maximum concentrated current distribution. It is mainly used to reveal the preferable MIMO antenna configuration by the usage of the model significant and model current distribution property. To reduce the mutual coupling between the radiating elements, a complementary dumbbell-structure Metamaterial (MTM) unit cell is etched in the ground plane to block the coupling mode without any affection on the dominant mode. The preferred MIMO configuration gives high isolation of −55 dB between the radiating patches. The fundamental characteristics have been discussed in detail. The proposed MIMO design offers several attractive features such as large bandwidth of 0.6 THz, low envelope correlation coefficient (ECC) of 0.000168, compact size, stable radiation, high gain of 7.23 dB, and low channel capacity loss (CCL) of 0.006. The proposed MIMO design is suitable for different applications in the THz band according to the high-performance parameters such as biomedical applications, security scanning, sensing, IoT, and 6G high-speed wireless communication systems.

Keywords: 6G communications; characteristic mode analysis (CMA); dumbbell-structure; graphene; metamaterial; MIMO antenna; IoT

Citation: Khaleel, S.A.; Hamad, E.K.I.; Parchin, N.O.; Saleh, M.B. MTM-Inspired Graphene-Based THz MIMO Antenna Configurations Using Characteristic Mode Analysis for 6G/IoT Applications. *Electronics* **2022**, *11*, 2152. https://doi.org/10.3390/electronics11142152

Academic Editors: Faisel Tubbal, Ladislau Matekovits and Raad Raad

Received: 9 June 2022
Accepted: 6 July 2022
Published: 9 July 2022

1. Introduction

The new era of the 6 G wireless communication system has gained significant attention in recent years as a result of the growing need for greater data rates and ultra-high-speed connectivity. Around 2030, it is anticipated that the 6G wireless communication technology will be operational [1,2]. Terahertz (THz) frequency bands will be critical components of the 6G wireless communication infrastructure [3–5]. The evolution of the 6 G in terahertz frequency was the least studied band in the electromagnetic spectrum, owing to the unavailability of needed materials, sources, measurements, equipment, and detectors that operate inside the terahertz band [6]. The THz spectrum lies between the microwave and infrared in the range of 0.1–10 THz. Recently, the terahertz band has been emersed in different applications such as high data rate transmission in IoT applications [7], biomedical applications [8,9], material characterizations [10], and high-speed secure data transmission [11]. Numerous antennas have been developed for terahertz spectrum uses, including the Leaky-Wave antenna [12], Yagi-Uda antenna [13,14], Bow-Tie antenna [15], Monopole Antenna [16], and log-periodic antenna [17]. The planar microstrip patch antenna possesses numerous advantages over the other types due to the low cost of manufacturing, low profile, simple structure, and Monolithic Microwave Integrated Circuit (MMIC) compatibility. Although, the metallic microstrip antenna suffers from low antenna performance such as gain, bandwidth, and radiation efficiency, especially on the scale of micrometer due to the low conductivity and mobility [18]. The limitation of the metallic planar microstrip patch antenna due to its drawbacks pushes researchers to the usage of 2D organic materials such as graphene, which is extracted from graphite [19–21]. The graphene utilization as a radiating patch treats the shortage of metallic materials. Graphene as a single-layer honeycomb lattice structure has excellent electrical, mechanical, thermal, and optical properties [22]. The most important property of graphene is the ability to support the propagation of the surface plasmon polariton (SPP) waves at the THz window. Additionally, graphene conductivity can be adjusted dynamically by applying an electric field via an external gate voltage, which alters the chemical potential and hence the conductivity [23]. As of today, research in millimeter-wave technology is advancing toward commercial implementation and is still spurred by the scarcity of bandwidth; the terahertz (THz) band is envisioned as the near future frontier for communication. In the present day, improved channel capacity and a high data rate are required for wireless communication to fulfill requirements. So, MIMO-communication systems operating in the THz range with data transfer rates of up to Tera-bit/sec (Tbps) are required [24]. Moreover, there are always some obstacles to the THz wireless communication frequency band such as high attenuation, Multipath fading, absorptions, and path loss. So, the MIMO system is a good candidate to face these problems.

In the MIMO antenna, the distance between elements has a very strong impact on the mutual coupling between the antenna elements. Greater mutual coupling occurs as a result of closer separation, which raises the correlation coefficients and reduces efficiency. Thus, it is required to have enough space between elements, which should be taken as $\lambda/2$, or for even better isolation it can be taken as $\lambda/4$ of the frequency to have a mutual coupling less than -15 dB [25]. Reduction in mutual coupling leads to reducing the Envelop Correlation Coefficient (ECC), increasing diversity gain (DG), and lowering the Channel Capacity Loss (CCL). It is crucial to note that graphene has inherent benefits in reducing mutual coupling effects because of its short SPP wavelength resulting from the slow propagation speed of SPP waves. As a result, graphene-based antenna arrays with the same spacing between components display fewer mutual coupling effects than metallic antenna arrays of the same distance. Different techniques and strategies are presented for the development of the decoupling methods. A neutralization line is used for isolation improvement between MIMO antenna elements, defected ground structure (DGS), electromagnetic bandgap (EBG), frequency selective surfaces (FSS), and artificial magnetic conductor (AMC) metamaterial structures that are considered effective methods to reduce the isolation between the MIMO antenna elements. The MTM structure is used as a band stop filter to reduce the coupling. Furthermore, the main parameters that the MTM used to enhance the isolation process are the structure form, the position of the unit cell, and the number of elements used. All of these parameters can be analyzed using the characteristic mode analysis (CMA) technique. CMA has become a popular technique for studying and constructing antennas, especially owing to the physical understanding acquired regarding antenna working principles without regard for any specific feeding method. Characteristic mode theory is a method of modal analysis for antennas of any form. In addition, it clearly describes the resonating frequency of certain modes, radiation patterns, and the associated mode current. Fundamental modes are extracted from an eigenvalue issue using the CMA.

Calculating Z, the Method of Moments (MOM) impedance matrix, is often required to solve the following eigenvalue equation.

$$X(J_n) = \lambda_n R(J_n) \quad (1)$$

where X and R are the imaginary and real parts, respectively, of the MOM impedance matrix Z, $J_n's$ are the modes or eigen-currents, and λ_n are the eigenvalues. The eigenvalue problem is solved, for each frequency, using the procedure detailed by Harrington and Mautz [26]. Antenna analysis using the MOM approach requires an understanding of the antenna's structure and its excitation field. In the meantime, just the antenna structure is needed as a known requirement in CMA. The characteristic modes are obtained by solving the following mathematical equations:

$$\vec{J} = \sum_n a_n J_n \quad (2)$$

where J_n is the eigen-current and a_n is the modal weighting coefficient

$$a_n = \frac{V_n^i}{1 + j\lambda_n} \quad (3)$$

where λ_n is the eigenvalue and V_n^i is the modal excitation coefficient. The modal significance, MS_n represented by the normalized amplitude of the current modes is calculated as follows:

$$MS_n = \left| \frac{1}{1 + j\lambda_n} \right| \quad (4)$$

Model significance demonstrated how each mode is near to resonance at each frequency. It reaches the maximum value of 1 at its resonance frequency (when $\lambda_n = 0$). The modal significance ranges from $0 < MS_n \leq 1$. The resonance mode is associated with a model significance of unity. In other words, the characteristic angle β_n can be calculated as in Equation (5). The characteristic angle has the advantage that its value changes more quickly making it ideal for analysis.

$$\beta_n = 180 \quad tan^{-1}(\lambda_n) \quad (5)$$

The CMA analysis is carried out using CST microwave studio using two different solvers, which are a surface integral solver and a multi-layer solver. The model significant (MS_n), characteristic angle (β_n), and the eigenvalue λ_n, all indicate the behavior of the concerned structure. At resonance mode, the value of λ_n is equal to 0, while (MS_n) is equal to 1 and the value of the characteristic angle (β_n) is equal to 180°. The inductive mode that stores the magnetic energy occurs when $\lambda_n > 0$ or $\beta_n < 180^\circ$. In contrast, the capacitive mode stores the electric energy when the value of $\lambda_n < 0$ or $\beta_n > 180^\circ$.

Herein, in this paper, a graphene plasmonic microstrip nano-antenna is proposed to be used in the 6G wireless communication system applications to operate at a 3.5 THz frequency band. A three configurations graphene plasmonic two-ports MTM microstrip patch antenna is reported based on the theory of characteristic mode analysis (CMA), which gives us the opportunity to reach the optimum design of the antenna without wasting a lot of time in the try and error iteration process. The CMA is also used to detect the optimal configuration that gives high isolation and good performance to withstand the current distribution and the model significance. In order to reduce the mutual coupling between the two-port graphene plasmonic microstrip nano-antenna, a metamaterial unit cell of Dumbbell-structure is introduced. The diversity performance is verified by the usage of the envelope correlation coefficient (ECC), channel capacity loss (CCL), and diversity gain (DG). The strategy of the proposed design is demonstrated in the following order: firstly, the design procedure of a single element graphene plasmonic nano-antenna at the resonance frequency of 3.5 THz by the usage of CMA to obtain the optimum dimensions then a verification of the desired is carried out by another software such as HFSS and ADS to invoke the equivalent circuit model to verify the intended performance. Secondly, the graphene plasmonic two-port MIMO array antenna is used to increase the gain with three configurations. Thirdly, the characteristics of the Dumbbell-structure MTM unit cell is investigated and etched in the antenna to reduce the mutual coupling between the two elements. Finally, the diversity performance of the graphene-MIMO configuration is also carried out with and without the MTM, and this design is validated using the HFSS and ADS software.

2. Graphene Patch Antenna Design

2.1. Graphene's Conductivity and Dispersion Relation

Graphene, as a carbon-based material, has been prepared with different techniques. A single layer of graphene was extracted from the bulk graphite crystal structure using a mechanical exfoliation approach. Chemical vapor deposition (CVD), thermal exfoliation, and carbon segregation [27] are other preparation methods for the graphene layers. At the THz frequency band, this exceptional material exhibits excellent mobility, transparency, flexibility, and environmental stability. As a result, it is the best choice that can be used in antenna applications. In comparison to other nano-plasmonic materials such as gold and silver, graphene is an excellent contender for promoting the propagation of SPP waves. The dispersion relation of Transverse Magnetic (TM) of SPP depends on the conductivity of graphene given by [28].

$$-i\frac{\sigma_s}{\omega\varepsilon_0} = \frac{\varepsilon_{r1} + \varepsilon_{r2}\coth\left(K_{eff}t_o\right)}{K_{eff}} \tag{6}$$

where ε_{r1} and ε_{r2} are the relative permittivity of air over the graphene layer and the dielectric constant of the substrate materials, respectively. K_{eff}, is the wavenumber of the *SPP* wave which can be expressed by:

$$K_{eff} = K_0\left(\eta_{eff}\right) \tag{7}$$

where η_{eff} is the effective refractive index of SPP and K_0 is the free space wavenumber $K_0 = \frac{2\pi}{\lambda}$. According to Kubo's formula [29,30]. The graphene conductivity is the sum of both intraband and interband conductivity, and can be described as:

$$\sigma_s = \sigma_{inter} + \sigma_{intra} \tag{8}$$

$$\sigma_s(\omega, \mu_c, \tau, T) = \frac{je^2(\omega - j\tau^{-1})}{\pi\hbar^2} \times \left[\begin{array}{c} \frac{1}{(\omega - j\tau^{-1})^2} \\ \int_0^\infty \varepsilon\left(\frac{\partial f_d(\varepsilon)}{\partial\varepsilon} - \frac{\partial f_d(-\varepsilon)}{\partial\varepsilon}\right) d\varepsilon - \\ \int_0^\infty \frac{f_d(-\varepsilon) - f_d(\varepsilon)}{(\omega - j\tau^{-1})^2 - 4(\varepsilon/\hbar)^2} d\varepsilon \end{array} \right] \tag{9}$$

$$f_d(\varepsilon) = \frac{1}{\left[e^{(\varepsilon - \mu_c)/K_BT} + 1\right]} \tag{10}$$

where T is the temperature, K_B is the Boltzmann constant, $\hbar$ is the reduced Plank's constant, μ_c is the chemical potential, e is the charge of electron τ is the relaxation time and $f_d(\varepsilon)$ is the Fermi-Dirac distribution. The first term in the preceding equation denotes graphene's intraband conductivity, whereas the second term denotes graphene's interband conductivity. Intraband conductivity dominates in the MW and THz bands, whereas inter-band conductivity dominates in the near-infrared and visible spectrums [31]. Graphene's intraband and interband conductivities can be expressed as:

$$\sigma_{intra}(\omega, \mu_c, \tau, T) = -j\frac{e^2 K_B T}{\pi\hbar^2(\omega - j\tau^{-1})}\left[\frac{\mu_c}{K_B T} + 2\ln\left(e^{-\mu_c/K_B T} + 1\right)\right] \quad (11)$$

$$\sigma_{inter}(\omega, \mu_c, \tau, T) = -\frac{je^2}{4\pi\hbar} ln\left[\frac{2|\mu_c| - \hbar(\omega - j\tau^{-1})}{2|\mu_c| + \hbar(\omega - j\tau^{-1})}\right] \quad (12)$$

2.2. Graphene Modeling

The graphene layer is simulated and optimized using CST Microwave Studio, which is based on the Finite Element Method (FEM). This program has two built-in graphene models, namely *"graphene"* and *"graphene-Eps"*. The first model uses a single layer of graphene with a thickness of 0.345 nm, whereas the second uses a multilayer of graphene sheets with varying thicknesses. In this paper, we proposed an application of a single layer of graphene sheet of 0.34 nm thickness for our presented graphene plasmonic nanoantenna. Figure 1a shows the graphical representation for the surface conductivity of the graphene layer at zero value of chemical potential and the relaxation time is 0.1 p sec at room temperature which is identical to the one presented by Llatser et al. [32].

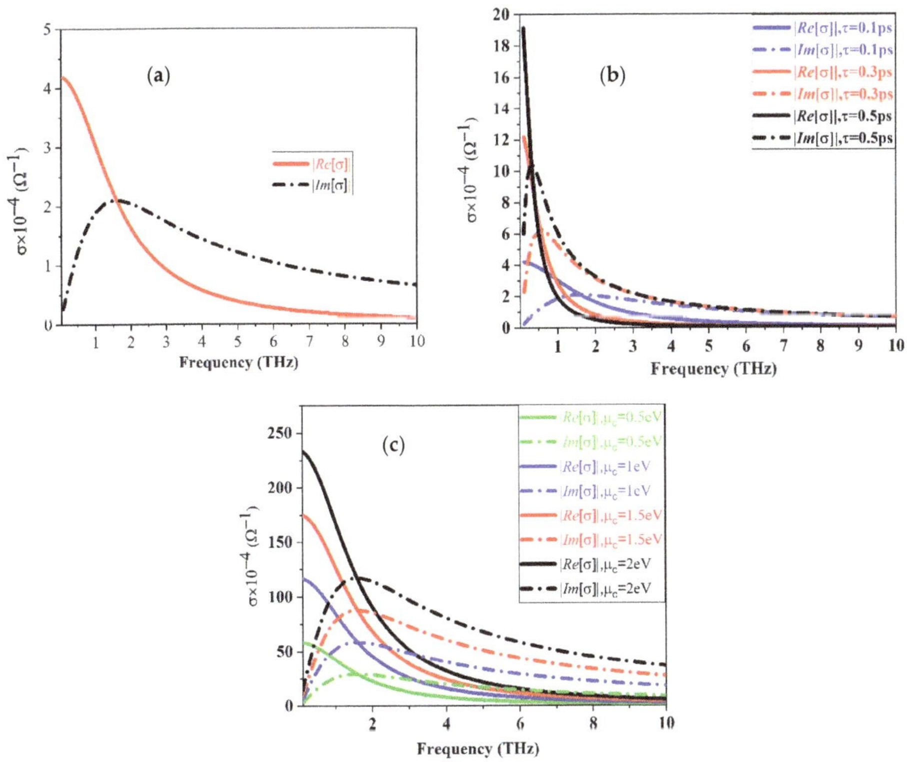

Figure 1. Surface conductivity of graphene for (**a**) μc = 0 eV and τ = 0.1 ps, (**b**) different relaxation times at μc = 0 eV, and (**c**) different chemical potential values at τ = 0.1 ps T = 300 K.

The real and imaginary parts of our simulated graphene are presented in Figure 1b,c for various relaxation time (τ) and chemical potential (μc) values, respectively, using the

MATLAB program. This figure demonstrates that the overall conductivity of a graphene slab is strongly reliant on the chemical potential and the relaxation time. Chemical potential is proportional to carrier density, which can be varied via an external gate voltage or chemical doping:

$$n = \frac{2}{\pi \hbar^2 v_f^2} \int_0^\infty \mathcal{E}[f_d(\mathcal{E} - \mu_c) - f_d(\mathcal{E} + \mu_c)] d\mathcal{E} \quad (13)$$

The relationship between the chemical potential and gate voltage is explained by [33]. Tunability is required for the wideband operation of the proposed antennas which can prevent path loss and enhance data rates in THz communication systems. Tunable antennas have been considered the best tools for manipulating the interaction of light with electrons present in materials. The fundamental capabilities of graphene antennas are vastly utilized in a broad scope of practices, including high-speed communication with high data rate, terahertz detection energy harvesting, and optical light emission.

$$V_g = \left[\frac{e \mu_c{}^2 h}{\pi \hbar^2 v_f{}^2 \varepsilon_0 \varepsilon_r} \right] \quad (14)$$

where h is the substrate thickness and ε_r is the relative permittivity of the substrate material. The conductivity of graphene can be changed with the relaxation time (τ) as follows [34].

$$\tau \simeq \mu_g \hbar \sqrt{\frac{n\pi}{e v_f}} \quad (15)$$

where μ_g is the electron mobility of graphene and v_f is the Fermi velocity. The relaxation time increases with the chemical potential values. Long relaxation time supports the propagation of the SPP wave.

3. Antenna Design

3.1. Single Element Patch Antenna

Initially, a conventional microstrip patch antenna with a graphene plasmonic layer is engraved on the top side of Teflon with a relative permittivity of 2.1, and a thickness of 10 μm. The proposed antenna is designed to resonate at 3.5 THz. The dimensions of the patch are recognized by the usage of the characteristic mode analysis (CMA) using the integral equation solver in the CST microwave studio. The optimum width and length of the patch are 32 μm and 25 μm, respectively. The single element of the patch antenna is matched to a transmission line of 50 Ω characteristic impedance using a quarter-wavelength impedance transformation. The model significance and the characteristic angel curves are shown in Figure 2a,b. Mode 1 is the resonance mode (desired mode) at a frequency of 3.5 THz in which the characteristic angle is 180° and the model significance in this frequency is the highest value (closely to unity). The validation process of the CST result can be achieved by the usage of HFSS and the equivalent circuit model using ADS software. The values of the electrical RLC components are calculated, and the electric circuit model of the antenna is developed using the circuit theory approach [35].

The desired current distribution specifies the optimum position for the excitation process so, the proposed structure of the single-element antenna is shown in Figure 3. The calculated value of the circuit parameters is listed in Table 1. Figure 4 shows the circuit model connection diagram of the proposed antenna. The results obtained from the HFSS and ADS circuit theory approach are in good agreement with those obtained using CST software. The $|S_{11}|$ curve of the graphene plasmonic antenna is illustrated in Figure 5. There is a small deviation in the $|S_{11}|$ curves, which may be due to the different numerical approaches used in different software. The 3D radiation pattern of the proposed graphene plasmonic antenna is depicted in Figure 6. Figure 7 shows the graphene's frequency-reconfigurable behavior of the graphene plasmonic antenna when the chemical potential applied varied between 0.5 and 2.5 eV. This figure reveals that the plasmonic resonance frequency of the proposed antenna increases via the chemical potential and can be dynamically controlled in a wide frequency range from 2.75 to 4 THz. So, this design has dynamic reconfigurability by applying an external electric field.

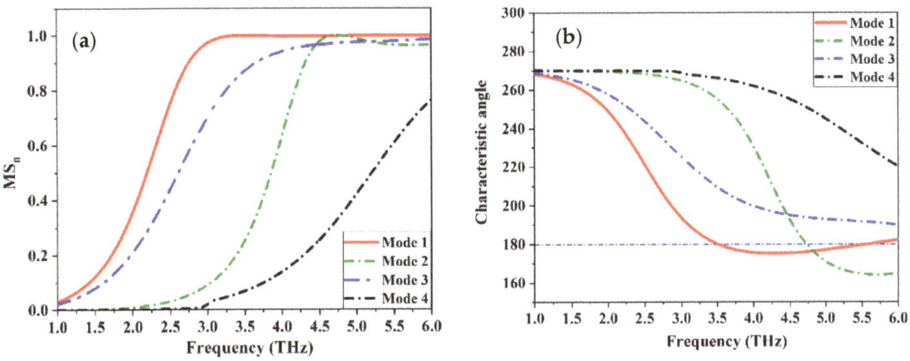

Figure 2. (**a**) Significant and (**b**) characteristic angle of the first four modes of the proposed patch antenna.

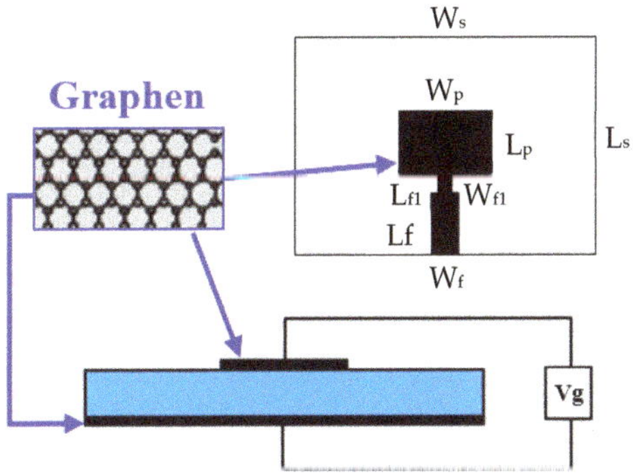

Figure 3. Schematic of the graphene plasmonic patch antenna.

Table 1. The proposed antenna parameters.

Par.	Value (μm)	Par.	Value (μm)
W_s	130	L_s	85
W_p	35	L_p	25
W_f	12	L_f	25
W_{f1}	5	L_{f1}	10
Single Element Parameters of ADS Circuit			
L_1	5.25 pH	L_2	0.6499 pH
C_1	1.544 fF	C_2	3.5 fF
R_1	200.23 Ω		
Two Element Parameters of ADS Circuit			
$L_3 = L_5$	1.5 pH	$L_6 = L_6$	1.8 pH
$C_3 = C_5$	1.1 fF	$C_4 = C_6$	1.8 fF
$R_2 = R_3 = R_4$	100 Ω	C_7	0.4 fF
L_7	0.6 pH	-	-
Dumbbell-Structure MTM parameters			
r_o	12	X	7
y	25	-	-

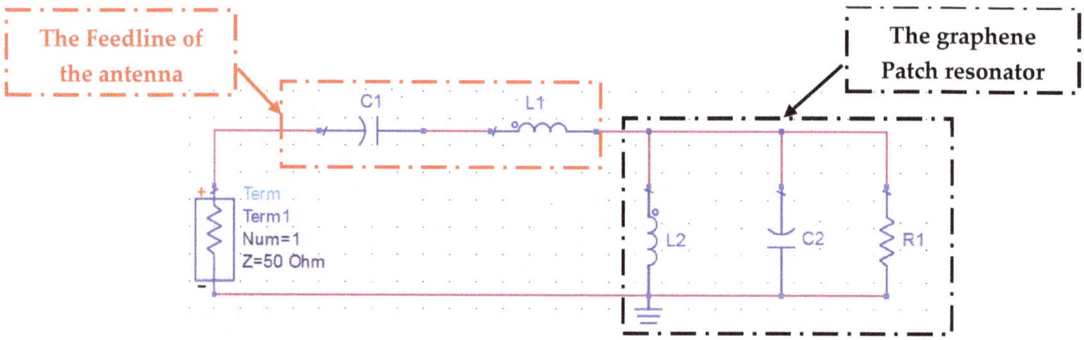

Figure 4. ADS equivalent circuit model of the proposed graphene antenna.

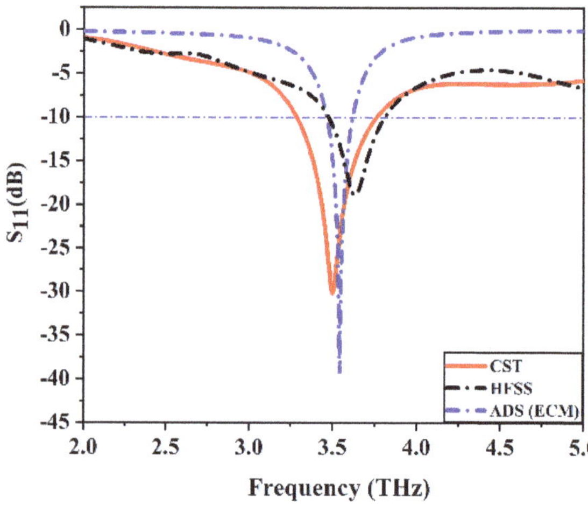

Figure 5. $|S_{11}|$ of the single element graphene plasmonic patch antenna with different simulation programs.

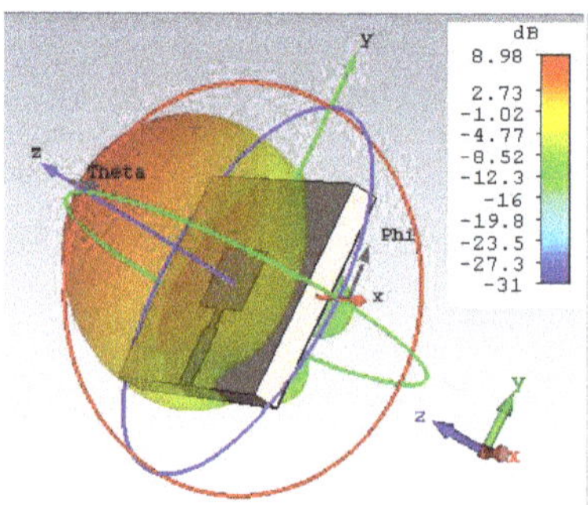

Figure 6. 3D radiation pattern of the proposed antenna.

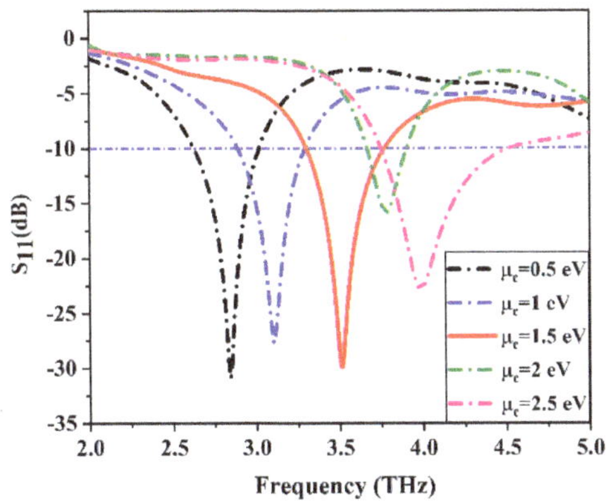

Figure 7. $|S_{11}|$ parameters of the graphene plasmonic antenna for different chemical potential.

3.2. *Two Elements Graphene Plasmonic MIMO Antenna Configurations*

The MIMO system's advantage is that it provides data rate augmentation even in the presence of signal fading, multipath fading, and interference. The MIMO system was designed to meet the requirements of acceptable MIMO parameters because of the need to transmit more data over greater distances. Mutual coupling between elements in a MIMO array could alter the performance of a single-element antenna. The distance between the MIMO units regulates this mutual coupling, which is formed via electromagnetic interaction between them. The optimum separation distance between adjacent MIMO elements is the system's most important goal. An antenna with a large distance between elements ensures non-correlation but increases the overall antenna size, whereas a small distance will increase the mutual coupling between the elements. So, mutual coupling is an important parameter that should be studied that faces the MIMO antenna systems. The two-element graphene plasmonic microstrip MIMO patch antennas are illustrated in the following part. In order to create the optimal MIMO antenna configuration, it is necessary to first determine the orientation of the optimal elements to achieve the desired high level of isolation. Figure 8 demonstrates the proposed three configurations that are used in the antenna design. The first configuration (Ant. A) is shown in Figure 8a the two elements have a 0° orientation side by side with the first element. The second configuration (Ant. B) is shown in Figure 8b where the two elements are orthogonally oriented with a 90° between them. In the last configuration (Ant. C), the two elements have a 180° orientation as shown in Figure 8c. The simulated reflection coefficients $|S_{11}|$ and transmission coefficients $|S_{21}|$ versus frequency of the three MIMO configurations are investigated in Figure 9. The transmission coefficients $|S_{21}|$ are an important parameter while designing the MIMO system. It is the measurable parameter of the mutual coupling between the MIMO elements.

These are three possible orientations that can be used using two elements MIMO antenna. Figure 9 illustrates the $|S_{11}|$ parameters of the three MIMO configurations that are designed to resonate at 3.5 THz. The $|S_{21}|$ curves reveal that Ant. C has the lowest mutual coupling of −25.6 dB within the entire interested frequency band rather than the other configurations of Ant. B that has a −22.4 dB in addition, the Ant. A has a mutual coupling of −18 dB The three graphene patch MIMO configurations are printed on Teflon substrate with a relative permittivity (ε_r) of 2.1 and a thickness of 10 μm. The CMA is used to describe the behavior of the graphene plasmonic MIMO antenna configurations with the help of model significant (Ms), and the model current distribution for the three configurations of Ant. A, Ant. B and Ant. C.

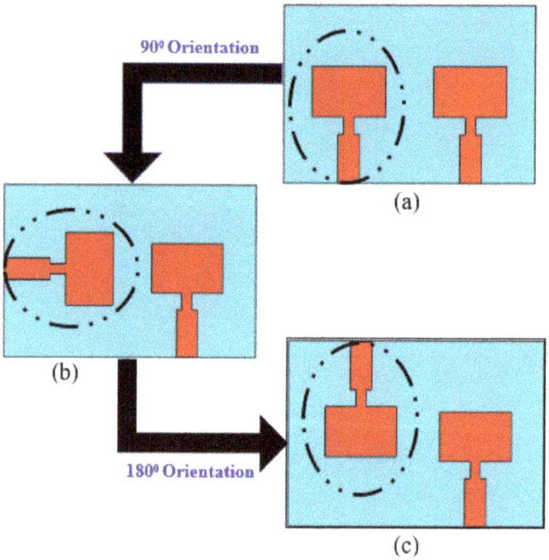

Figure 8. The proposed three different MIMO antenna configurations (**a**) 0° orientation, (**b**) 90° orientation, (**c**) 180°orientation.

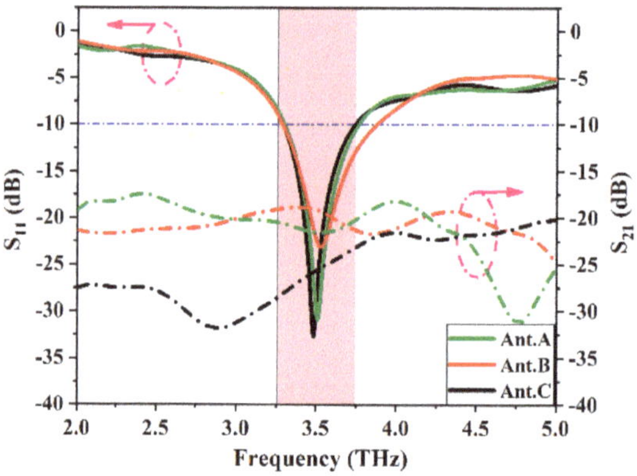

Figure 9. S–parameters (S_{11} and S_{21}) of the three different MIMO configurations.

3.3. *Analysis of the Graphene Plasmonic MIMO Configurations*

The evaluation of the two-element graphene plasmonic MIMO antenna structure is carried out and optimized using CMA for three different configurations. Firstly, in Ant. A structure, the model significance, and the model current distribution curves at the resonance frequency of 3.5 THz are investigated in Figures 10a and 11a, respectively. The model significance curve indicates the first four modes of operation through the entire band. It can be seen that the four modes play a significant role in the desired bandwidth ($M_s \geq 0.707$), but modes 1 and 2 give a high contribution in the desired bandwidth rather than modes 3 and 4. Moreover, modes 1 and 2, have the highest values (close to unity). So, it is expected that one of these modes is responsible for the antenna resonation at 3.5 THz and the other one is responsible for coupling between the two-elements' graphene plasmonic MIMO antenna. The model current distribution makes the divergence between the first two modes. From the current distribution depicted in Figure 11a; mode (J_{n1}) is the dominant mode that is responsible for the antenna resonation and mode (J_{n2}) causes the coupling between the two-radiating elements of the MIMO antenna because of the high current density between the elements which, in turn, affects the overall antenna performance negatively. Therefore, it is imperative that the characteristics of this mode need to be tuned again with a certain approach. In addition, mode (J_{n3}) and mode (J_{n4}) have no impact on the coupling between the MIMO elements. Secondly, in Ant. B, the model significant curve shown in Figure 10b illustrates the first four modes. It can be seen that all modes have the opportunity to contribute to the radiating bandwidth ($M_s \geq 0.707$) with different levels of acceptance such as modes 1 and 3, which have the highest values of M_s, while modes 2 and 4, have the lowest one. So, the model significant property is not sufficient to specify all the modes and clarify which of them is responsible for the coupling, non-coupling, and resonance effect. The model current distribution in Figure 11b is a good candidate that is used to illustrate this point. This figure reveals that modes (J_{n3}) and, (J_{n4}), have a current null between the antenna elements, and mode (J_{n1}) is the resonance one while mode (J_{n2}) is the coupling mode that needs to be characterized and tuned again to reduce the mutual coupling between the graphene plasmonic MIMO antenna elements. Finally, in Ant. C configuration, the model significant curve depicted in Figure 10c demonstrates the first four modes that participate in the desired bandwidth ($M_s \geq 0.707$). It is interesting to note that, modes 1,2, and 3 only contribute to the desired bandwidth of the antenna as they have the highest value of M_s. However, mode 4 can be neglected because of its low value of M_s within interested bandwidth. This result can be assured by introducing the model current distribution in Figure 11c. From this curve, we can overlook mode (J_{n3}) and (J_{n4}) as the current density concentration is very weak between the patches. So, mode (J_{n1}) is the resonance mode, and mode (J_{n2}) is the coupling one. The model significance and model current distribution curves of the proposed three configurations of the MIMO antenna reveal that the configuration of the designed antenna is responsible for the degree of the coupling between the two elements. The model current distribution indicates that the best configuration is Ant. C which gives a low coupling between the patches and low current density concentration in comparison with Ant. A and Ant. B.

3.4. De-Coupling Structure of Dumbbell Shape MTM Unit Cell

As the distance between the elements of the MIMO antenna is responsible for the mutual coupling issue, the performance of the antenna seriously deteriorates as this distance decreases. The three configurations of the MIMO antenna introduced above suffer from mutual coupling problems, so, a de-coupling structure is needed to overcome this problem. A complimentary new shape of Dumbbell-structure MTM unit cell is etching in the ground plane with a strong band rejection characteristic (Band-stop Filter) illustrated in Figure 12 and its dimensions listed in Table 1. The S-parameter curve depicted in Figure 13 illustrates that the Dumbbell-structure MTM unit cell has a high band-stop characteristic $|S_{21}|$. The permittivity and permeability of the unit cell in Figure 14a,b reveal that this structure is an MTM unit cell with a negative value if (ε_r and μ_r) in the desired bandwidth. The MTM unit cell of the Dumbbell-structure that is etched in the ground plane will reduce the mutual coupling between the two elements of the graphene MIMO configurations due to its band rejection characteristic at the desired frequency band. As mentioned before, in the proposed three configurations there is only one mode responsible for the coupling between the two elements (mode 2) in all configurations. So, this mode should be returned to improve the isolation. The MTM unit cell is designed to produce a current flow distribution in the opposite direction of the coupling mode current (mode 2). The characteristic mode theory (CMA) has a vital role in illustrating the behavior of the graphene MIMO systems with and without introducing the Dumbbell-structure unit cell by the usage of model significant and model current distribution. The Ant. C is the best choice to invoke the MTM unit cell on its ground plane as it has a good characteristic rather than the other configurations.

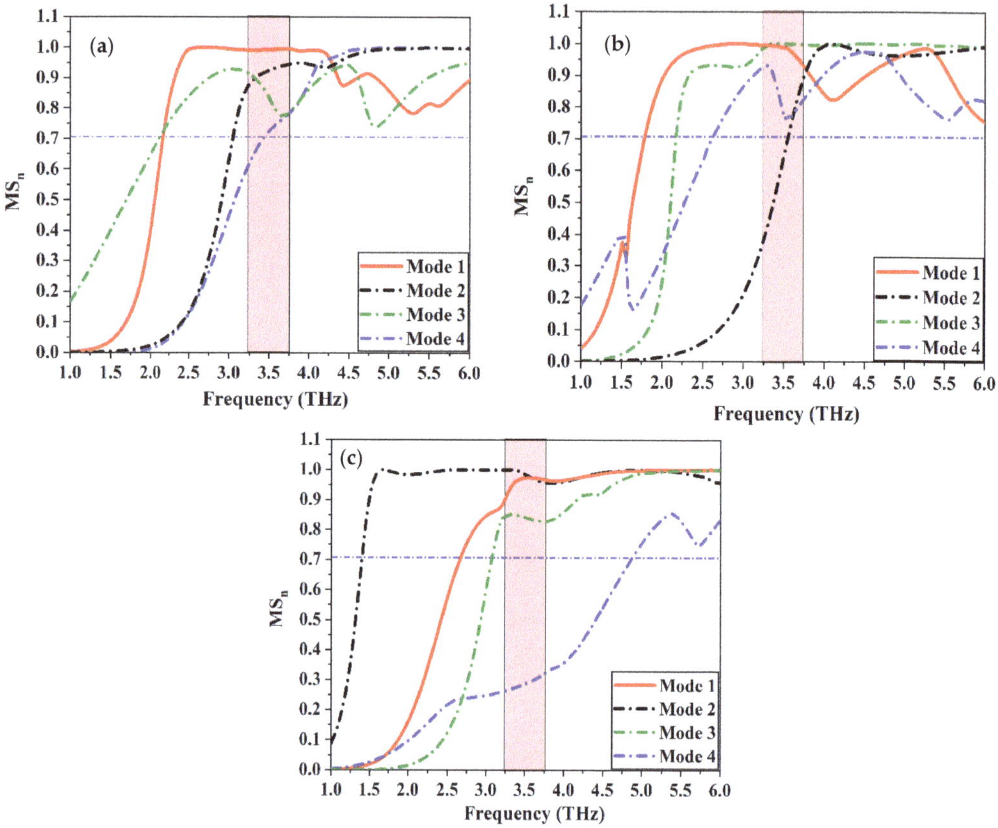

Figure 10. Model Significant of the proposed MIMO configurations (**a**) Ant. A, (**b**) Ant. B, and (**c**) Ant. C.

In Ant. C configuration, with the presence of an MTM unit cell as depicted in Figure 15, the model current distribution of the first four modes is shown in Figure 16. It can be seen that the mutual coupling is reduced and the path of the current through the MTM design is orthogonal to the path of the current that occurred due to the coupling mode (mode 2). So, the net interference between the two patches is remarkably reduced. This result can be assured from the model significance curve illustrated in Figure 17, we can note that the first four modes of operations and only mode 1 affects the desired bandwidth ($M_s \geq 0.707$). Mode 2, which is responsible for the coupling is moved away from the bandwidth in the condition that the resonance mode 1 does not affect and stays the same. Moreover, modes 3 and 4 do not affect the desired bandwidth which can be seen from the model current distribution and model significant curves. At last, from the study of the CMA analysis in the three proposed configurations of graphene MIMO antenna, introducing the Dumbbell-structure MTM unit cell in the ground plane has a good impact on the MIMO antenna parameters, especially in the interference problem between the patches that enhance the overall performance of the antenna. Ant. C, in the presence of an MTM unit cell, is the preferable configuration of the two-element graphene plasmonic MIMO structure, which gives high isolation between the patches as shown in Figure 18, which presents the $|S_{11}|$ and $|S_{21}|$ of the proposed antenna with and without the presence of the MTM unit cell.

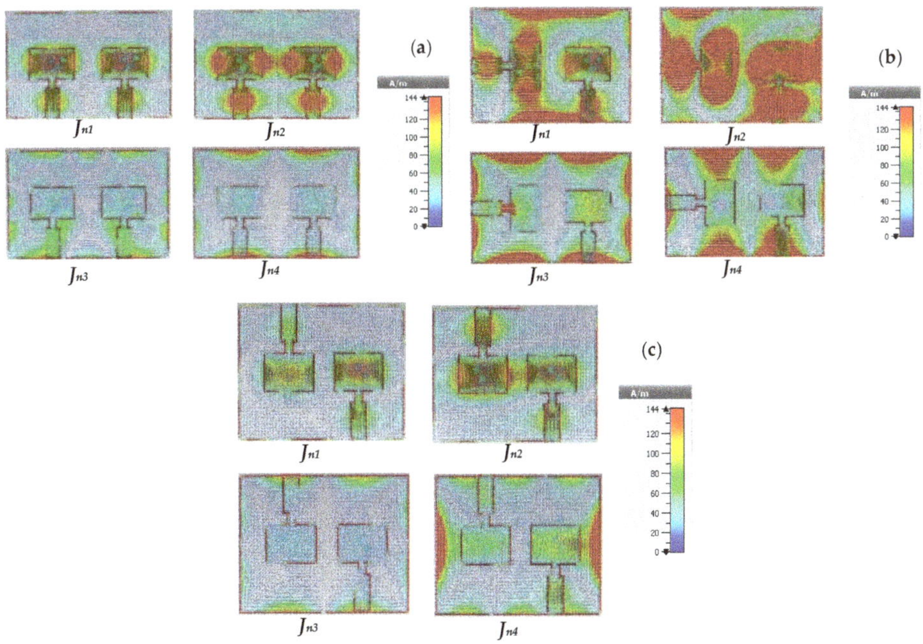

Figure 11. Model current distribution (J_{n1}–J_{n4}) of the proposed MIMO configurations (**a**) Ant. A, (**b**) Ant. B, and (**c**) Ant. C.

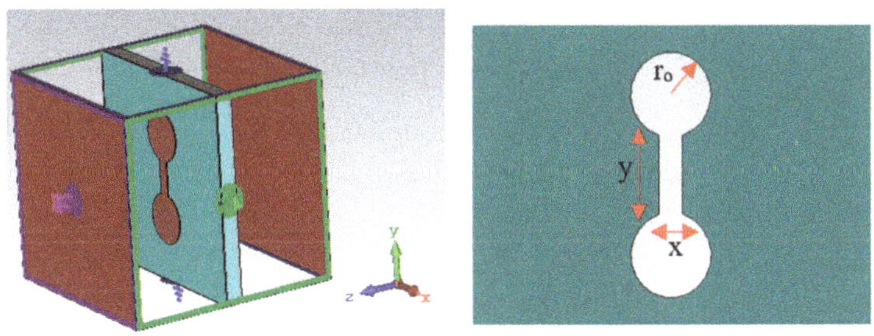

Figure 12. Simulation set up and dimensions for the Dumbbell-structure MTM unit cell.

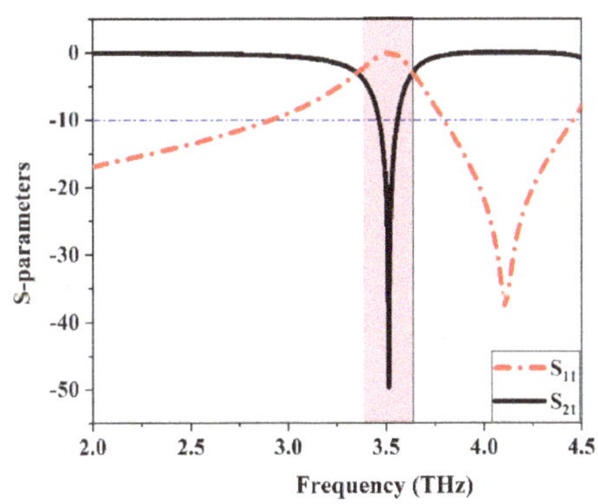

Figure 13. S–Parameters of Dumbbell-structure MTM unit cell.

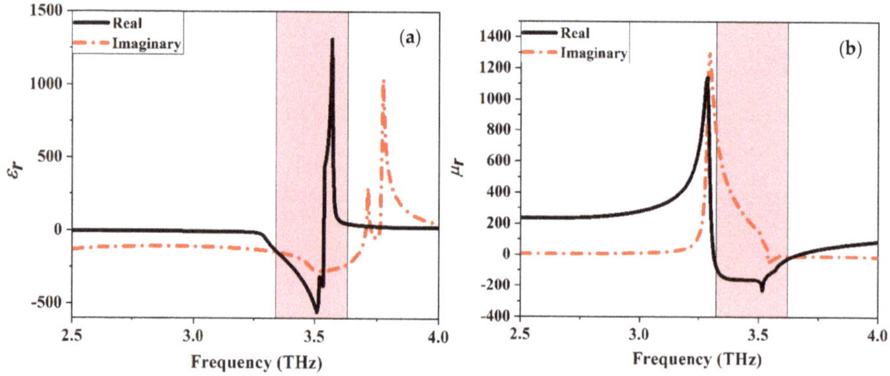

Figure 14. Extracted (**a**)permittivity and (**b**) permeability of Dumbbell-structure MTM unit cell.

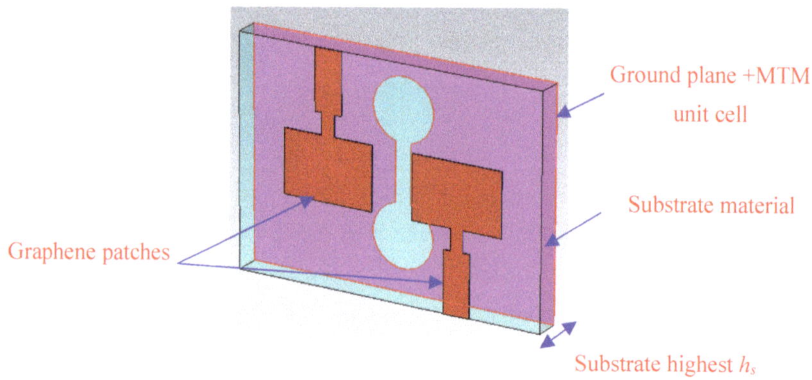

Figure 15. Ant. C structure in the presence of MTM unit cell.

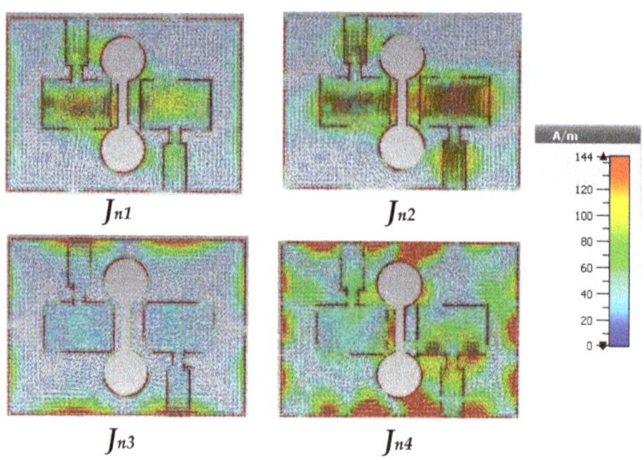

Figure 16. Model current distribution of Ant. C with the MTM unit cell.

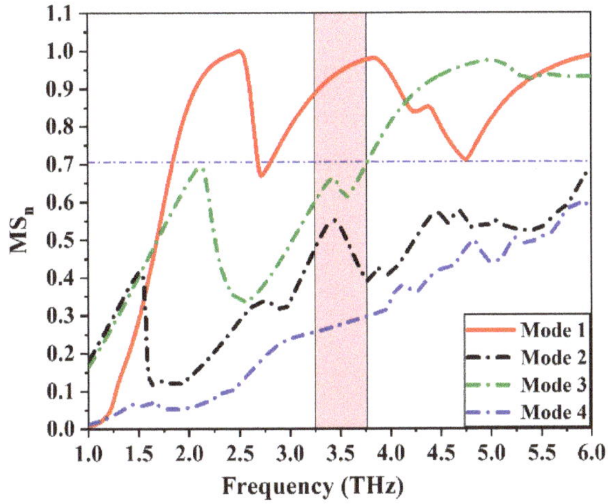

Figure 17. Model significant of Ant. C with the MTM unit cell.

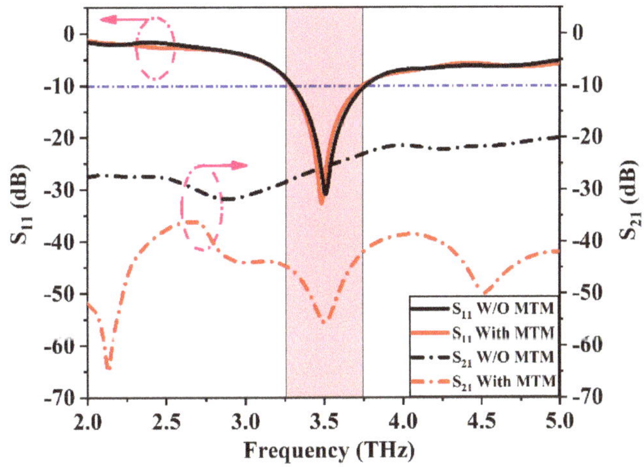

Figure 18. S–Parameters (S_{11} and S_{21}) of Ant. C with and without the MTM unit cell.

The S–Parameters and the ADS for the equivalent circuit models of the proposed design are listed in Figures 19 and 20 respectively. The isolation coefficient $|S_{21}|$ has been reduced from −25 dB to −55 dB because of the introduction of the Dumbbell-structure MTM unit cell in the ground. The ADS software is used to validate the results by using the equivalent circuit model (ECM) analysis. The ECM consists of a two-input port with 50 Ω input impedance and is separated by the resonator circuit that represents the Dumbbell-structure MTM unit cell. The calculated values of the circuit elements are listed in Table 1. The S–Parameters of the ECM analysis is compared with the CST microwave studio. It can be seen that the ADS results are in good agreement and coincide with the CST result with a little change due to the values of the RLC components.

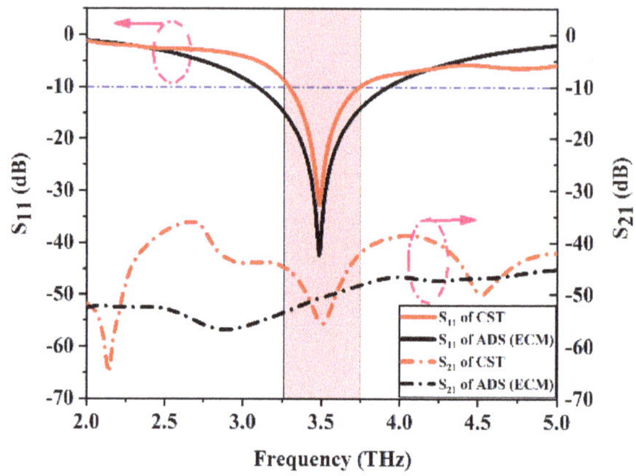

Figure 19. S–Parameters of Ant. C with CST and ADS equivalent circuit model.

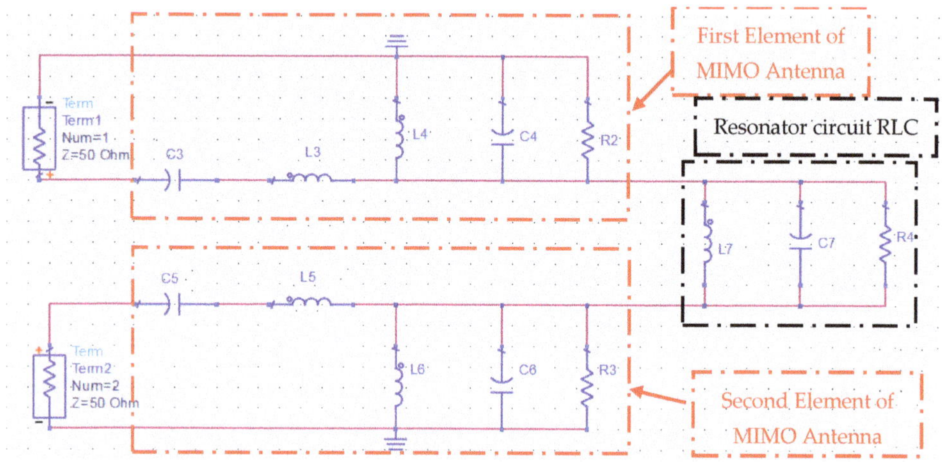

Figure 20. Equivalent circuit model of the proposed structure of two-element MIMO configurations with the MTM unit cell.

3.5. Analysis of MIMO Antenna Parameters

In this section, we discuss the most important parameters of the MIMO antenna structure. These parameters are very crucial in the MIMO systems. As the demand for a higher data rate for long-distance transmission is one of the major motivations behind the MIMO configuration to ensure the compatibility of the graphene plasmonic MIMO antenna in the THz frequency band. The envelope correlation coefficient (ECC), diversity gain (DG), and channel capacity loss (CCL) are evaluated and discussed [36,37]. The ECC is defined as the correlation between the patches in the chosen MIMO configuration. The lower amount of ECC between the MIMO elements means higher MIMO performance. There are two methods to extract the ECC parameters, the first one is by extracting from the S–Parameters as in [38].

$$ECC = \rho_{ij} = \frac{\left|S_{11}^{*}S_{12} + S_{21}^{*}S_{22}\right|^2}{\left(1 - \left(|S_{11}|^2 + |S_{21}|^2\right)\right)\left(1 - \left(|S_{22}|^2 + |S_{12}|^2\right)\right)} \quad (16)$$

where S_{11}^* and S_{21}^* represent the conjugate of S_{11} and S_{21}, respectively. The second one is by extracting from the 3D radiation pattern as illustrated in [25].

$$ECC = \rho_{ij} = \frac{\left| \iint_{4\pi} \left[F_i(\theta, \varphi) \bullet F_j(\theta, \varphi) \right] d\Omega \right|^2}{\iint_{4\pi} \left| F_i(\theta, \varphi) \right|^2 d\Omega \ \iint_{4\pi} \left| F_j(\theta, \varphi) \right|^2 d\Omega} \tag{17}$$

where $F_n(\theta, \varphi)$ is the 3D field radiation pattern of the elements associated with the nth port, • denotes the Hermitian product operator and Ω represents the solid angle. The evaluation of the ECC using a far-field radiation pattern is a complex method, so the S–Parameters method is very convenient to calculate the ECC between the two antenna elements as it is faster and easy to be used as it depends only on the port parameters of the antenna system. The agreeable limit of the ECC is lower than 0.5. The proposed graphene plasmonic MIMO antenna is simulated and ECC is calculated and displayed in Figure 21a. This figure reveals the comparisons between the MIMO elements with and without introducing the complementary Dumbbell-structure MTM unit cell in the ground plane. The ECC without MTM has a value of 0.0189, while the ECC with MTM is reduced to 0.000168 through the entire bandwidth of operation. The lower value of ECC indicates a lower correlation between antenna elements. This value guarantees good MIMO performance of the proposed antenna. Another way to describe and evaluate the performance of the MIMO antenna structure is the diversity gain (DG), which describes the losses that occur in the transmitted power. The formula given in [39] is used to calculate the DG value:

$$DG = 10\sqrt{1 - (ECC)^2} \tag{18}$$

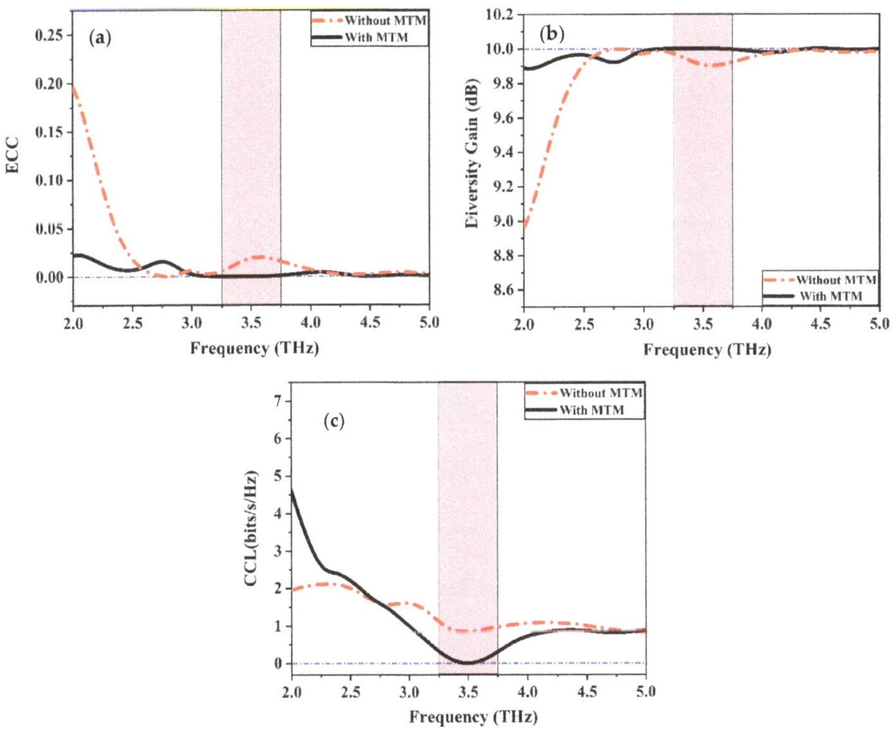

Figure 21. (**a**) Envelop correlation coefficient (ECC), (**b**) Diversity gain (DG), and (**c**) channel capacity loss (CCL) of Ant. C with and without the presence of the MTM unit cell.

It can be seen that the DG depends on the value of ECC, moreover, equation (18) states that the small value of ECC produces a higher value of DG and makes the MIMO antenna suitable for diverse applications. Figure 21b depicts the DG value of the proposed MIMO structure in the presence and absence of the MTM structure it has a value of 9.999 and 9.904, respectively.

Channel capacity loss (CCL) is one of the most significant MIMO parameters that describe the potential loss of channel capacity due to the correlation between the MIMO antenna elements, in other words, CCL defines the maximum limit of information transmission rate without significant loss. It should be less than 0.5 bits/s/Hz for a well-designed MIMO configuration. The CCL is indicated in (19) as reported in [39].

$$CCL = -\log_2[\det(\Phi_k)], \tag{19}$$

$$\Phi_k = \begin{bmatrix} \rho_{11} & \rho_{12} \\ \rho_{21} & \rho_{22} \end{bmatrix} \tag{20}$$

$$\rho_{ii} = \left(1 - \left(|S_{ii}|^2 + |S_{ij}|^2\right)\right), \quad and \tag{21}$$

$$\rho_{ij} = -\left(S_{ii}^* S_{ij} + S_{ij}^* S_{jj}\right), \tag{22}$$

where Φ_k indicates the correlation matrix at the receiving antenna. Figure 21c depicts the simulated CCL of the proposed MIMO antenna. It recorded the specified limit for the proposed MIMO configuration. A comparison between the CCL curves with and without introducing the MTM unit cell. So, it has a value of 0.006 and 0.840, respectively. It can be confirmed that the proposed design of graphene plasmonic MIMO configuration delivers a higher transmission data rate in any scattering environment. Figure 22 illustrates the representation of the far-field radiation pattern component of the E and H plane at 3.5 THz. The E and H plane radiation patterns are represented in terms their components. The proposed graphene plasmonic MIMO antenna radiation patterns are illustrated in the presence and the absence of the MTM unit cell. The proposed MIMO configuration exhibits a little change in the E and H planes as shown in Figure 22. At last, the E and H patterns indicate that no significant difference between the main lobe pattern with and without the Dumbbell-structure MTM unit cell.

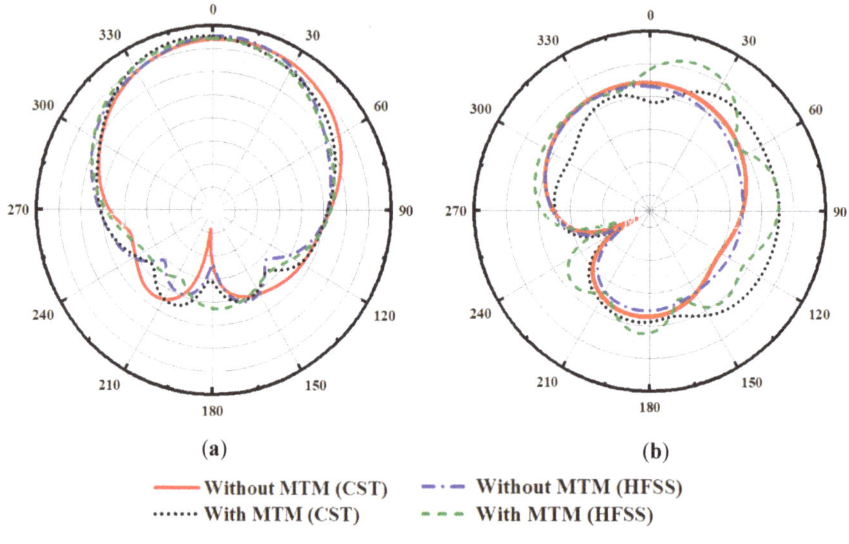

Figure 22. Far-field radiation of Ant. C with and without MTM unit cell (**a**) E-plane and (**b**) H-Plane.

3.6. Performance Analysis with State-of-the-Art of Printed THz Antennas

In this part, comparisons between proposed optimized Ant. C design with previously reported antennas has been performed for the terahertz frequency band. The comparison is conducted based on performance parameters such as gain, size of the antenna elements, and resonance frequency. Moreover, the MIMO parameters such as ECC, DG, and CCL are tabulated in Table 2. Based on the presented data, we can conclude that the proposed graphene plasmonic MIMO antenna has the best isolation and ECC. Although the proposed design presents a high gain with all compared references, the design in [40] has a higher gain, but gives low isolation and CCL. Finally, the proposed antenna demonstrates a superior MIMO performance to that of the other listed designs in Table 2. So, it is a good candidate to be used in 6G for mobile generation and medical applications. It has a suitable compact size to easily integrate with on-chip integrated circuits.

Table 2. Comparison of the proposed antenna performance with the state-of-the-art works.

Ref.	Freq. (THz)	Antenna Size (μm^2)	Isolation Technique	Mutual Coupling (dB)	Gain (dB)	ECC/DG (dB)	CCL
[14]	(1–10) (15–30)	620 × 800	Separated Distance	−20: −45	7	0.01/10	0.5
[40]	0.472	600 × 300	Separated Distance	−52	4.27	0.0458/9.99	0.00018
[41]	1.82	60 × 40	Separated Distance	−25	4.45	0.0372/9.99	-
[42]	0.2–10	125 × 125	elliptical fractal slots	−20	8.2	0.0022/9.98	0.31
[43]	0.33–10	1000 × 1400	DGS	−25	19	0.0015/9.99	0.25
[44]	1.68 and 1.81	60 × 40	Slots	−40	-	0.17/9.99	-
[45]	2.3, 3.2 and 4.5	50 × 40	Decoupling structure	−17, −30 and −23	5	0.2/10	-
[46]	1.1–1.7	84 × 84	FSS	−25	15	0.01/-	-
[47]	1.1	380 × 380	MTM	−20	8.28	-	-
[48]	1.76	56 × 56	Slots	−50	4.02	0.0086/9.989	-
Proposed Work	3.5	130 × 85	MTM+CMA+ Elements Arrangement	−55	7.23	0.000168/9.999	0.006

4. Conclusions

In this paper, a high-performance graphene plasmonic two-port MIMO antenna has been introduced to be used in terahertz wireless communication applications. In the theory of characteristic mode analysis (CMA) with the help of a complementary Dumbbell-structure MTM unit cell, the mutual coupling between the antenna elements has been significantly reduced. The MTM structure acts as a band stop filter in the desired bandwidth. It is etched in the ground plane so as to reduce the mutual coupling from −25 dB to −55 dB in the presence of the MTM unit cell. The CMA has been used to study the behavior of the antenna modes of operations and clarify the mode responsible for the coupling between the elements and detect the optimum position of this mode which can be illuminated by the MTM structure without any distortion in the main antenna resonance mode (dominant mode). Furthermore, the MIMO antenna parameters such as ECC, DG coefficients, and CCL are investigated and presented to confirm the suitability of the proposed graphene MIMO antenna for different THz applications. The proposed MIMO antenna resonates at 3.5 THz with a return loss of −35 dB and exhibits a wide impedance characteristic bandwidth covering 3.2 to 3.8 THz frequency spectrum in the THz regime (BW = 600 GHz). The designed MIMO antenna maintains maximum isolation of −55 dB at the resonance frequency of 3.5 THz. The good MIMO performance is ensured by achieving the MIMO performance parameters (ECC = 0.000168, DG = 9.999, and CCL = 0.006) within acceptable limits. The proposed MIMO antenna design in the future can be used as a massive MIMO and used the reconfigurable intelligent surfaces to be suitable for the 6G communication technology The proposed MIMO antenna is suitable for high-speed short-distance indoor communication applications such as video-rate imaging, biomedical imaging, sensing, and security scanning in the terahertz frequency band.

Author Contributions: S.A.K. raised the idea, wrote the manuscript, and perform the numerical simulations, E.K.I.H. and N.O.P., revised the manuscript and perform the theoretical analysis, N.O.P. and M.B.S., discussed the numerical results in the main manuscript; S.A.K. and E.K.I.H. contribute to the simulation results and the validation, improved the revised version. All authors have read and agreed to the published version of the manuscript.

Funding: This research received no external funding.

Institutional Review Board Statement: Not applicable.

Informed Consent Statement: Not applicable.

Data Availability Statement: All the data have been included in the study.

Conflicts of Interest: The authors declare no conflict of interest.

References

1. Giordani, M.; Polese, M.; Mezzavilla, M.; Rangan, S.; Zorzi, M. Toward 6G Networks: Use Cases and Technologies. *IEEE Commun. Mag.* **2020**, *58*, 55–61. [CrossRef]
2. Alsabah, M.; Naser, M.A.; Mahmmod, B.M.; Abdulhussain, S.H.; Eissa, M.R.; Al-Baidhani, A.; Noordin, N.K.; Sait, S.M.; Al-Utaibi, K.A.; Hashim, F. 6G Wireless Communications Networks: A Comprehensive Survey. *IEEE Access* **2021**, *9*, 148191–148243. [CrossRef]
3. Chowdhury, M.Z.; Shahjalal, M.; Ahmed, S.; Jang, Y.M. 6G Wireless Communication Systems: Applications, Requirements, Technologies, Challenges, and Research Directions. *IEEE Open J. Commun. Soc.* **2020**, *1*, 957–975. [CrossRef]
4. Akyildiz, I.F.; Jornet, J.M.; Han, C.J.P.C. Terahertz Band: Next Frontier for Wireless Communications. *Phys. Commun.* **2014**, *12*, 16–32. [CrossRef]
5. Tekbıyık, K.; Ekti, A.R.; Kurt, G.K.; Görçin, A.J.P.C. Terahertz Band Communication Systems: Challenges, Novelties and Standardization Efforts. *Phys. Commun.* **2019**, *35*, 100700. [CrossRef]
6. Hosseininejad, S.E.; Alarcon, E.; Komjani, N.; Abadal, S.; Lemme, M.C.; Bolívar, P.H.; Cabellos-Aparicio, A. Study of hybrid and pure plasmonic terahertz antennas based on graphene guided-wave structures. *Nano Commun. Netw.* **2017**, *12*, 34–42. [CrossRef]
7. Koenig, S.; Lopez-Diaz, D.; Antes, J.; Boes, F.; Henneberger, R.; Leuther, A.; Tessmann, A.; Schmogrow, R.; Hillerkuss, D.; Palmer, R.; et al. Wireless sub-THz communication system with high data rate enabled by RF photonics and active MMIC technology. In Proceedings of the IEEE Photonics Conference, San Diego, CA, USA, 12–16 October 2014; IEEE: Piscataway, NJ, USA, 2014; pp. 414–415.
8. Poorgholam-Khanjari, S.; Zarrabi, F.B. Reconfigurable Vivaldi THz antenna based on graphene load as hyperbolic metamaterial for skin cancer spectroscopy. *Opt. Commun.* **2021**, *480*, 126482. [CrossRef]
9. Geetharamani, G.; Aathmanesan, T. Split ring resonator inspired THz antenna for breast cancer detection. *Opt. Laser Technol.* **2020**, *126*, 106111. [CrossRef]
10. Naftaly, M.; Foulds, A.P.; Miles, R.E.; Davies, A.G. Terahertz Transmission Spectroscopy of Nonpolar Materials and Relationship with Composition and Properties. *Int. J. Infrared Millim. Waves* **2005**, *26*, 55–64. [CrossRef]
11. Wu, K.; Xiao, J.; Ni, M.L. Rethinking the architecture design of data center networks. *Front. Comput. Sci.* **2012**, *6*, 596–603. [CrossRef]
12. Gao, M.; Li, K.; Kong, F.; Zhuang, H.; Zhu, G. Graphene-Based Composite Right/Left-Handed Leaky-Wave Antenna at Terahertz. *Plasmonics* **2020**, *15*, 1199–1204. [CrossRef]
13. Poorgholam-Khanjari, S.; Zarrabi, F.B.; Jarchi, S. Compact and wide-band Quasi Yagi-Uda antenna based on periodic grating ground and coupling method in terahertz regime. *Optik* **2020**, *203*, 163990. [CrossRef]
14. Alharbi, A.G.; Sorathiya, V. Ultra-Wideband Graphene-Based Micro-Sized Circular Patch-Shaped Yagi-like MIMO Antenna for Terahertz Wireless Communication. *Electronics* **2022**, *11*, 1305. [CrossRef]
15. Runge, M.; Engel, D.; Schneider, M.; Reimann, K.; Woerner, M.; Elsaesser, T. Spatial distribution of electric-field enhancement across the gap of terahertz bow-tie antennas. *Opt. Express* **2020**, *28*, 24389. [CrossRef]
16. Bekasiewicz, A.; Koziel, S.; Plotka, P.; Zwolski, K. EM-Driven Multi-Objective Optimization of a Generic Monopole Antenna by Means of a Nested Trust-Region Algorithm. *Appl. Sci.* **2021**, *11*, 3958. [CrossRef]
17. Abdulnabi, H.A.; Hussein, R.T.; Fyath, R.S. 0.1–10 thz single port log periodic antenna design based on hilbert graphene artificial magnetic conductor. *ARPN J. Eng. Appl. Sci.* **2017**, *12*, 1189–1196.
18. Walther, M.; Cooke, D.G.; Sherstan, C.; Hajar, M.; Freeman, M.R.; Hegmann, F.A. Terahertz conductivity of thin gold films at the metal-insulator percolation transition. *Phys. Rev. B* **2007**, *76*, 125408. [CrossRef]
19. Razaq, A.; Bibi, F.; Zheng, X.; Papadakis, R.; Jafri, S.H.M.; Li, H. Review on Graphene-, Graphene Oxide-, Reduced Graphene Oxide-Based Flexible Composites: From Fabrication to Applications. *Materials* **2022**, *15*, 1012. [CrossRef]
20. Olabi, A.G.; Abdelkareem, M.A.; Wilberforce, T.; Sayed, E.T. Application of graphene in energy storage device—A review. *Renew. Sustain. Energy Rev.* **2021**, *135*, 110026. [CrossRef]

21. Zhang, H.; Yang, D.; Lau, A.; Ma, T.; Lin, H.; Jia, B. Hybridized graphene for supercapacitors: Beyond the limitation of pure graphene. *Small* **2021**, *17*, 2007311. [CrossRef]
22. Beiranvand, B.; Sobolev, A.S. A proposal for a multi-functional tunable dual-band plasmonic absorber consisting of a periodic array of elliptical grooves. *J. Opt.* **2020**, *22*, 105005. [CrossRef]
23. Dash, S.; Patnaik, A.; Kaushik, B.K. Performance enhancement of graphene plasmonic nanoantennas for THz communication. *IET Microw. Antennas Propag.* **2019**, *13*, 71–75. [CrossRef]
24. Mumtaz, S.; Jornet, J.M.; Aulin, J.; Gerstacker, W.H.; Dong, X.; Ai, B. Terahertz Communication for Vehicular Networks. *IEEE Trans. Veh. Technol.* **2017**, *66*, 5617–5625. [CrossRef]
25. Abdelaziz, A.; Hamad, E.K.I. Isolation enhancement of 5G multiple-input multiple-output microstrip patch antenna using metamaterials and the theory of characteristic modes. *Int. J. RF Microw. Comput.-Aided Eng.* **2020**, *30*, e22416. [CrossRef]
26. Harrington, R.; Mautz, J. Computation of characteristic modes for conducting bodies. *IEEE Trans. Antennas Propag.* **1971**, *19*, 629–639. [CrossRef]
27. Song, R.; Chen, X.; Jiang, S.; Hu, Z.; Liu, T.; Calatayud, D.G.; Mao, B.; He, D. A Graphene-Assembled Film Based MIMO Antenna Array with High Isolation for 5G Wireless Communication. *Appl. Sci.* **2021**, *11*, 2382. [CrossRef]
28. Dash, S.; Patnaik, A. Behavior of graphene based planar antenna at microwave and terahertz frequency. *Photonics Nanostructures-Fundam. Appl.* **2020**, *40*, 100800. [CrossRef]
29. Hanson, G.W. Dyadic Green's Functions for an Anisotropic, Non-Local Model of Biased Graphene. *IEEE Trans. Antennas Propag.* **2008**, *56*, 747–757. [CrossRef]
30. Beiranvand, B.; Sobolev, A.S.; Sheikhaleh, A. A proposal for a dual-band tunable plasmonic absorber using concentric-rings resonators and mono-layer graphene. *Optik* **2020**, *223*, 165587. [CrossRef]
31. Qin, X.; Chen, J.; Xie, C.; Xu, N.; Shi, J. A tunable THz dipole antenna based on graphene. In Proceedings of the 2016 IEEE MTT-S International Microwave Workshop Series on Advanced Materials and Processes for RF and THz Applications (IMWS-AMP), Chengdu, China, 20–22 July 2016; IEEE: Piscataway, NJ, USA, 2016; pp. 1–3.
32. Llatser, I.; Kremers, C.; Cabellos Aparicio, A.; Jornet, J.M.; Alarcón, E.; Chigrin, D.N. Graphene-based nano-patch antenna for terahertz radiation. *Photonics Nanostructures-Fundam. Appl.* **2012**, *10*, 353–358. [CrossRef]
33. Gatte, M.T.; Soh, P.J.; Rahim, H.A.; Ahmad, R.B.; Malek, F. The performance improvement of thz antenna via modeling and characterization of doped graphene. *Prog. Electromagn. Res. M* **2016**, *49*, 21–31. [CrossRef]
34. Jablan, M.; Buljan, H.; Soljačić, M. Plasmonics in graphene at infrared frequencies. *Phys. Rev. B* **2009**, *80*, 245435. [CrossRef]
35. Varshney, G.; Verma, A.; Pandey, V.S.; Yaduvanshi, R.S.; Bala, R. A proximity coupled wideband graphene antenna with the generation of higher order TM modes for THz applications. *Opt. Mater.* **2018**, *85*, 456–463. [CrossRef]
36. Abdelaziz, A.; Mohamed, H.A.; Hamad, E.K.I. Applying Characteristic Mode Analysis to Systematically Design of 5G Logarithmic Spiral MIMO Patch Antenna. *IEEE Access* **2021**, *9*, 156566–156580. [CrossRef]
37. Das, P.; Singh, A.K.; Mandal, K. Metamaterial loaded highly isolated tunable polarisation diversity MIMO antennas for THz applications. *Opt. Quantum Electron.* **2022**, *54*, 250. [CrossRef]
38. Hussain, N.; Awan, W.A.; Ali, W.; Naqvi, S.I.; Zaidi, A.; Le, T.T. Compact wideband patch antenna and its MIMO configuration for 28 GHz applications. *AEU-Int. J. Electron. Commun.* **2021**, *132*, 153612. [CrossRef]
39. Babu, K.V.; Das, S.; Varshney, G.; Sree, G.N.J.; Madhav, B.T.P. A micro-scaled graphene-based tree-shaped wideband printed MIMO antenna for terahertz applications. *J. Comput. Electron.* **2022**, *21*, 289–303. [CrossRef]
40. Fritz-Andrade, E.; Perez-Miguel, A.; Gomez-Villanueva, R.; Jardon-Aguilar, H. Characteristic mode analysis applied to reduce the mutual coupling of a four-element patch MIMO antenna using a defected ground structure. *IET Microw. Antennas Propag.* **2019**, *14*, 215–226. [CrossRef]
41. Varshney, G.; Gotra, S.; Pandey, V.S.; Yaduvanshi, R.S. Proximity-coupled two-port multi-input-multi-output graphene antenna with pattern diversity for THz applications. *Nano Commun. Netw.* **2019**, *21*, 100246. [CrossRef]
42. Das, S.; Mitra, D.; Chaudhuri, S.R.B. Fractal loaded planar Super Wide Band four element MIMO antenna for THz applications. *Nano Commun. Netw.* **2021**, *30*, 100374. [CrossRef]
43. Saxena, G.; Awasthi, Y.K.; Jain, P. High Isolation and High Gain Super-Wideband (0.33–10 THz) MIMO Antenna for THz Applications. *Optik* **2020**, *223*, 165335. [CrossRef]
44. Ali, M.F.; Bhattacharya, R.; Varshney, G. Graphene-based tunable terahertz self-diplexing/MIMO-STAR antenna with pattern diversity. *Nano Commun. Netw.* **2021**, *30*, 100378. [CrossRef]
45. Vijayalakshmi, K.; Selvi, C.S.K.; Sapna, B. Novel tri-band series fed microstrip antenna array for THz MIMO communications. *Opt. Quantum Electron.* **2021**, *53*, 395. [CrossRef]
46. Zhang, B.; Jornet, J.M.; Akyildiz, I.F.; Wu, Z.P. Mutual Coupling Reduction for Ultra-Dense Multi-Band Plasmonic Nano-Antenna Arrays Using Graphene-Based Frequency Selective Surface. *IEEE Access* **2019**, *7*, 33214–33225. [CrossRef]
47. Esfandiyari, M.; Jarchi, S.; Ghaffari-Miab, M. Channel capacity enhancement by adjustable graphene-based MIMO antenna in THz band. *Opt. Quantum Electron.* **2019**, *51*, 137. [CrossRef]
48. Ali, M.F.; Bhattacharya, R.; Varshney, G. *Tunable Four-Port MIMO/Self-Multiplexing THz Graphene Patch Antenna with High Isolation*; PREPRINT (Version 1); Research Square: Durham, NC, USA, 2022. [CrossRef]

electronics

MDPI

Article

Ultra-Wideband Pentagonal Fractal Antenna with Stable Radiation Characteristics for Microwave Imaging Applications

Muhammad Abbas Khan [1], Umair Rafique [2,*], Hüseyin Şerif Savci [3], Anis Nurashikin Nordin [4], Saad Hassan Kiani [5,6] and Syed Muzahir Abbas [7,*]

1 Department of Electrical Engineering, Balochistan University of Information Technology, Engineering and Management Sciences, Quetta 87300, Pakistan; muhammad.abbas@buitms.edu.pk
2 Department of Information Engineering, Electronics, and Telecommunications, Sapienza University of Rome, 00184 Rome, Italy
3 Electrical and Electronics Engineering Department, Istanbul Medipol University, Istanbul 34810, Turkey; hsavci@medipol.edu.tr
4 Department of Electrical & Computer Engineering, International Islamic University Malaysia, Kuala Lumpur 43200, Malaysia; anisnn@iium.edu.my
5 Department of Electrical Engineering, IIC University of Technology, Phnom Penh 121206, Cambodia; iam.kiani91@gmail.com
6 Smart Systems Engineering Laboratory, College of Engineering, Prince Sultan University, Riyadh 11586, Saudi Arabia
7 Faculty of Science and Engineering, School of Engineering, Macquarie University, Sydney, NSW 2109, Australia
* Correspondence: umair.rafique@uniroma1.it (U.R.); syed.abbas@mq.edu.au (S.M.A.)

Abstract: For microwave imaging applications, a design for an ultra-wideband (UWB) fractal antenna is presented. The antenna design is composed of a pentagonal fractal patch radiator fed by a modified co-planar waveguide (CPW) ground plane. It is built on a low-loss Rogers RT/Duroid 5880 dielectric substrate with a dimensions of $24 \times 30 \times 0.787$ mm^3. According to the measurements, the designed antenna offers a fractional bandwidth of 123.56% ranging from 3 GHz to 12.7 GHz. In addition, a maximum gain of 3.6 dBi is achieved at 8.5 GHz. From the results, it is also observed that the proposed antenna structure attains constant radiation characteristics in the operating bandwidth, which is useful for microwave imaging applications. The time domain analysis of the proposed design is also performed, and it is observed that the designed antenna offers a group delay of $\leq$1.5 ns, which ensures minimum pulse distortion.

Keywords: microwave imaging; ultra-wideband; fractal antenna; co-planar waveguide; fractional bandwidth; group delay

Citation: Khan, M.A.; Rafique, U.; Savci, H.Ş.; Nordin, A.N.; Kiani, S.H.; Abbas, S.M. Ultra-Wideband Pentagonal Fractal Antenna with Stable Radiation Characteristics for Microwave Imaging Applications. *Electronics* **2022**, *11*, 2061. https://doi.org/10.3390/electronics11132061

Academic Editor: Andrea Randazzo

Received: 9 June 2022
Accepted: 29 June 2022
Published: 30 June 2022

Publisher's Note: MDPI stays neutral with regard to jurisdictional claims in published maps and institutional affiliations.

1. Introduction

Near-field microwave imaging has gotten a lot of attention in recent years, especially for detecting and localizing cancer tissues in the human body [1]. Microwave imaging offers promising solutions for a number of biological applications [2] when compared to existing approaches such as ultrasound [3] and mammography [4]. In this technique, short pulses of low power are emitted by antennas towards the human body. The backscattered radiation is then collected by the antennas and processed to form an image. One of the examples is shown in Figure 1. From the figure, one can observe that multiple antennas operating in the near field are placed around the breast. The antennas forming the array are sequentially selected to transmit pulses into the breast. Then, the backscattered signals are collected by the receiving antennas. Finally, a suitable signal processing technique is applied to backscattered signals for image reconstruction.

Microwave imaging antennas should have a wide bandwidth and be able to provide constant radiation performance. Furthermore, the antenna's time domain performance should be sufficient to allow transmitted pulses to pass through human tissues with minimal signal distortion. Ultra-wideband (UWB) technology may be useful in meeting these criteria as the pulse duration in UWB systems is often on the scale of a few nanoseconds, resulting in a bandwidth of a few gigahertz. In addition, the antenna size should be small enough so that it can easily be integrated into microwave imaging systems. To accommodate this requirement, researchers presented a variety of antenna designs. From the presented designs, planar antennas received a lot of attention due to their small size and light weight.

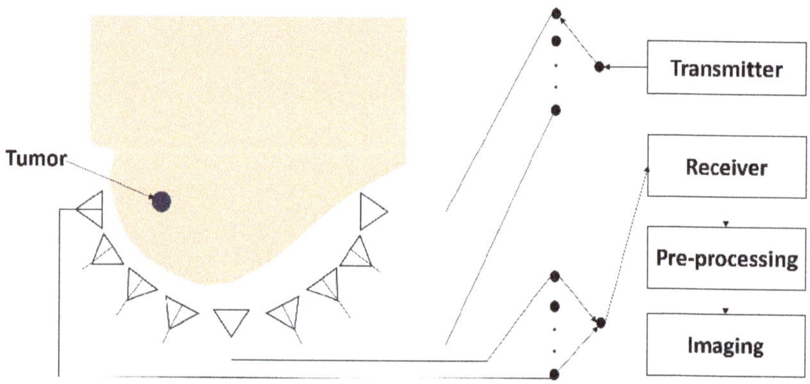

Figure 1. A typical near-field microwave imaging system for breast cancer detection [5] (Reprinted from an open access article published under Creative Commons Attribution 4.0 License).

For UWB microwave imaging systems, the authors in [6] presented a modified square monopole antenna design. In the ground plane, a pair of loop sleeves was designed, which tend to achieve an impedance bandwidth in the frequency range of 3.19–11.03 GHz. Although the designed antenna offers a wide impedance bandwidth, its efficiency is decreasing with an increase in frequency. For body-centric imaging applications, Danjuma et al. [7] designed a co-planar waveguide (CPW)-fed quasi-cross slot-based circular monopole antenna. The presented antenna can operate at frequencies ranging from 3 GHz to more than 11 GHz. A Hibiscus petal-like patch antenna was presented in [8] for microwave imaging systems. To obtain an impedance bandwidth in the frequency range of 3.04–11 GHz, a tapered microstrip feed line and a partial trapezoid ground plane were used. The performance of the antenna was evaluated in the time domain and a high value of group delay was observed for the side-by-side configuration, which is not useful for microwave imaging applications. In [9], an octagonal planar antenna was designed to detect malignant tissues in the human body. To achieve UWB response, the radiating element was fed by a 50 Ω microstrip feeding line and a step impedance transformer. For improved radiation characteristics, a circular slot including a symmetrical metal cross was designed in the octagonal radiator. From the presented configuration, the authors achieved an impedance bandwidth of 12 GHz from 3 GHz to 15 GHz.

Lee et al. [10] introduced a UWB-resistive dipole antenna for microwave imaging applications. They demonstrated that the designed dipole antenna offers minimal internal reflections compared to other UWB antenna structures. They employed chip resistors on both arms of the dipole antenna according to the technique presented in [11], and observed a UWB response ranging from 1 GHz to 10 GHz. However, this kind of configuration led to a large antenna size and was complex in nature. A square fractal monopole antenna with a slotted partial ground plane was presented in [12] for imaging purposes. The designed antenna-measured results demonstrate that it offers a fractional bandwidth (FBW) of 117.88% from 3.1 GHz to 12 GHz. A modified square monopole antenna design was presented in [13] for medical applications. With the use of modified square patch and inverted U-shaped parasitic elements, the authors achieved a 5.2 GHz impedance bandwidth from 3.8 GHz to 9 GHz. Another square monopole antenna with a T-slot-loaded ground plane was designed in [14] for UWB microwave imaging applications. In addition to the T-slot, a parasitic element with a π shape was placed beneath the microstrip line to achieve a UWB response in the frequency range of 2.91–14.72 GHz.

Ojaroudi and Ghadimi [15] utilized the same technique as presented in [14] to design a multi-resonance UWB antenna for microwave imaging applications. In this design, an E-shaped slot was etched in the ground plane, and an E-shaped parasitic structure was designed behind the microstrip feeding line. The presented configuration extends the upper-frequency limit of the antenna from 10.3 GHz to 15 GHz. The designs presented in [14,15] were able to provide a UWB response, but their radiation characteristics were not stable over the operating bandwidth. In [16], three T-shaped slots were etched in the ground plane to achieve resonance ranging from 2.96 GHz to 15.8 GHz. Lasemi and Atlasbaf [17] designed compact UWB fractal antennas for breast tumor detection. The Sierpinski gasket configuration was used for one antenna, and a ring-inscribed hexagonal fractal antenna was used for the other. According to the measurements, the designed antennas operate well from 2.95 GHz to 12 GHz. However, the time domain performance of the presented designs reveals that the transmission phase of the antennas was not stable over the operating bandwidth, which ultimately led to pulse distortion.

A negative index mematerial-based planar antenna was designed in [18] for microwave imaging applications. To achieve a compact size, the edges of the triangular patch were loaded with four left-handed metamaterial unit cells. To achieve negative permittivity and permeability, each metamaterial unit cell was composed of a modified split-ring resonator (SRR) and a capacitance-loaded strip. Measurement results show that the presented antenna operates well from 3.4 GHz to 12.5 GHz and offers a maximum gain of 5.16 dBi at 10.15 GHz. The same technique was utilized in [19] to design a compact planar metamaterial antenna. In this design, a trapezoidal patch radiator was used, which was fed through a stepped tapered microstrip feed line. The presented antenna was able to operate in the frequency range of 4.23–14 GHz. In [20], a wearable textile antenna with a full ground plane was designed for breast cancer detection. Photonic bandgap (PBG) and substrate-integrated waveguide (SIW) technology were utilized for antenna design. The antenna was fed using a grounded CPW (GCPW) through a SIW transition. The results demonstrate that the antenna offered an impedance bandwidth of 21 GHz from 7 GHz to 28 GHz. The presented antenna operates at higher frequencies that are not useful for biomedical applications. In addition, the use of PBG and SIW technologies offers complexities in the fabrication process.

In this article, a CPW-fed pentagonal fractal antenna is designed for near-field microwave imaging applications, especially for breast cancer detection. The results demonstrate that the antenna ranges from 3 GHz to 12.7 GHz, with a peak gain of 3.6 dBi at 8.5 GHz. Furthermore, from the time domain analysis, it is noted that a group delay for both face-to-face and side-by-side configurations is $\leq$1.5 ns, ensuring minimum pulse distortion, which is a fundamental criterion of near-field microwave imaging systems. To further demonstrate the suitability of the proposed antenna, a comparison is presented among proposed and previously published UWB antennas meant for microwave imaging applications (see Table 1).

From Table 1, it can be noted that all the antennas have fractional bandwidth >100%. The designs presented in [9,10,20] have maximum and approximately equal fractional bandwidth, but their electrical dimensions are large in comparison to the proposed design. The rest of the antennas exhibit low fractional bandwidth compared to the proposed antenna design. Furthermore, the proposed fractal antenna offers a high impedance bandwidth compared to the designs presented in [6–8,10,12,17,18].

Table 1. Comparative analysis among proposed and previously presented planar monopole antennas for microwave imaging applications.

Ref.	Dimensions (mm^2)	Dimensions (λ^2)	Dielectric Material	Dielectric Constant	Frequency Band (GHz)	Fractional Bandwidth (%)	Impedance Bandwidth (GHz)	Peak Gain (dBi)	Average Efficiency (%)
[6]	20 × 25	0.47 × 0.59	FR-4	4.3	3.19–11.03	110.26	7.84	6	80
[7]	33.14 × 14.90	0.77 × 0.34	FR-4	4.4	3–11	114.28	8	4.74	–
[8]	31 × 31	0.72 × 0.72	Rogers RT/Duroid 5870	2.33	3.04–11	113.39	7.96	5	–
[9]	29 × 27	0.87 × 0.81	FR-4	4.4	3–15	133.33	12	–	–
[10]	51 × 95	0.93 × 1.74	FR-4	4.4	1–10	163.63	9	–	–
[12]	23.1 × 32	0.58 × 0.8	FR-4	4.6	3.1–12	117.88	8.9	3.54	–
[17]	20 × 28 16 × 22	0.49 × 0.69 0.39 × 0.54	FR-4	4.4	2.95–12	121	9.05	4 3.7	– –
[18]	16 × 21	0.42 × 0.55	FR-4	4.4	3.4–12.5	114.5	9.1	5	85
[19]	10.2 × 15.5	0.3 × 0.47	FR-4	4.3	4.23–14	107	9.77	5	85
[20]	50 × 60	2.9 × 3.5	Denim substrate	1.4	7–28	120	21	10	88.5
Proposed	24 × 30	0.62 × 0.78	Rogers RT/Duroid 5880	2.2	3–12.7	123.56	9.7	3.6	88

2. Pentagonal Fractal Antenna Design

Figure 2 illustrates the geometry of the proposed UWB antenna. The proposed design is composed of a modified CPW-fed pentagonal fractal radiator, printed on a low-loss Rogers RT/Duroid 5880 dielectric substrate with a relative permittivity (ε_r) of 2.2. The thickness of the dielectric substrate for the antenna design is chosen to be 0.787 mm. The overall dimensions of the antenna are 24 × 30 mm^2. From Figure 2, one can note that a trapezoidal ground plane is used for better impedance matching in the band of interest. Furthermore, the CPW feed technique is utilized because it is easy to design and fabricate and it provides low losses [21]. The rest of the design parameters' values are as follows: W_S = 24 mm, L_S = 30 mm, W_f = 2.5 mm, L_f = 12 mm, t = 1.5 mm, g = 0.3 mm, L_g = 5.5 mm, W_g = 4.2 mm, R_b = 3 mm, R_e = 1 mm, S_1 = 9.6 mm, S_2 = 7.6 mm, S_3 = 5.4 mm, S_4 = 3 mm, R_1 = 7 mm, R_2 = 5 mm, and R_3 = 2.8 mm.

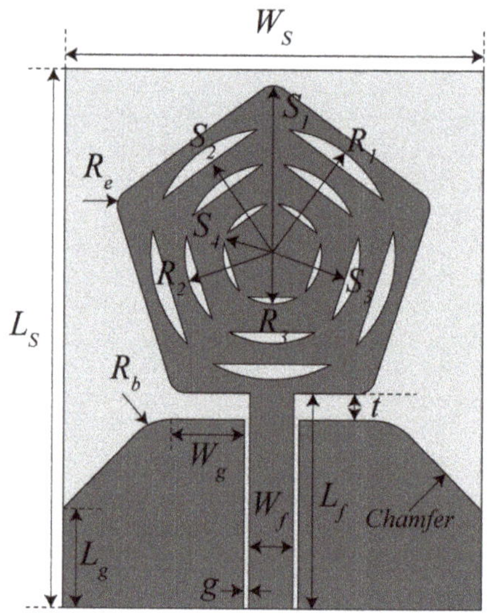

Figure 2. Geometry of the proposed CPW-fed pentagonal fractal antenna.

The design steps of the proposed CPW-fed pentagonal fractal antenna are shown in Figure 3a–d. In the first step, a CPW-fed circle-inscribed pentagonal radiator is designed, as shown in Figure 3a. The pentagonal radiator's dimensions are calculated using the following expression [22].

$$f_l = \frac{7.2}{(L + r + p) \times k} \quad \text{GHz} \tag{1}$$

where

$$L = 2A, \quad r = \frac{A}{4} \tag{2}$$

where A denotes the radius of the of the pentagonal radiator, which is equal to 9.6 mm. The parameters f_l and k in Equation (1) correspond to the lowest operating frequency of the antenna and the effective relative permittivity (ε_{reff}) of the dielectric substrate, respectively.

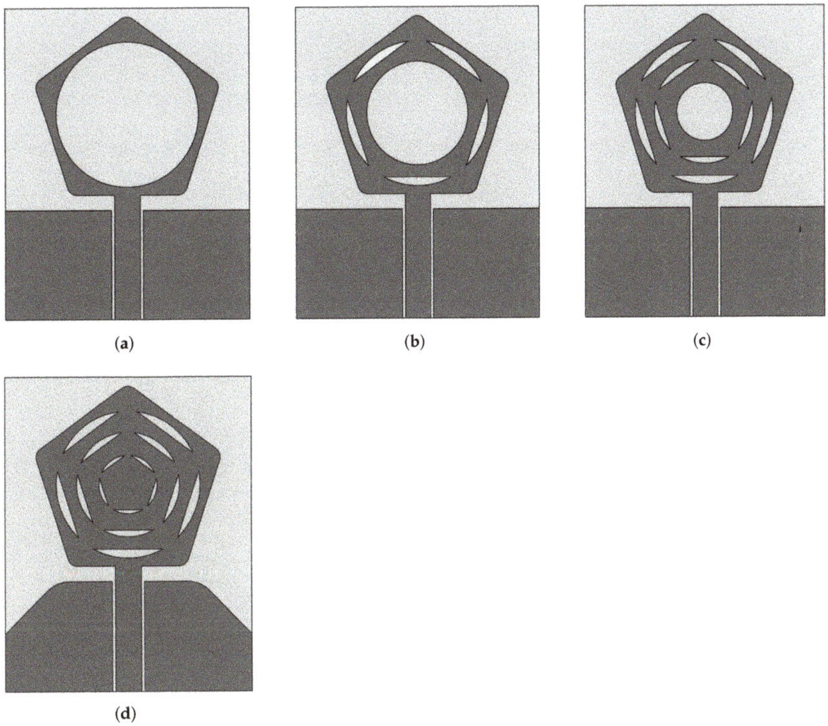

Figure 3. Design evolution of the proposed CPW-fed pentagonal fractal antenna (**a**) Step 1, (**b**) Step 2, (**c**) Step 3, and (**d**) Step 4 (proposed).

The antenna design shown in Figure 3a is simulated in the CST Microwave Studio and the respective reflection coefficient (S_{11}) is given in Figure 4. From the figure, it is observed that the design shown in Figure 3a provides a dual-band response at 4 GHz and around 12 GHz. In the second step, another circle-inscribed pentagonal patch is added in the Step-1 design, as shown in Figure 3b. This modification tends to excite higher order modes and ultimately leads to a UWB response in the band of interest (see Figure 4, Step-2). For improved matching, another pentagonal radiator is added in the Step-2 design, as depicted in Figure 3c. This design offers improved impedance matching in the band of interest, as shown in Figure 4. To further optimize the response, another pentagonal patch is designed with the Step-3 design, as shown in Figure 3d. In this step, the ground plane shape is changed to a trapezoidal shape (see Figure 3d). This modification led to a UWB response in the frequency range of 3.38–12.89 GHz, as shown in Figure 4.

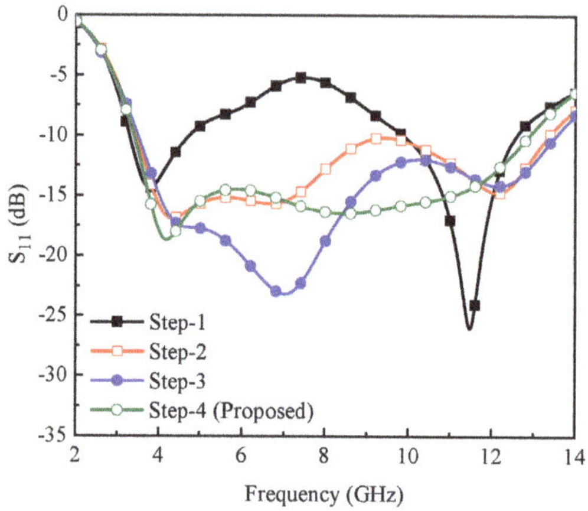

Figure 4. Reflection coefficient (S_{11}) of different antenna designs shown in Figure 3.

The simulated surface current distribution plots for the proposed antenna are shown in Figure 5. The surface current is plotted for three different frequencies, i.e., 4 GHz, 8 GHz, and 12 GHz. It is evident from the figure that at 4 GHz, the dense current is distributed on the feed line, on the edges of the ground plane, and in the lower portion of the patch radiator, as shown in Figure 5a. For higher frequencies, the current is uniformly distributed on the surface of the radiator and modified the CPW ground plane (see Figure 5b,c).

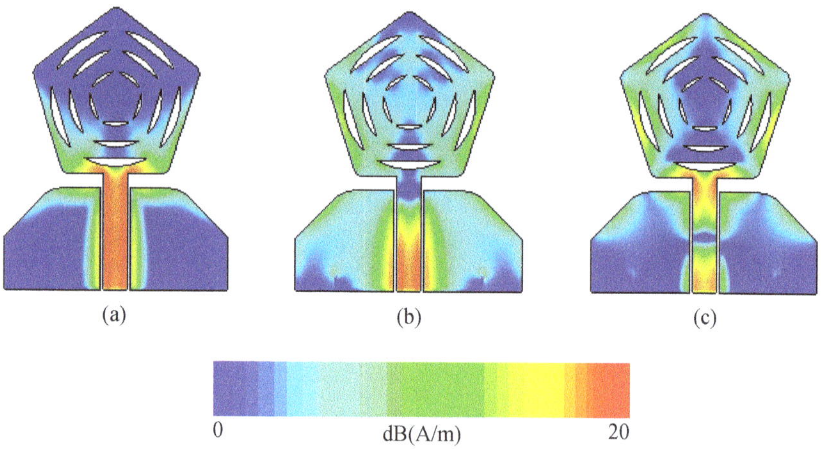

Figure 5. Simulated surface current distribution of the proposed CPW-fed pentagonal fractal antenna at (**a**) 4 GHz, (**b**) 8 GHz, and (**c**) 12 GHz.

A parametric study was performed to understand the behavior of the proposed UWB antenna by varying the gap between the microstrip feeding line and the ground plane, denoted as g, and the gap between the ground plane and the main radiator, denoted as t. From Figure 6a, it is observed that when the value of g is increased from 0.3 mm to 0.9 mm, the bandwidth of the antenna decreases. The maximum impedance matching is obtained for g = 0.3 mm (see Figure 6a). On the other hand, when the value of t is changed from 0.5 mm to 1.5 mm, the impedance bandwidth of the antenna improves, as shown in Figure 6b. For t >1.5 mm, the bandwidth of the antenna is reduced (see Figure 6b).

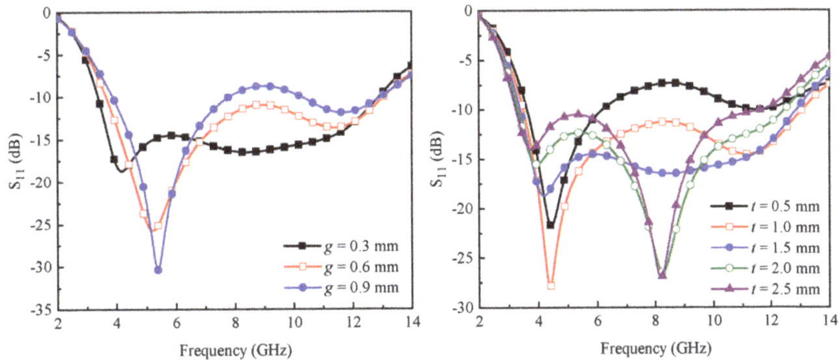

(a) (b)

Figure 6. Effect of parameter (**a**) *g* and (**b**) *t* on the antenna's performance.

3. Fabricated Prototype and Measurements

To verify the simulation results, the designed antenna was fabricated using the miling process. The prototype of the fabricated antenna is shown in Figure 7a. The S_{11} of the proposed antenna was measured by Rohde Schwarz (R&S) using the vector network analyzer (VNA) ZVA-67. The simulated and measured S_{11} characteristics are shown in Figure 7b. The antenna has a computed impedance bandwidth of 9.51 GHz in the frequency range of 3.38–12.89 GHz and a measured impedance bandwidth of 9.7 GHz in the frequency range of 3–12.7 GHz (see Figure 7b). Some discrepancies are observed between the simulated and measured results, which could arise due to fabrication tolerances, SMA connector losses, and the scattering measurement environment.

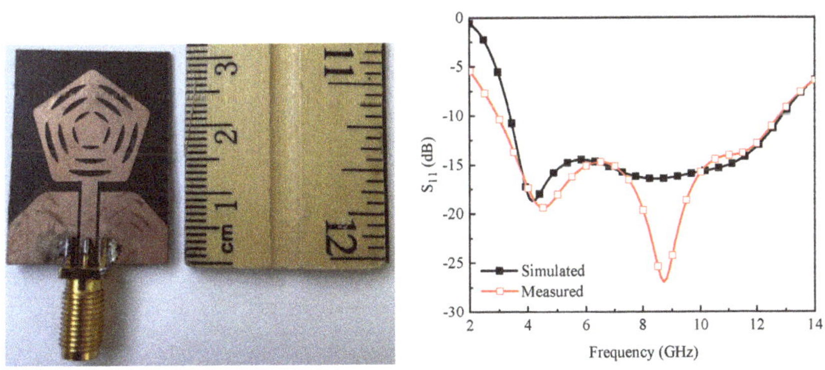

(a) (b)

Figure 7. (**a**) Fabricated prototype. (**b**) Simulated and measured S_{11} of the proposed CPW-fed pentagonal fractal antenna.

The simulated and measured radiation characteristics of the proposed fractal antenna are shown in Figure 8. For both the YZ and XZ planes, radiation characteristics are observed at three different frequencies: 4 GHz, 8 GHz, and 12 GHz. From the figure, it is observed that the designed antenna exhibits bi-directional radiation properties for the YZ plane and omnidirectional characteristics for the XZ-plane. It is also observed from Figure 8 that the radiation patterns are almost constant for the entire operating bandwidth, which is a useful property for microwave imaging applications.

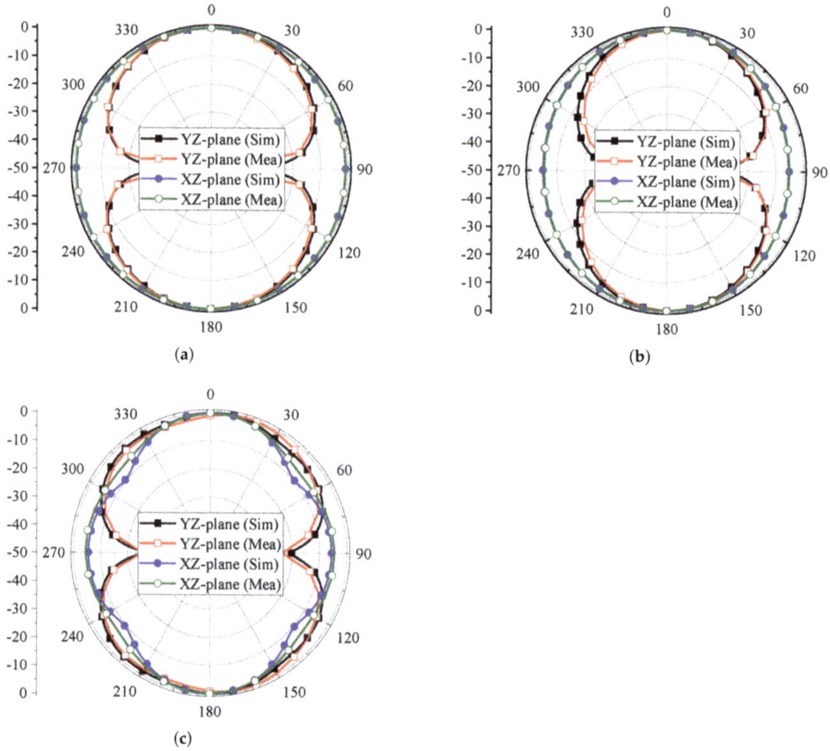

Figure 8. Simulated and measured radiation characteristics of the proposed CPW-fed pentagonal fractal antenna at (**a**) 4 GHz, (**b**) 8 GHz, and (**c**) 12 GHz.

Figure 9 shows the realized gain and efficiency of the proposed antenna. The peak gain of the antenna is noted to be 3.6 dBi at 8.5 GHz, while the minimum gain is 0.6 dBi at 3 GHz, as shown in Figure 9. Furthermore, the efficiency of the antenna is noted to be >80% for the entire operating bandwidth (see Figure 9).

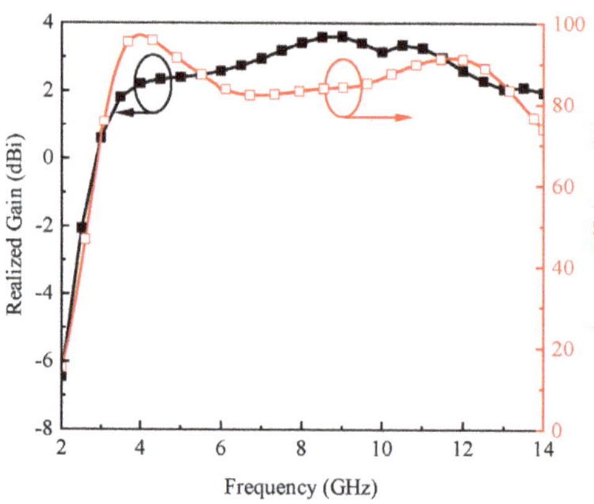

Figure 9. Realized gain and efficiency of the proposed CPW-fed pentagonal fractal antenna.

4. Time Domain Analysis

For near-field microwave imaging applications, it is necessary to assess the performance of the antenna in the time domain [23]. For this purpose, two identical antennas are placed 30 cm apart from each other in two different configurations, i.e., face to face and side by side, respectively, as shown in Figure 10a,b, in which both the antennas work in the transceiver mode.

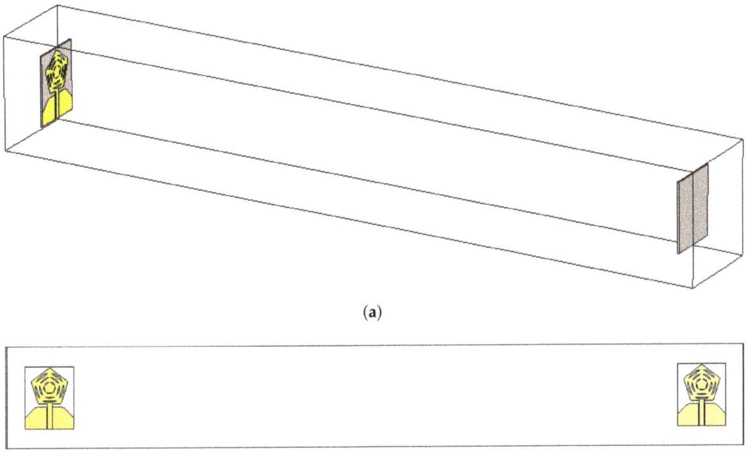

(a)

(b)

Figure 10. (**a**) Face-to-face and (**b**) side-by-side configuration for the evaluation of time domain performance of the proposed antenna CPW-fed pentagonal fractal antenna.

To excite the antennas, a Gaussian pulse, centered at 8 GHz with a frequency band of 1–15 GHz, is utilized. Figure 11a shows the normalized amplitudes of both input and output signals for both configurations. The cross-correlation between transmitted and received signals is calculated using Equation (3), which is known as the fidelity factor (FF) [24].

$$FF = \max\left[\frac{\int_{-\infty}^{\infty} Y_t(t)Y_r(t+\tau)d\tau}{\int_{-\infty}^{\infty}|Y_t(t)|^2dt\int_{-\infty}^{\infty}|Y_r(t)|^2dt}\right] \quad (3)$$

where $Y_t(t)$ and $Y_r(t)$ represent transmitted and received signals, respectively; and τ represents the group delay. The FF values for face-to-face and side-by-side configurations are noted to be 96.85% and 88.52%, respectively. The high value of FF for the face-to-face configuration shows that the transmitted signal will be less distorted.

The group delay for both the configurations, shown in Figure 10, is presented in Figure 11b. The group delay is defined as *the negative rate of change of the transfer function phase, $\phi(\omega)$ with respect to frequency*. The group delay quantifies the transition time taken by the signal to travel through a device [23]. Mathematically, it can be calculated as:

$$\tau(\omega) = -\frac{d\phi(\omega)}{d\omega} = -\frac{d\phi(\omega)}{2\pi df} \quad (4)$$

For UWB applications, $|\tau| \leq 2$ ns is desirable to ensure linearity of the phase in the far-field region. From Figure 11b, one can observe that the $|\tau|$ value is ≤ 1.5 ns for the entire frequency range [25].

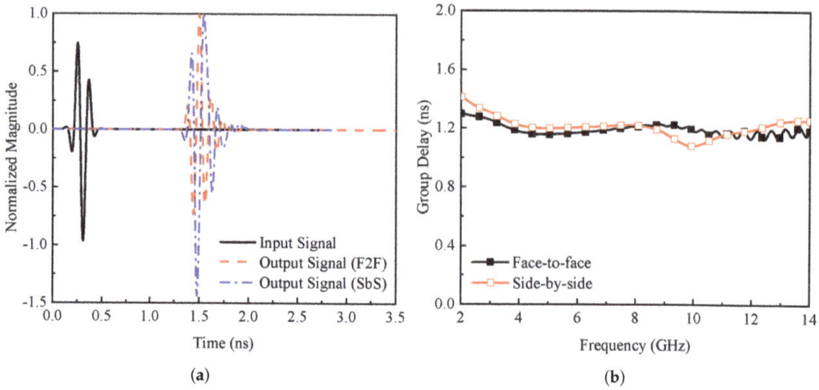

Figure 11. (**a**) Pulse response and (**b**) group delay of the proposed antenna CPW-fed pentagonal fractal antenna.

5. Conclusions

A compact UWB planar antenna is reported for near-field microwave imaging applications. The radiator geometry consists of a modified CPW-fed pentagonal fractal structure. To improve the impedance matching, a trapezoidal-shaped ground plane is utilized. The proposed design has an impedance bandwidth of 9.7 GHz, ranging from 3 GHz to 12.7 GHz, and offers a fractional bandwidth of 123.56%. It is also observed from the results that the proposed antenna exhibits constant radiation characteristics in the operating bandwidth, which is useful for near-field microwave imaging. The performance of the proposed antenna is also assessed in the time domain, and acceptable characteristics are observed.

Author Contributions: Conceptualization, M.A.K., U.R. and S.H.K.; methodology, M.A.K., A.N.N. and S.H.K.; software, U.R., H.Ş.S., U.R., S.H.K. and S.M.A.; validation, M.A.K., U.R., S.H.K. and S.M.A.; formal analysis, H.Ş.S. and A.N.N.; investigation, U.R., A.N.N. and S.H.K.; resources, H.Ş.S., A.N.N. and S.M.A.; writing—original draft preparation, M.A.K. and S.H.K.; writing—review and editing, U.R., H.Ş.S., A.N.N. and S.M.A.; visualization, U.R.; supervision, A.N.N. and S.H.K.; project administration, M.A.K., U.R. and A.N.N.; funding acquisition, S.M.A. All authors have read and agreed to the published version of the manuscript.

Funding: This research received no external funding.

Conflicts of Interest: The authors declare no conflict of interest.

References

1. Fear, E.C.; Hagness, S.C.; Meaney, P.M.; Okoniewski, M.; Stuchly, M.A. Enhancing breast tumor detection with near-field imaging. *IEEE Microw. Mag.* **2002**, *3*, 48–56. [CrossRef]
2. Fear, E.; Stuchly, M. Microwave detection of breast cancer. *IEEE Trans. Microw. Theory Tech.* **2000**, *48*, 1854–1863.
3. Wells, P.N. Ultrasound imaging. *Phys. Med. Biol.* **2006**, *51*, R83. [CrossRef] [PubMed]
4. Nass, S.J.; Henderson, I.C.; Lashof, J.C. *Mammography and Beyond: Developing Technologies for the Early Detection of Breast Cancer*; National Academy Press: Washington, DC, USA, 2001.
5. El Misilmani, H.M.; Naous, T.; Al Khatib, S.K.; Kabalan, K.Y. A survey on antenna designs for breast cancer detection using microwave imaging. *IEEE Access* **2020**, *8*, 102570–102594. [CrossRef]
6. Ojaroudi, M.; Civi, Ö.A. High efficiency loop sleeve monopole antenna for array based UWB microwave imaging systems. In Proceedings of the 2016 IEEE International Symposium on Antennas and Propagation (APSURSI), Fajardo, PR, USA, 26 June–1 July 2016; pp. 1781–1782.
7. Danjuma, I.M.; Akinsolu, M.O.; See, C.H.; Abd-Alhameed, R.A.; Liu, B. Design and optimization of a slotted monopole antenna for ultra-wide band body centric imaging applications. *IEEE J. Electromagn. Microw. Med. Biol.* **2020**, *4*, 140–147. [CrossRef]
8. Mahmud, M.; Islam, M.T.; Samsuzzaman, M. A high performance UWB antenna design for microwave imaging system. *Microw. Opt. Technol. Lett.* **2016**, *58*, 1824–1831. [CrossRef]

9. Subramanian, S.; Sundarambal, B.; Nirmal, D. Investigation on simulation-based specific absorption rate in ultra-wideband antenna for breast cancer detection. *IEEE Sens. J.* **2018**, *18*, 10002–10009. [CrossRef]
10. Lee, D.; Nowinski, D.; Augustine, R. A UWB sensor based on resistively-loaded dipole antenna for skull healing on cranial surgery phantom models. *Microw. Opt. Technol. Lett.* **2018**, *60*, 897–905. [CrossRef]
11. Wu, T.; King, R. The cylindrical antenna with nonreflecting resistive loading. *IEEE Trans. Antennas Propag.* **1965**, *13*, 369–373. [CrossRef]
12. Islam, M.; Islam, M.; Samsuzzaman, M.; Faruque, M.; Misran, N. Microstrip line-fed fractal antenna with a high fidelity factor for UWB imaging applications. *Microw. Opt. Technol. Lett.* **2015**, *57*, 2580–2585. [CrossRef]
13. Ahadi, M.; Isa, M.B.M.; Saripan, M.I.B.; Hasan, W.Z.W. Square monopole antenna for microwave imaging, design and characterisation. *IET Microw. Antennas Propag.* **2014**, *9*, 49–57. [CrossRef]
14. Halili, K.; Ojaroudi, M.; Ojaroudi, N. Ultrawideband monopole antenna for use in a circular cylindrical microwave imaging system. *Microw. Opt. Technol. Lett.* **2012**, *54*, 2202–2205. [CrossRef]
15. Ojaroudi, N.; Ghadimi, N. Omnidirectional microstrip monopole antenna design for use in microwave imaging systems. *Microw. Opt. Technol. Lett.* **2015**, *57*, 395–401. [CrossRef]
16. Abdollahvand, A.; Pirhadi, A.; Ebrahimian, H.; Abdollahvand, M. A compact UWB printed antenna with bandwidth enhancement for in-body microwave imaging applications. *Prog. Electromagn. Res.* **2014**, *55*, 149–157. [CrossRef]
17. Lasemi, Z.; Atlasbaf, Z. Impact of Fidelity Factor on Breast Cancer Detection. *IEEE Antennas Wirel. Propag. Lett.* **2020**, *19*, 1649–1653. [CrossRef]
18. Islam, M.; Islam, M.; Samsuzzaman, M.; Faruque, M. A negative index metamaterial antenna for UWB microwave imaging applications. *Microw. Opt. Technol. Lett.* **2015**, *57*, 1352–1361. [CrossRef]
19. Afifi, A.; Abdel-Rahman, A.B.; Allam, A.; Abd El-Hameed, A.S. A compact ultra-wideband monopole antenna for breast cancer detection. In Proceedings of the 2016 IEEE 59th International Midwest Symposium on Circuits and Systems (MWSCAS), Abu Dhabi, United Arab Emirates, 16–19 October 2016; pp. 1–4.
20. Mahmood, S.N.; Ishak, A.J.; Saeidi, T.; Soh, A.C.; Jalal, A.; Imran, M.A.; Abbasi, Q.H. Full Ground Ultra-Wideband Wearable Textile Antenna for Breast Cancer and Wireless Body Area Network Applications. *Micromachines* **2021**, *12*, 322. [CrossRef]
21. Rafique, U.; ud Din, S.; Khalil, H. Compact CPW-fed super wideband planar elliptical antenna. *Int. J. Microw. Wirel. Technol.* **2021**, *13*, 407–414. [CrossRef]
22. Ray, K.P. Design aspects of printed monopole antennas for ultra-wide band applications. *Int. J. Antennas Propag.* **2008**, *2008*. [CrossRef]
23. Wiesbeck, W.; Adamiuk, G.; Sturm, C. Basic properties and design principles of UWB antennas. *Proc. IEEE* **2009**, *97*, 372–385. [CrossRef]
24. Singhal, S.; Singh, A.K. CPW-fed hexagonal Sierpinski super wideband fractal antenna. *IET Microw. Antennas Propag.* **2016**, *10*, 1701–1707. [CrossRef]
25. Kwon, D.H. Effect of antenna gain and group delay variations on pulse-preserving capabilities of ultrawideband antennas. *IEEE Trans. Antennas Propag.* **2006**, *54*, 2208–2215. [CrossRef]

electronics

MDPI

Article

A Novel and Compact Slotted-Decahedral Antenna for 5G Devices

Karen Nallely Olan Nuñez and Roberto S. Murphy Arteaga *

Electronics Department, Instituto Nacional de Astrofísica, Óptica y Electrónica (INAOE), Tonantzintla, Puebla 72840, Mexico; karen.olan@inaoep.mx
* Correspondence: rmurphy@ieee.org; Tel.: +52-222-736-9044

Abstract: In this paper, a compact SISO (Single-Input Single-Output) antenna with a novel slotted-decahedral geometry is presented. The design was performed using a full-wave simulator (FWS). A decahedral patch with an eight-pointed star-shaped slot and two rectangular grooves is the main characteristic of the proposed novel radiator. The decagon shape, eight-pointed star-shaped slot, rectangular grooves, and separate section give rise to radiation. The SISO antenna operates from 23.1 to 29.94 GHz covering the proposed frequency bands for 5G wireless communication systems. The radiation pattern stability, moderate to high gain values (6.5 dBi, average), good radiation efficiency (higher than 89.4%), wide impedance bandwidth (6.84 GHz), compactness ($13 \times 13 \times 0.787$ mm^3) and ease of manufacture are the principles advantages of this novel geometry. The experimental validation of the SISO antenna was performed, and good agreement was obtained with simulation results, and an equivalent circuit model was proposed. The proposed SISO antenna can be easily integrated into wireless sensors, drones, backhaul, and 5G devices which support IoT, working in the n257, n258 and n261 frequency bands.

Keywords: 5G; antennas; SISO antennas; n257; n258; n261

Citation: Olan Nuñez, K.N.; Murphy Arteaga, R.S. A Novel and Compact Slotted-Decahedral Antenna for 5G Devices. *Electronics* **2022**, *11*, 1813. https://doi.org/10.3390/electronics11121813

Academic Editors: Faisel Tubbal, Ladislau Matekovits and Raad Raad

Received: 19 May 2022
Accepted: 30 May 2022
Published: 7 June 2022

Publisher's Note: MDPI stays neutral with regard to jurisdictional claims in published maps and institutional affiliations.

1. Introduction

Nowadays, the 5G standard of communications is ubiquitous and pervasive, taking every day more importance for our daily lives. And we can be sure that its impact will continue to grow, as there are more devices and applications that require the signal transfer quality that this network can offer. Moreover, human endeavor is shifting toward the Internet of Things (IoT), also called the Internet of Everything (IoE), to provide, control, and monitor a slew of routinely activities in the fields of personal and commercial communications, industry, health services, banking, safety, education, and entertainment, to mention just a few [1,2].

The initial 5G network operating bands are divided into two sub-bands, called FR1 and FR2 (for Frequency Range). The first consists of a host of assigned frequencies between 450 MHz and 6 GHz, thus also dubbed "the sub 6 GHz band". The second one comprises allocated frequencies between 24.25 to 52.6 GHz [3–7]. In this one, the frequencies that are being exploited cover the n257 (26.50 GHz to 29.50 GHz), n258 (24.25 GHz to 27.50 GHz), and the n261 (27.50 GHz to 28.35 GHz) sub-bands. Some of the antennas designed for FR1, however, are the basis for those used for FR2 [8–10]. In this frequency range, special attention has been placed for applications around 28 GHz, since it can operate with a broader bandwidth [11].

This paper focuses on the design of a Single Input Single Output (SISO) antenna to cover the needs for devices using the frequency bands from 24.25 GHz to 29.50 GHz. In fact, the antenna was designed, simulated, fabricated, and measured, demonstrating a very good response from 23.1 GHz to 29.94 GHz.

Moreover, this antenna can be used as the basis for Multiple Input Multiple Output (MIMO) systems. MIMO allows for size optimization, broader bandwidth, stable radiation patterns and higher gains, among other fundamental figures of merit [12–24]. These, however, require careful design considerations, especially related to coupling. This has become a very important field of endeavor, tackled by many researchers around the world [25–33].

The paper is organized as follows. Section 2 details the antenna design process, based on full wave simulations. Experimental results are presented and discussed in Section 3. Section 4 presents a discussion of the results, compared to similar recently reported structures. The paper closes with general conclusions, presented in Section 5.

2. Materials and Methods

The proposed novel geometry, named "slotted-decahedral antenna", is shown in Figure 1, which includes its geometrical design parameters.

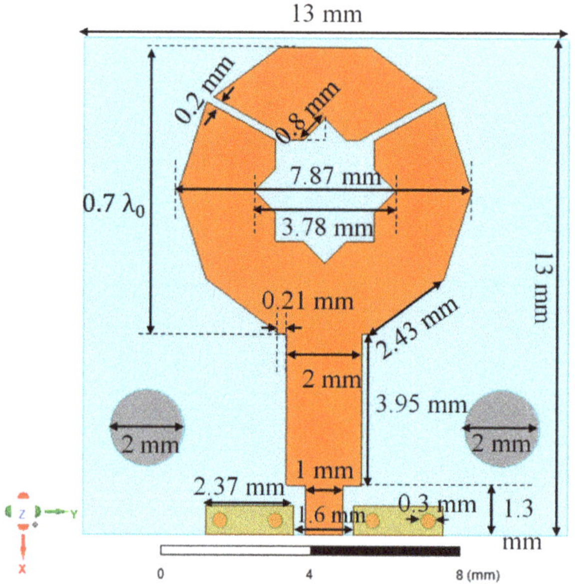

Figure 1. Proposed novel geometry and design parameters.

The geometry is based on a decahedral patch with an eight-pointed star-shaped slot at its center, and two rectangular grooves at the end of the radiator, which enhance stability in the shape of the radiation pattern across the impedance bandwidth (from 23 GHz to 29.9 GHz) through its resonating response. A commercial Rogers 4350B substrate was used for the proposed antenna design, which has a thickness of 0.787 mm, a loss tangent δ of 0.004 and a relative dielectric constant (ε_r) of 3.66.

The width of the feed line (1 mm) and the spacing between the feed line and top ground (0.3 mm) were chosen considering connector dimensions. The 2 mm wide line has a characteristic impedance of 50 Ω to achieve impedance matching.

The structure was defined on a full-wave simulator, including as many material, geometrical and process-related parameters as possible. This structure is shown in Figure 2, a render that is very close to the real structure.

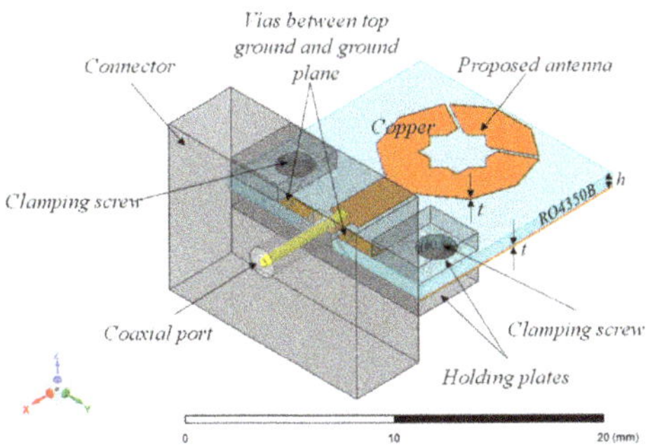

Figure 2. Antenna structure defined in the full-wave simulator, including the connector (h = 0.787 mm, t = 17.5 µm).

The design evolution and a comparison of four important figures of merit during the design process is shown in Figure 3, step #5 being the final design. Figure 3a illustrates the five design steps; (b) the reflection coefficient at the input port, using a reference impedance of 50 Ω; (c) the behavior of peak gain and radiation efficiency for every design step; and (d), the radiation patterns obtained for each design step. Something that is noteworthy in this figure is the change in shape of the radiation pattern, that is, each design step causes the pattern to become more directive and more symmetrical.

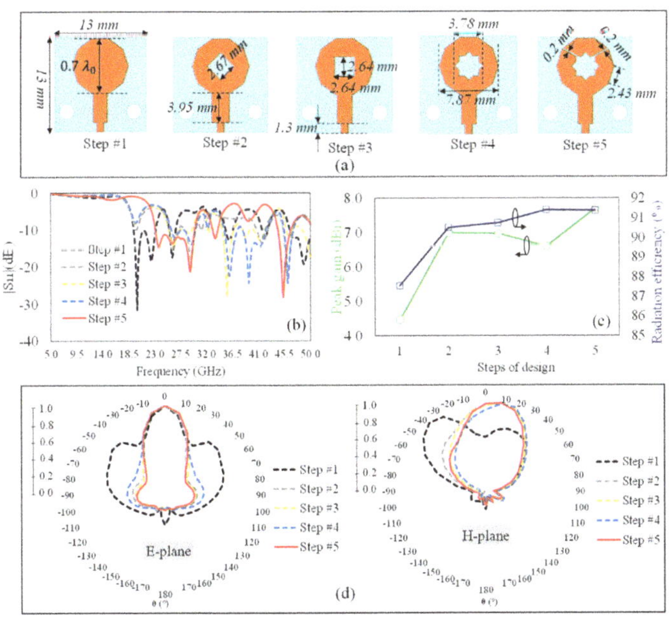

Figure 3. Design evolution and comparison of (**a**) Steps of the design process and antenna dimensions; (**b**) Return loss; (**c**) Peak gains and radiation efficiencies at 28 GHz; and (**d**) 2D—normalized radiation patterns at 28 GHz.

During the design process different shapes for the antenna were tested, as well as different shapes for the slots. Figure 4 shows six 3D-radiation patterns from the full-wave simulator. The same figure contains squares and ellipses highlighting the points where the radiation is the least intense.

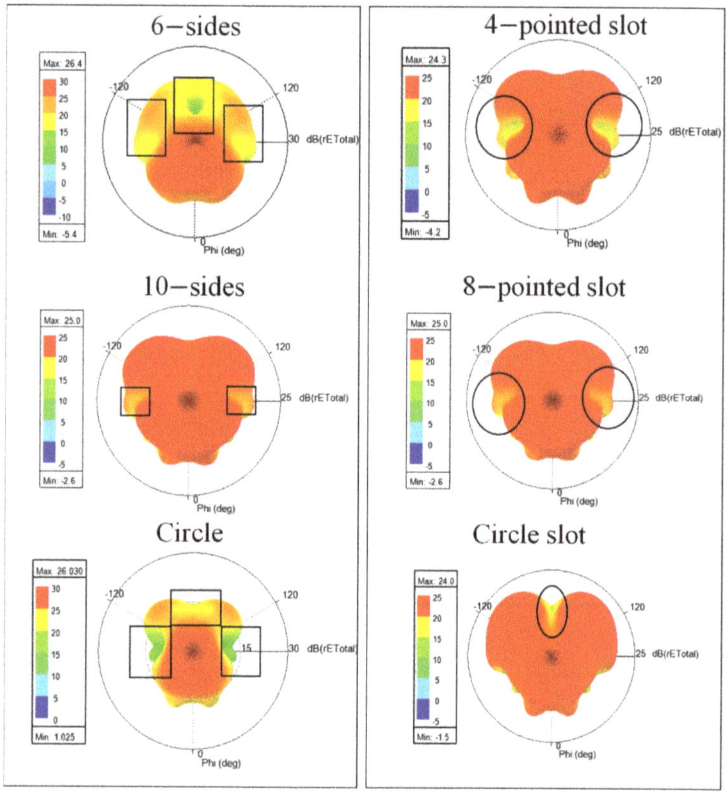

Figure 4. 3D—radiation patterns (in dB) from simulation, at the left—side of different antenna shape, and at the right-side of different slot shape.

The best performance, considering the radiation pattern shape, gain, radiation efficiency and impedance bandwidth, is obtained when the antenna shape is 10 sided and the slot is 8-pointed, herein named "eight-pointed star-shaped slot".

To verify the working principle, the electric and magnetic fields were plotted; these graphs are shown in Figure 5. At the lower frequency, 23 GHz, resonance is mainly due to the decagon shape and rectangular grooves.

In the medium frequency range and beyond—that is 28 GHz to 30 GHz—some other effects occur. When the fields reach the rectangular grooves, Figure 6, the magnetic field (B_i) induces an electromotive force (induced EMF) in the separate section. The induced EMF gives rise to a magnetic field (B_{EMF}) going in the opposite direction, propagating in this small separate section of the antenna. This occurs when the phase of the magnetic field reaches its first positive half-cycle maximum (phase = 90°). On the contrary, when the second maximum of the negative half cycle occurs (phase = 270°) the magnetic field produced in the separate section now goes in the opposite direction than that generated with the 90° phase. This change in the direction of magnetic field happens continuously, since the phase of the electromagnetic wave changes as it propagates. Moreover, the electric field (associated to the induced EMF, E_{EMF}) in the separate section also causes this section to enter a state of resonance. Furthermore, the initial electric field (from the excitation pathway, E_i) is cancelled at the rectangular opening. At the same time, around 28 GHz and 30 GHz, the eight-pointed star-shaped slot at the center, becomes resonant.

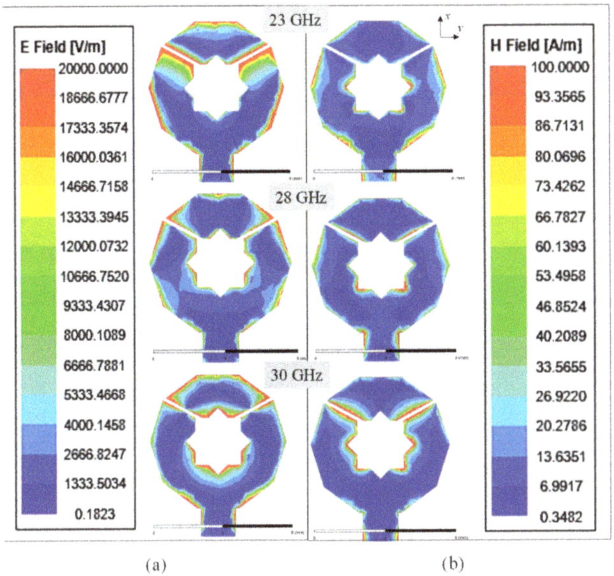

Figure 5. (**a**) Electric field (**b**) magnetic field (in magnitude) plotted in the upper layer of the metal at 23 GHz, 28 GHz, and 30 GHz.

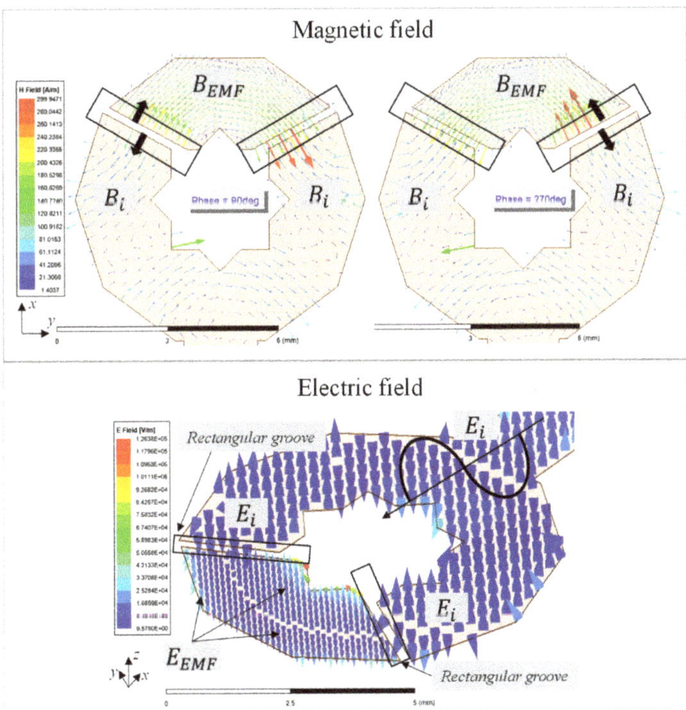

Figure 6. Magnetic and electric fields around the rectangular grooves.

On the other hand, the 33 GHz and 45 GHz frequency bands from $|S_{11}|$ in Figure 3a (step#5), are generated by the 8-pointed star slot, at its center, and by the two rectangular grooves. However, when analyzing some radiation parameters, such as, radiation efficiency, gain, and radiation pattern, the structure presents some nulls and secondary lobes, such as Figure 7 shows. Besides, the radiation efficiency is about 84% and 91% at 33 and 45 GHz, respectively. In consequence, these bands are not considered in the bandwidth of the novel proposed SISO antenna.

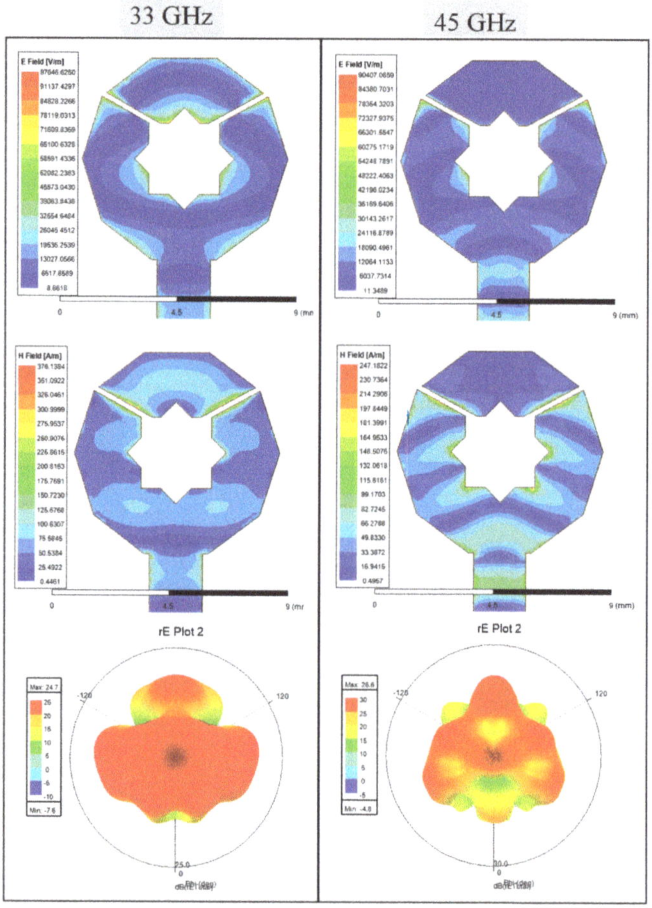

Figure 7. Magnetic and electric fields (in magnitude) and 3D—radiation patterns (in dB) at 33 and 45 GHz, respectively.

3. Results

To verify that the proposed design performs correctly in the frequency band from 23.1 to 29.94 GHz, apt for 5G technologies, the antenna was fabricated and measured. Its performance was then assessed from measured and simulated results.

A photograph of the fabricated antenna, including connectors, is presented as Figure 8a,b. The challenge in the manufacture of this prototype was making the perforations in the substrate to be able to use the connector, as well as the paths from the upper ground to the lower ground plane.

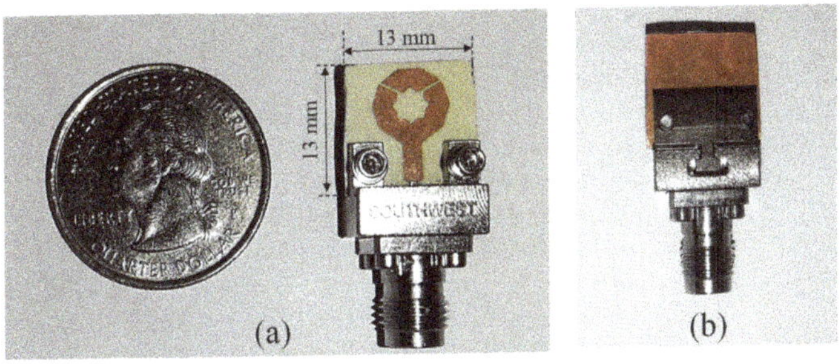

Figure 8. Fabricated antenna; (**a**) Front view, and (**b**) back view.

The measurements were performed using an Anritsu Vector Network Analyzer (VNA), in conjunction with an automatic calibration device (Autocal 36585 V). Figure 9 shows the experimental setup.

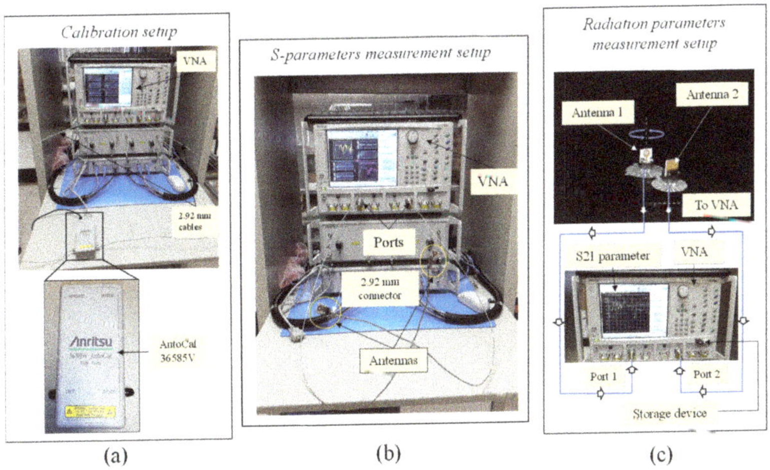

Figure 9. Measurement setup to (**a**) calibration, (**b**) S-parameters, and (**c**) radiation test.

The return loss at the input port from simulation and measurement is shown in Figure 10a. Moreover, the same figure shows a small frequency shift (600 MHz and 550 MHz) between measurement and simulation, which does affect the operating bandwidth.

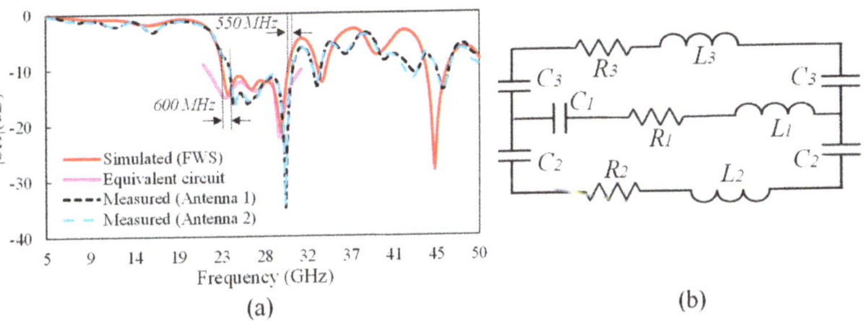

Figure 10. (**a**) Comparison of theoretical (using a full—wave simulator, FWS), equivalent circuit, and experimental reflection coefficient at the input port of the proposed design; and (**b**) the proposed equivalent circuit suitable for the 23 to 30 GHz frequency range.

In order to measure the radiation pattern, two antennas were built to determine the appropriate parameters following the methodology described in [34], which requires the two antennas to be identical. As can be seen on Figure 10a, the response of the two antennas is practically identical in the frequency range of interest, and thus we can conclude that they behave as identical electrically.

Additionally, a proposed equivalent circuit is shown in Figure 10b. The lumped elements represent the multiple phenomena occurring in this frequency range, and the model is seen to have a good correlation with full-wave simulation (FWS) results, carried out in Ansys Electronics' HFSS. The values of the proposed model are $R_1 = 10.29\ \Omega$, $R_2 = 29\ \Omega$, $R_3 = 30.2\ \Omega$, $L_1 = 3$ nH, $L_2 = 2.22$ nH, $L_3 = 1.45$ nH, $C_1 = 0.01$ pF, $C_2 = 0.04$ pF, and $C_3 = 0.049$ pF. Such values were obtained through an optimization process using ADS (Advance Design System, version 2022) simulations. Here, C_2 and C_3 represent the rectangular grooves. R_1-L_1-C_1 correspond to the response around 23 GHz, R_2-L_2-C_2 to that around 26.5 GHz, and R_3-L_3-C_3 account for the response about 29.7 GHz. All together, they model the behavior of the antenna from 23 to 30 GHz.

The measured and simulated 2D-radiation patterns are graphed in Figure 11, while the gain values across the impedance bandwidth are shown in Figure 12.

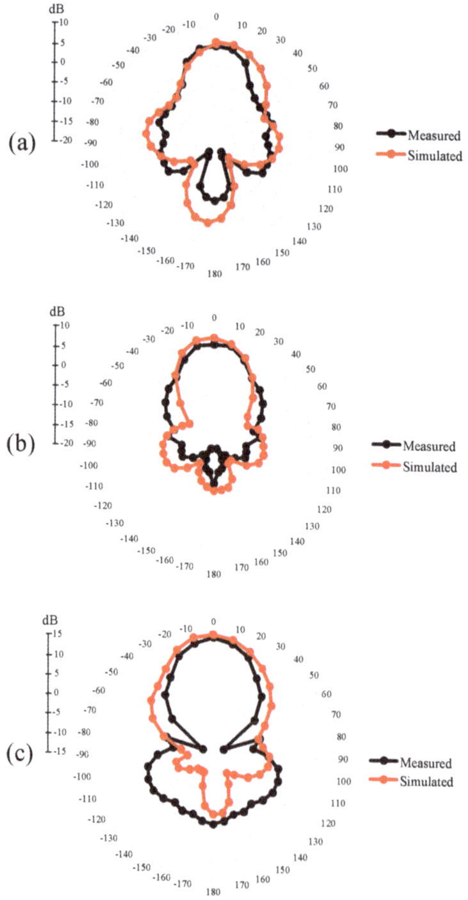

Figure 11. 2D—Normalized radiation patterns at (**a**) 23 GHz, (**b**) 28 GHz, and (**c**) 30 GHz from simulation and measurement.

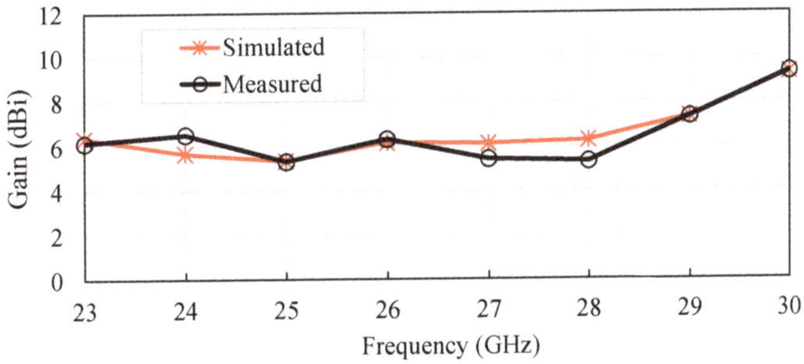

Figure 12. Measured and simulated gains from 23 to 30 GHz at $\theta = 0°$.

Once the performance of the proposed SISO antenna has been tested and validated, its principal figures of merit were determined to compare it with related works. This comparison is summarized in Table 1.

The proposed design has a compact size, which makes it suitable for integration into compact 5G devices that support IoT, and presents a wide impedance bandwidth covering the n257, n258, and n261 bands.

The gain values in Figure 12 are above 6 dBi at 24 GHz and up to 9.2 dBi at 29.9 GHz, from simulation, and above 5 dBi and up to 9 dBi from measurement. It is important to note that these values are high considering they were obtained with a single radiating element (SISO antenna).

The cross-polarization and co-polarization, in dB, are shown in Figure 13, from the full-wave simulator, the co-pol values are higher than 0 dB and the cross-pol values are around −40 dB.

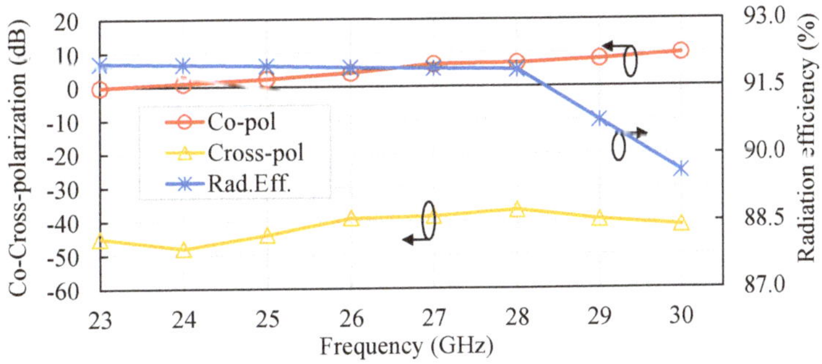

Figure 13. Co-polarization, cross-polarization, and radiation efficiency over the bandwidth of operation.

The radiation efficiency is also shown in Figure 13; it was obtained from the full-wave simulation, considering finite conductivity, the loss coefficient of the substrate, and other effects, such as mismatching of impedances of the proposed antenna and the connector's impedance across the impedance bandwidth, as well as the coupling between the holding plates and the antenna. The values are better than 89% (at 29.9 GHz) and up to 92% (at 24 GHz).

The radiation efficiency is reduced when including the connector during the simulation: 8% less at 23.1 GHz, 7% less at 28 GHz, and 9% less at 29.9 GHz. Besides, the shape of the radiation pattern is affected by the connector (holding plates and backplane).

Although during the simulations many parameters were considered, some others cannot be predicted and added to the simulation, such as imperfections in the metal of the radiator element, reduction of the metal layer during manufacture, small differences in the thickness of the substrate, imperfections inherent to the substrate, reflections due to the size of the connector being comparable to the size of the antenna, as well as random errors due to setup calibration, cable attenuation (1.96–2.32 dB/m [35]), and misalignment of the device under test.

We thus attribute the difference in gain between experimental and simulation results, shown in Figure 12, to these effects.

Despite obtaining lower gains during the experimental validation, we believe that the attained gain values are good, since they have a similar behavior to those of simulation, and they are higher than 5 dBi throughout the impedance bandwidth. The average gain from measurement is 6.44 dBi whereas the simulated one is 6.56 dBi.

The radiation patterns show good correlation with the simulated patterns. These have one only beam, normal to the decahedral geometry, and an HPBW close to 40°.

The back lobes (in the three patterns) are apparently large, but this is only an appreciation, these lobes have lower values than the principal beam.

4. Discussion

Considering the recent 5G developments regarding antennas, a comparison with reported works for the same frequency band for 5G technology above of 6 GHz (5G mm-wave frequency band) is shown in Table 1.

It should be noted that this table contains two types of SISO (Single-Input Single-Output) antennas, for a fair comparison; SISO antennas of a single element [36,37] and SISO antennas of multiple elements [38–40].

Moreover, is important to highlight that the listed designs are the most recent developments (2020–2021).

The antenna herein proposed has many advantages when compared with [36], such as higher gain, higher radiation efficiency, higher bandwidth, and compactness.

The designs presented by [38–40] have higher gains than our proposal, but it is important to note that those works are arrays of elements. Thus, it is reasonable to expect higher gains.

Our design bests that of [38] in impedance bandwidth, HPBW, and above all, it is much more compact.

With respect to [39], the proposed design in this work is better in impedance bandwidth, HPBW, radiation efficiency, and volume (considering that it requires two substrate layers).

Furthermore, the design presented by [40] requires two substrate layers, and the total area is large compared with our work. That work, however, attains good values of gain, impedance bandwidth, and polarization.

A single element antenna is reported in [37], but it has a narrow bandwidth of just 1.5 GHz. Despite having a high gain value this design is large (19 times larger than ours), and its HPBW is narrow.

Table 1. Comparison with Related Works Operating at 28 Ghz (SISO Antennas).

Parameter \ Work	[36]	[38]	[39]	[40]	[37]	This Work
Substrate and thickness (h)	Rogers 5880 h = 0.787 mm	Rogers RT/Duroid 5880 h = 0.52 mm	Rogers Duroid RT5880 h1 = 0.5 mm h2 = 0.8 mm	Rogers 5880 h1 = 1 mm h2 = 0.508 mm	Rogers RT/Duroid 5880 h = 0.127 mm	Rogers 4350 B h = 0.787 mm
Total area (mm^2) (λ_0 at 28 GHz)	21.9 × 5.64 2.05 λ_0 × 0.53 λ_0	75 × 100 7 λ_0 × 9.34 λ_0	34 × 36 3.18 λ_0 × 3.36 λ_0	~ 40 × 40 3.74 λ_0 × 3.74 λ_0	48 × 67 4.48 λ_0 × 6.26 λ_0	13 × 13 1.21 λ_0 × 1.21 λ_0
Design type	Antipodal Vivaldi antenna	2 × 2 Patch array	2 × 2 Dual off-center-fed dipoles (array)	Magneto-electric dipole 1 × 2 array	Reconfigurable Leaky-wave antenna based on a HMSIW	Single element antenna based on decahedral geometry
Reflection coefficient	−25 dB	~−18 dB	~−15 dB	~−15 dB	~−15 dB	−21.5 dB

Table 1. *Cont.*

Parameter \ Work	[36]	[38]	[39]	[40]	[37]	**This Work**
Impedance bandwidth	4.66 GHz (28 GHz band)	720 MHz (27.75–28.47 GHz)	3 GHz (27.2–30.2 GHz)	7 GHz (24.4–31.4 GHz)	1.5 GHz (28–29.5 GHz)	6.84 GHz (23.1–29.94 GHz)
Gain	3.4 dBi 3.6 dBi 7.4 dBi	9.97 dBi 12.3 dBi	13.1 dBi 13.2 dBi	10 dBic	8.2 ± 0.6 dBi	6.56 dBi *Average*
Polarization	N.A.	Linear	Dual linear	Circular	N.A.	Linear
Radiation efficiency	>86% (28 GHz band)	>96% (28 GHz band)	88% (28 GHz band)	N.A.	N.A.	>89.4%
HPBW	266° (E-plane) 160° (H-plane)	29.9° (E-plane) 60.0° (H-plane)	12° (xz-plane) 11° (yz-plane)	N.A.	29°	40° (E-plane) 65° (H-plane)
Operation frequencies	14.44–20.98 GHz 24.34–29 GHz 33–40 GHz	5.9 GHz 28 GHz	28 GHz 38 GHz	28 GHz bands	28 GHz band	28 GHz bands
Design complexity	high	low	high	high	medium	medium
Fabrication	Easy	Easy	Difficult	Difficult	Easy	Easy

N.A. is Not Available. HMSIW is Half-Mode Substrate Integrated Waveguide. HPBW is Half Power Beam Width.

5. Conclusions

In this paper we present a novel antenna design based on a decahedral patch with an eight-pointed star-shaped slot and two rectangular grooves at the end of the radiator.

This design has a very compact size ($13 \times 13 \times 0.787$ mm^3), wide impedance bandwidth (6.84 GHz) covering the proposed bands for 5G technology, above 6 GHz (n257, n258, and n261), specifically from 23.1 GHz to 29.9 GHz, stable radiation pattern, average gain of 6.56 dBi (values from 5.3 dBi to 9.28 dBi), and radiation efficiency better than 89%, from simulation.

The performance of the proposed novel geometry (SISO antenna) is suitable for compact 5G devices that support IoT. The experimental validation was performed, showing a very good agreement with simulation results, especially in return loss, gain and radiation pattern.

Author Contributions: K.N.O.N. conceived the idea, designed the structure, and wrote the principal ideas. R.S.M.A. revised the manuscript, contributed some ideas, and supervised the work. All authors have read and agreed to the published version of the manuscript.

Funding: This research was funded by CONACyT, México, under grant numbers 852217 and 285199.

Institutional Review Board Statement: Not applicable.

Informed Consent Statement: Not applicable.

Data Availability Statement: Not applicable.

Conflicts of Interest: The authors declare no conflict of interest.

References

1. Andrews, J.G.; Buzzi, S.; Choi, W.; Hanly, S.V.; Lozano, A.; Soong, A.C.K.; Zhang, J.C. What will 5G be? *IEEE J. Sel. Areas Commun.* **2014**, *32*, 1065–1082. [CrossRef]
2. Attiah, M.L.; Isa, A.; Zakaria, Z.; Abdulhameed, M.; Mohsen, M.K.; Ali, I. A survey of mmWave user association mechanisms and spectrum sharing approaches: An overview, open issues and challenges, future research trends. *Wirel. Netw.* **2020**, *26*, 2487–2514. [CrossRef]
3. Kumar, S.; Dixit, A.S.; Malekar, R.R.; Raut, H.D.; Shevada, L.K. Fifth Generation Antennas: A Comprehensive Review of Design and Performance Enhancement Techniques. *IEEE Access* **2020**, *8*, 163568–163593. [CrossRef]
4. *Guidelines for Evaluation of Radio Interface Technologies for IMT-2020*; Document ITU-R M.2412-0; International Telecommunication Union: Geneva, Switzerland, 2017; pp. 1–144.
5. *Requirements, Evaluation Criteria and Submission Templates for the Development of IMT-2020*; Document Report ITU-R M.2411-0; International Telecommunication Union: Geneva, Switzerland, 2017; pp. 1–32.

6. Zhao, A.; Ren, Z. Wideband MIMO antenna systems based on coupled-loop antenna for 5G N77/N78/N79 applications in mobile terminals. *IEE Access* **2019**, *7*, 93761–93771. [CrossRef]
7. *Technical Feasibility of IMT in Bands Above 6 GHz*; Document ITU-R M.2376-0; International Telecommunication Union: Geneva, Switzerland, 2015; pp. 1–134.
8. Abdullah, M.; Altaf, A.; Anjum, M.R.; Arain, Z.A.; Jamal, A.A.; Alibakhshikenari, M.; Falcone, F.; Limiti, E. Future Smartphone: MIMO Antenna System for 5G Mobile Terminals. *IEEE Access* **2021**, *9*, 91593–91603. [CrossRef]
9. Kiani, S.H.; Altaf, A.; Anjum, M.R.; Afridi, S.; Arain, Z.A.; Anwar, S.; Khan, S.; Alibakhshikenari, M.; Lalbakhsh, A.; Khan, M.A.; et al. MIMO Antenna System for Modern 5G Handheld Devices with Healthcare and High Rate Delivery. *Sensors* **2021**, *21*, 7415. [CrossRef]
10. Khalid, H.; Awan, W.A.; Hussain, M.; Fatima, A.; Ali, M.; Hussain, N.; Khan, S.; Alibakhshikenari, M.; Limiti, E. Design of an Integrated Sub-6 GHz and mmWave MIMO Antenna for 5G Handheld Devices. *Appl. Sci.* **2021**, *11*, 8331. [CrossRef]
11. Rappaport, T.S.; Xing, Y.; MacCartney, G.R.; Molisch, A.F.; Mellios, E.; Zhang, J. Overview of millimeter wave communications for fifth generation (5G) wireless networks with a focus on propagation models. *IEEE Trans. Antennas Propag.* **2017**, *65*, 6213–6230. [CrossRef]
12. Xi, L. A wideband planar filtering dipole antenna for 5G communication applications. *Microw. Opt. Technol. Lett.* **2019**, *61*, 2746–2751. [CrossRef]
13. Hussain, N.; Jeong, M.J.; Abbas, A.; Kim, T.J.; Kim, N. A metasurface-based low-profile wideband circularly polarized patch antenna for 5G millimeter-wave systems. *IEEE Access* **2020**, *8*, 22127–22135. [CrossRef]
14. Zeng, J.; Luk, K.-M. Single-layered broadband magnetoelectric dipole antenna for new 5G application. *IEEE Antennas Wirel. Propag. Lett.* **2019**, *18*, 911–915. [CrossRef]
15. Karthikeya, G.S.; Abegaonkar, M.P.; Koul, S.K. A wideband conformal antenna with high pattern integrity for mmWave 5G smartphones. *Prog. Electromagn. Res. Lett.* **2019**, *84*, 1–6. [CrossRef]
16. Desai, A.; Upadhyaya, T.; Patel, R. Compact wideband transparent antenna for 5G communication systems. *Mirow. Opt. Technol. Lett.* **2019**, *61*, 781–786. [CrossRef]
17. Alkaraki, S.; Andy, A.S.; Gao, Y.; Tong, K.-F.; Ying, Z.; Donnan, R.; Parini, C. Compact and low-cost 3-D printed antennas metalized using spray-coating technology for 5G mm-Wave communication systems. *IEEE Antennas Wirel. Propag. Lett.* **2018**, *17*, 2051–2055. [CrossRef]
18. Yin, J.; Wu, Q.; Yu, C.; Wang, H.; Hong, W. Broadband endfire magnetoelectric dipole antenna array using SICL feeding network for 5G millimeter-wave applications. *IEEE Trans. Antennas Propag.* **2019**, *67*, 4895–4900. [CrossRef]
19. Mujammami, E.H.; Sebak, A.B. Wideband high gain printed quasi-Yagi diffraction gratings-based antenna for 5G applications. *IEEE Access* **2019**, *7*, 18089–18100. [CrossRef]
20. Dzagbletey, P.A.; Jung, Y.-B. Stacked microstrip linear array for millimeter-wave 5G baseband communication. *IEEE Antennas Wireless Propag. Lett.* **2018**, *17*, 780–783. [CrossRef]
21. Goel, T.; Patnaik, A. Novel broadband antennas for future mobile communications. *IEEE Trans. Antennas Propag.* **2018**, *66*, 2299–2308. [CrossRef]
22. Kim, E.; Ko, S.T.; Lee, Y.J.; Oh, J. Millimeter-wave tiny lens antenna employing U-shaped filter arrays for 5G. *IEEE Antennas Wireless Propag. Lett.* **2018**, *17*, 845–848. [CrossRef]
23. Wen, B.J.; Peng, L.; Li, X.F.; Mo, K.S.; Jiang, X.; Li, S.M. A low profile and wideband unidirectional antenna using bandwidth enhanced resonance-based reflector for fifth generation (5G) systems applications. *IEEE Access* **2019**, *7*, 27352–27361. [CrossRef]
24. Feng, B.; Li, L.; Zeng, Q.; Chung, K.L. A wideband antenna using metasurface for the 2G/3G/LTE/5G communications. *Microw. Opt. Technol. Lett.* **2018**, *60*, 2482–2487.
25. Nadeem, I.; Choi, D.-Y. Study on mutual coupling reduction technique for MIMO antennas. *IEEE Access* **2019**, *7*, 563–586. [CrossRef]
26. Sharawi, M.S.; Podilchak, S.K.; Khan, M.U.; Antar, Y.M. Dual frequency DRA-based MIMO antenna system for wireless access points. *IET Microw., Antennas Propag.* **2017**, *11*, 1174–1182. [CrossRef]
27. Zhang, Y.; Deng, J.-Y.; Li, M.-J.; Sun, D.; Guo, L.-X. A MIMO dielectric resonator antenna with improved isolation for 5G mm-Wave applications. *IEEE Antennas Wireless Propag. Lett.* **2019**, *18*, 747–751. [CrossRef]
28. Kowalewski, J.; Eisenbeis, J.; Jauch, A.; Mayer, J.; Kretschmann, M.; Zwick, T. A mmW broadband dual-polarized dielectric resonator antenna based on hybrid modes. *IEEE Antennas Wirel. Propag. Lett.* **2020**, *19*, 1068–1072. [CrossRef]
29. Alibakhshikenari, M.; Khalily, M.; Virdee, B.S.; See, C.H.; Abd-Alhameed, R.A.; Limiti, E. Mutual Coupling Suppression between Two Closely Placed Microstrip Patches Using EM-Bandgap Metamaterial Fractal Loading. *IEEE Access* **2019**, *7*, 23606–23614. [CrossRef]
30. Alibakhshikenari, M.; Khalily, M.; Virdee, B.S.; See, C.H.; Abd-Alhameed, R.A.; Limiti, E. Mutual-Coupling Isolation Using Embedded Metamaterial EM Bandgap Decoupling Slab for Densely Packed Array Antennas. *IEEE Access* **2019**, *7*, 51827–51840. [CrossRef]
31. Jamshidi, M.B.; Roshani, S.; Talla, J.; Roshani, S.; Peroutka, Z. Size reduction and performance improvement of a microstrip Wilkinson power divider using a hybrid design technique. *Sci. Rep.* **2021**, *11*, 7773. [CrossRef]
32. Roshani, S.; Roshani, S. A compact coupler design using meandered compact microstrip resonant cell (MLCMRC) and bended lines. *Wirel. Netw.* **2021**, *27*, 677–684. [CrossRef]

33. Li, Y.; Sim, C.-Y.-D.; Luo, Y.; Yang, G. 12-port 5G massive MIMO antenna array in sub-6 GHz mobile handset for LTE bands 42/43/46 applications. *IEEE Access* **2018**, *6*, 344–354. [CrossRef]
34. Medina, J.L.; Díaz, E.; Olera, J.L.; Chávez, R.A.; Velázquez, A. Análisis y Comparación de Metodologías Para Determinar Experimentalmente la Ganancia de Antenas de RF Y Microondas. In Proceedings of the Metrology Symposium, Querétaro, Mexico, 22 October 2008.
35. Huber+Suhner. Formable Microwave Cable: SR_118_TP. Data Sheet. Available online: https://ecatalog.hubersuhner.com/media/documents/datasheet/en/pdf/22810073 (accessed on 26 May 2022).
36. Ullah, R.; Ullah, S.; Faisal, F.; Ullah, R.; Choi, D.Y.; Ahmad, A.; Kamal, B. High-Gain Vivaldi Antenna with Wide Bandwidth Characteristics for 5G Mobile and Ku-Band Radar Applications. *Electronics* **2021**, *10*, 667. [CrossRef]
37. Govindarajulu, S.R.; Hokayem, R.; Tarek, M.N.A.; Guerra, M.R.; Alwan, E.A. Low Profile Dual-Band Shared Aperture Array for Vehicle-to-Vehicle Communication. *IEEE Access* **2021**, *9*, 147082–147090. [CrossRef]
38. Hu, H.; Lai, F.; Chen, Y. Dual-Band Dual-Polarized Scalable Antenna Subarray for Compact Millimeter-Wave 5G Base Stations. *IEEE Access* **2020**, *8*, 129180–129192. [CrossRef]
39. Wang, Y.; Wu, B.; Zhang, N.; Zhao, Y.; Su, T. Wideband Circularly Polarized Magneto-Electric Dipole 1x2 Antenna Array for Millimeter-Wave Applications. *IEEE Access* **2020**, *8*, 27516–27523. [CrossRef]
40. Javanbakht, N.; Amaya, R.E.; Shaker, J.; Syrett, B. Fixed Frequency Beam-Scanning HMSIW-Based Leaky-Wave Antenna Composed of Circular Slots in V-Shape Configuration. *IEEE Access* **2021**, *9*, 52891–52901. [CrossRef]

electronics

MDPI

Article

A New mm-Wave Antenna Array with Wideband Characteristics for Next Generation Communication Systems

Mehr E Munir [1], Abdullah G. Al Harbi [2], Saad Hassan Kiani [3], Mohamed Marey [4], Naser Ojaroudi Parchin [5,*], Jehanzeb Khan [1], Hala Mostafa [6], Javed Iqbal [7,8], Muhammad Abbas Khan [9], Chan Hwang See [5] and Raed A. Abd-Alhameed [10]

1 Department of Electrical Engineering, Iqra National University, Peshawar 25000, Pakistan; mehre.munir@inu.edu.pk (M.E.M.); jehanzeb.khan@inu.edu.pk (J.K.)
2 Department of Electrical Engineering, Faculty of Engineering, Jouf University, Sakaka 42421, Saudi Arabia; a.g.alharbi@ieee.org
3 Department of Electrical Engineering, IIC University of Technology, Phnom Penh 121206, Cambodia; iam.kiani91@gmail.com
4 Smart Systems Engineering Laboratory, College of Engineering, Prince Sultan University, Riyadh 11586, Saudi Arabia; mfmmarey@psu.edu.sa
5 School of Engineering and the Built Environment, Edinburgh Napier University, Edinburgh EH10 5DT, UK; c.see@napier.ac.uk
6 Department of Information Technology, College of Computer and Information Sciences, Princess Nourah bint Abdulrahman University, P.O. Box 84428, Riyadh 11671, Saudi Arabia; hfmostafa@pnu.edu.sa
7 Department of Electrical Engineering, Faculty of Engineering and Technology, Gomal University, D.I. Khan 29050, Pakistan; javediqbal.iet@gu.edu.pk
8 School of Electrical and Electronic Engineering, Engineering Campus, Universiti Sains Malaysia, Nibog Tebal 14300, Penang, Malaysia
9 Department of Electrical Engineering, Balochistan University of Information Technology, Engineering and Management Sciences, Quetta 87300, Pakistan; muhammad.abbas@buitms.edu.pk
10 Faculty of Engineering and Informatics, University of Bradford, Bradford BD7 1DP, UK; r.a.a.abd@bradford.ac.uk
* Correspondence: n.ojaroudiparchin@napier.ac.uk

Citation: Munir, M.E.; Al Harbi, A.G.; Kiani, S.H.; Marey, M.; Parchin, N.O.; Khan, J.; Mostafa, H.; Iqbal, J.; Khan, M.A.; See, C.H.; et al. A New mm-Wave Antenna Array with Wideband Characteristics for Next Generation Communication Systems. *Electronics* **2022**, *11*, 1560. https://doi.org/10.3390/electronics11101560

Academic Editors: Dimitra I. Kaklamani and Manuel Arrebola

Received: 3 April 2022
Accepted: 11 May 2022
Published: 13 May 2022

Abstract: This paper presents a planar multi-circular loop antenna with a wide impedance bandwidth for next generation mm-wave systems. The proposed antenna comprises three circular rings with a partial ground plane with a square slot. The resonating structure is designed on a 0.254 mm thin RO5880 substrate with a relative permittivity of 2.3. The single element of the proposed design showed a resonance response from 26.5 to 41 GHz, with a peak gain of 4 dBi and radiation efficiency of 96%. The proposed multicircular ring antenna element is transformed into a four-element array system. The array size is kept at $18.25 \times 12.5 \times 0.254$ mm^3 with a peak gain of 11 dBi. The antenna array is fabricated and measured using the in-house facility. The simulated and measured results are well agreed upon and are found to be suitable for mm-wave communication systems.

Keywords: circular rings; dual-beam; mm-wave; efficiency; gain; linear array

1. Introduction

With the evolution of communication systems in modern telecommunication infrastructure, 5G is standardized throughout the world, offering high channel capacity characteristics with an increased number of users connected and low latency over the communication channel [1–4]. The 5G spectrum is categorized into two broad regions as Sub6 GHz, ranging under 6 GHz frequency, and mm-wave region, in which frequencies above 24 GHz and above are adopted. The bands of 28 GHz and 37 GHz and 39 GHz are licensed mm-wave bands in the mm-wave spectrum, for which in order to overcome high attenuation and propagation losses, high gain antenna arrays are required [5–8]. These antenna arrays include higher numbers of radiating elements with low-loss feeding networks [7,8]. There are several antenna systems reported in the literature [9–15]. These antenna systems include substrate integrated waveguide (SIW) antenna systems, air filled gap structures (AFGs), and planar structures. In [9], a multilayer antenna array with wideband and high gain characteristics is presented. The frequency range of the reported antenna is 57–71 GHz, with a peak gain of 13.9 dBi and minute sidelobe levels, with the number of Vias included. However, due to the nature of bonding films and asymmetric Vias inside SIW, the reported structure is complex to build and integrate into RF circuits. Waveguide antennas usually accomplish high gain, but the large volume of the metallic waveguide with a high manufacturing price is the main hitch for bulk production. In [10], an 8 × 8 element SIW array is presented with a bandwidth range from 35.4–41.7 GHz, total peak radiation efficiency of 82% and realized gain of 26.1 dBi. The array is difficult to assemble due to the higher numbers of Vias included. Planar antenna arrays are simple to assemble compared to SIWs. The transmission line losses are high compared to SIW, but they can be neglected with careful feed network modeling. A four-element T shape planar antenna array is presented in [11]. The peak gain achieved is 11.5 dBi and the bandwidth attained is 8 GHz, approximately. A modified four-element hook shape antenna array in [12] exhibited a dual-beam radiation response, with a small angular beamwidth of twenty-four Degree. A four-element planar antenna in [13] shows a dual-band response at the central frequency of 28 GHz and 38 GHz. The reported antenna showed a peak gain of 7 dBi, but the bandwidth at both resonances is extremely low. Metamaterial-induced mm-wave antennas have also been reported, but they are difficult to assemble and the cost and complexity generally increase [14,15].

In [16], a circular shape antenna with a slotted loop structure is presented for Bluetooth and Wi-Fi applications, covering resonance from 2.4 to 3.1 GHz. The antenna is fabricated on an FR4 substrate with a thickness of 1.6 mm. The size of the single element is 35 × 35 mm^2. This design is then transformed into a four-element linear array system using the Wilkinson power divider. In the four-element array, the radiating elements are separated at 40 mm apart and the total dimension of the proposed array is 155 × 75 mm^2. The peak gain achieved is 8.39, while total efficiency is 77%.

Similarly, Ref. [17] presents a single element with a super bandwidth operating from 6.5 to 100 GHz. The proposed antenna offers stable radiation patterns at lower resonance frequencies but at higher frequencies, the radiation patterns deviate. The size of the antenna is 20 × 20 mm^2 only while the maximum gain achieved is 10 dB.

Therefore, to address those limitations, in this paper, a three circular novel shape four-element antenna array is presented with dual-beam characteristics. The proposed antenna is simple in structure and provides high gain and dual beam characteristics with a narrow beamwidth. The proposed antenna offers a wide frequency bandwidth of 15.5 GHz and radiation efficiency above 90%. This paper is organized as follows. Section 1 covers the introduction. Section 2 shows the complete design analysis and array transformation. Section 3 shows the results and discussions from the simulated and obtained measured results with a comparison table. In the Section 4, the conclusion of the proposed work is provided.

2. Antenna Design

The proposed multi circular loop ring mm-wave antenna is designed on an ultra-thin Ro5880 substrate, with a relative permittivity of 2.3. The substrate is 0.254 mm in height. Figure 1 shows the proposed antenna's front and back view. The dimensions of the proposed antenna in mm are as follows: $A = 0.3$, $E = 6.75$, $M = 9$, $N = 11.5$, $DD = 1$, $O = 4.75$. The wide bandwidth characteristic of the proposed antenna is achieved using several numbers of the conducted parametrical studies. These studies included ring interspacing, transmission line feed modeling, and square slot on the ground plane. Figure 2 shows the overall response of the parametrical studies.

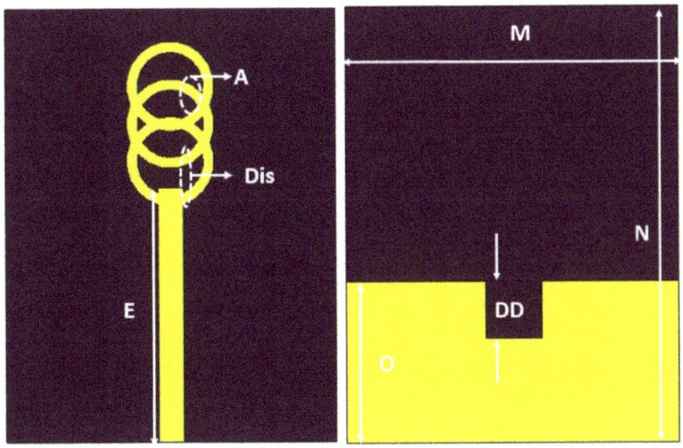

(a) (b)

Figure 1. (**a**) Front (**b**) back views of the proposed antenna.

As observed in Figure 2a, the parameter DD has a significant impact on the resonance response attained. The ground slot parametrical studies were conducted with a 0.1 mm apart value. As shown, at 0.8 mm, the antenna showed resonance from 27 to 42.6 GHz, while for 0.9 mm, the bandwidth ranged from 26.7 to 42.0 GHz. The optimum response was attained at a 1 mm ground slot value, exhibiting bandwidth characteristics from 26 to 41 GHz. It was observed that the ground slot affects higher frequency. In addition, the reflection coefficient response overall moves to lower values. Next, the parametric response concerning the distance between the circular loops is analyzed. The distance impact is analyzed at a 0.1 mm value. As observed in Figure 2b, the lower frequency range is affected by the starting values of 0.65 and 0.7 mm. The optimum response was achieved at 0.75 mm, as after this value, the reflection coefficient response is degraded. Figure 2c shows the feed line modeling effect for the proposed structure. As observed in Figure 2c, at the initial value of 6.55 mm, the resonance response stretches up to 42 GHz but the starting frequency response also moves forward. With the gradual increase of 0.05 mm, the resonance response moved backwards and the optimum response was achieved at 6.75 mm. Figure 2d shows the single element reflection coefficient obtained.

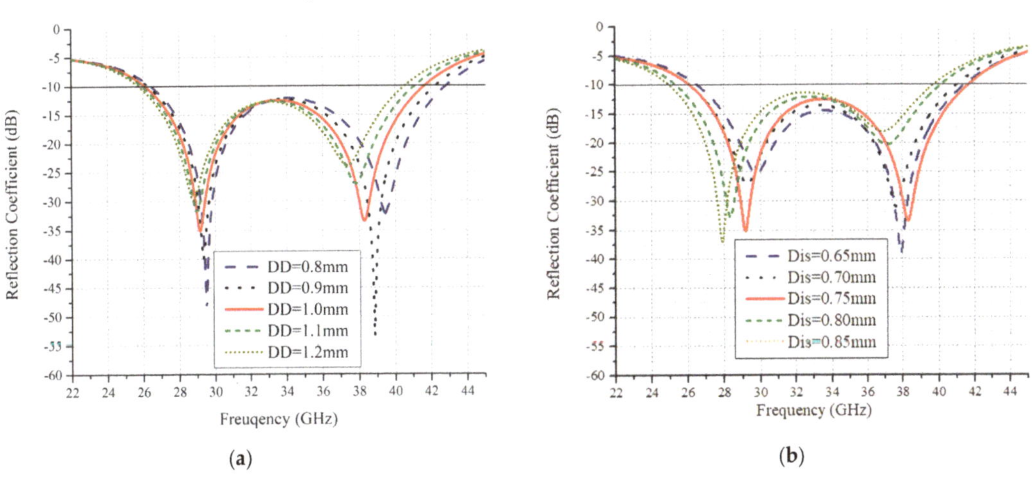

(a) (b)

Figure 2. *Cont.*

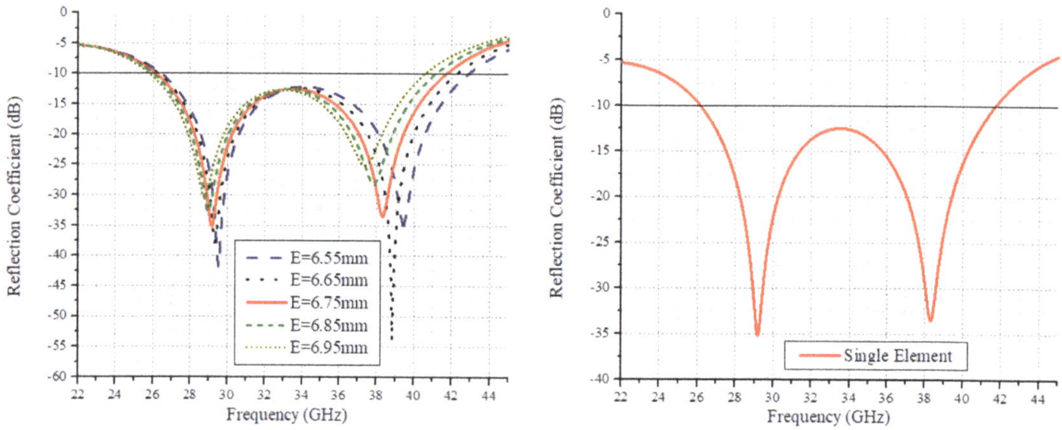

(**c**) (**d**)

Figure 2. Parametrical studies for (**a**) slot length (DD), (**b**) distance among circular loops (**c**) feedline length (E) and (**d**) single element.

3. Results and Discussions

The proposed multi circular is fabricated and tested using the in-house facility. Figures 3 and 4 show the simulated and measured reflection coefficient and efficiency with the measured gain. As observed in the figure, the reflection coefficient response is in good agreement. The measured gain differs 0.2 dB from the simulated gain at 28 GHz. The simulated radiation and total efficiency are above 95% at 28 GHz.

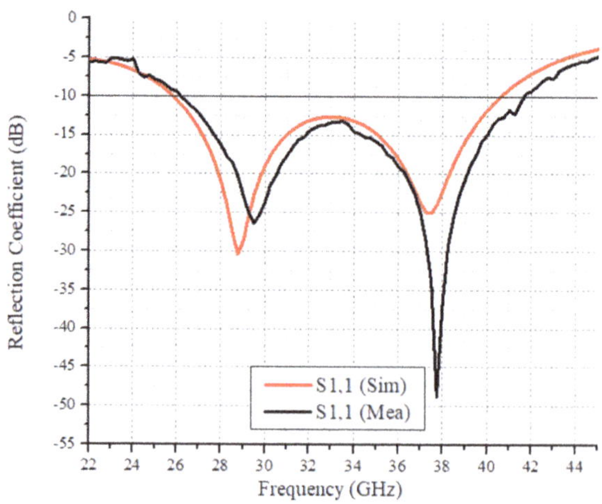

Figure 3. Single element simulated and measured reflection coefficient.

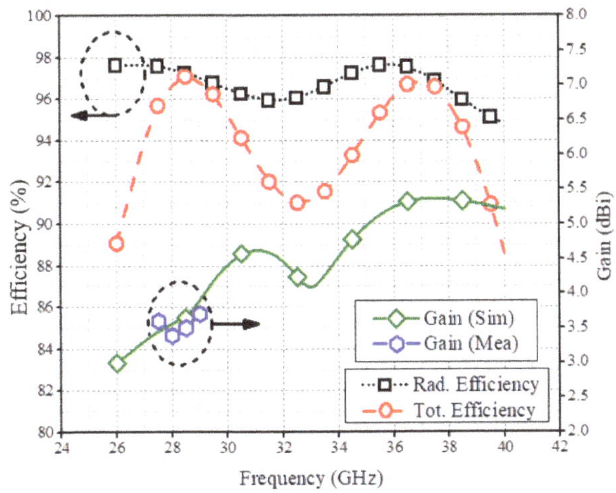

Figure 4. Single element gain and efficiency.

Array Transformation

The proposed design is transformed into a four-element linear array to achieve high gain characteristics. The array is shown in Figure 5. As observed in the figure, the array total length and width are 18.5 × 22.5 mm, respectively. The ground plane length is kept at 12.5 mm. The feed network is composed of a 50 Ω feed line, divided into a 100 Ω feedline, which is further subdivided into 70.7 Ω. The end feedlines connecting each radiating elements are 100 Ω feed lines. The widths of the feedlines are as follows: 50 Ω = 0.7 mm, 100 Ω = 0.28 mm and 70.7 Ω = 0.36 mm. Figure 5c shows the surface current distributions of the feed network at 28 GHz. As observed, the array network distributes power equally among the radiating elements. Figure 6 shows the simulated and measured reflection coefficient response of the proposed four-element array system. From the figure, it is clear that the simulated and measured results are in good agreement. The minute disruptions are the result of fabrication and measurement errors.

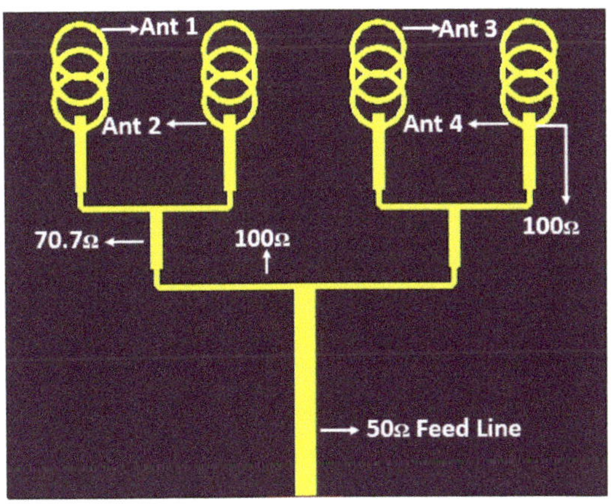

(a)

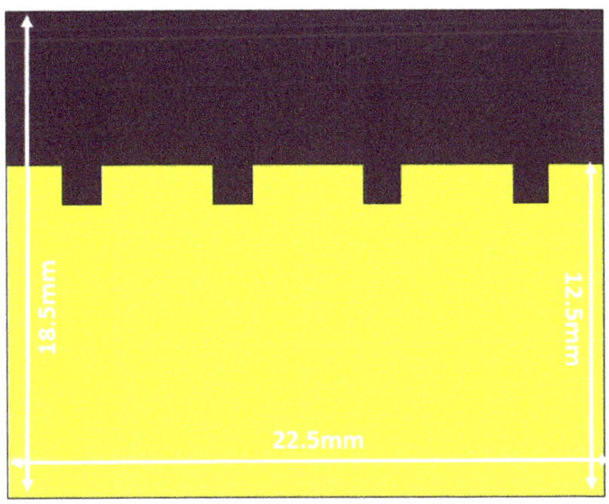

(b)

Figure 5. *Cont.*

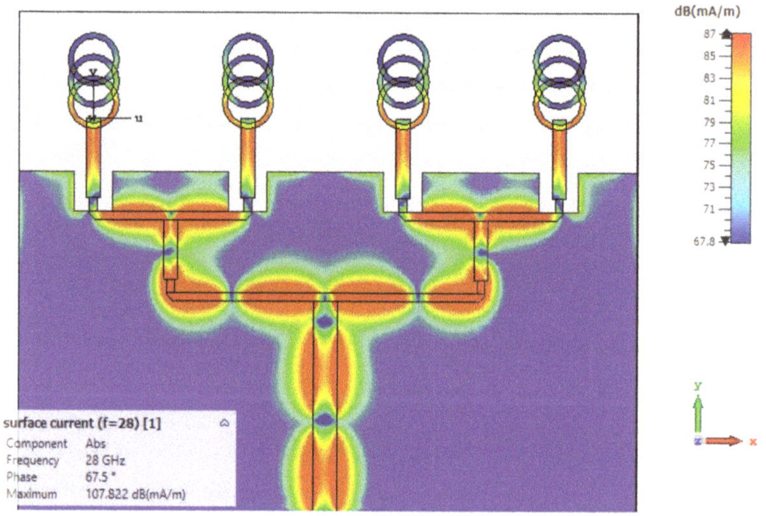

(c)

Figure 5. (**a**) Front layer, (**b**) back layer, and (c) surface current distributions of the proposed four-element antenna.

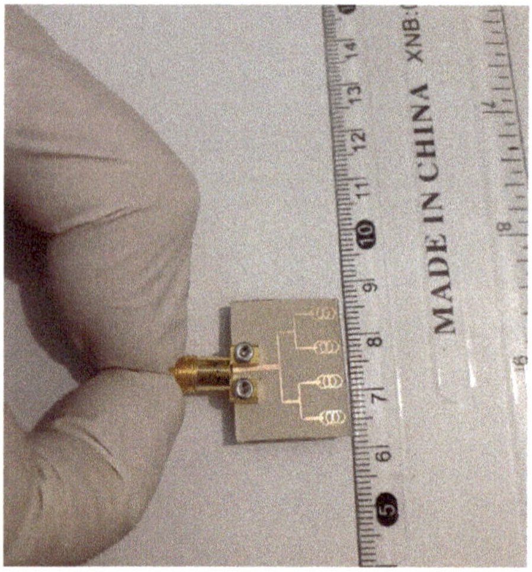

(**a**)

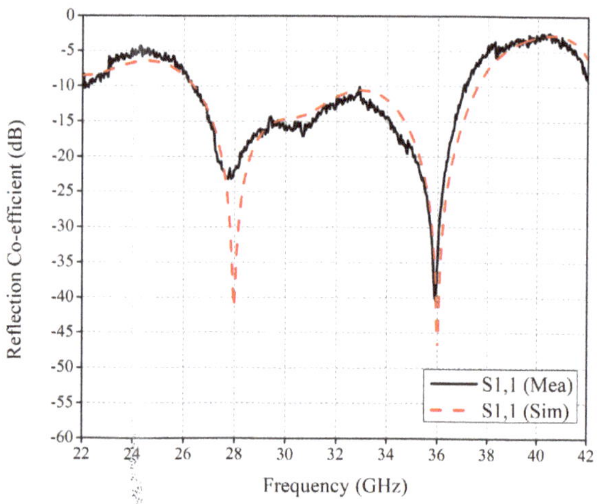

(**b**)

Figure 6. (**a**) Fabricated prototype (**b**) simulated and measured S-parameters.

The performance parameters of the proposed antenna array are shown in Figure 7. The radiation efficiency throughout the band of interest is above 89% and the measured efficiency is above 84%. At 28 GHz, the radiation and total efficiency is 94% and 93%, respectively. The gain of the proposed array ranges from 9 dBi to 11.5 dBi. The gain of the antenna array becomes low at the higher frequency end. The simulated gain of the array at 28 GHz is 10.7 dBi and the measured gain is 10.5 dBi. The peak gain achieved is 11.5 dB at 30.5 GHz. Figure 8 shows the proposed antenna array radiation patterns at 28 GHz resonance. The patterns show the far-field characteristics for Phi 90 and Phi = 0. As observed in the figure, the array exhibits a dual-beam characteristic with a very narrow beamwidth, as required for mm-wave systems. The simulated and measured radiation patterns are in good agreement. Figure 9 shows the 3D antenna radiation pattern of the 28 GHz system.

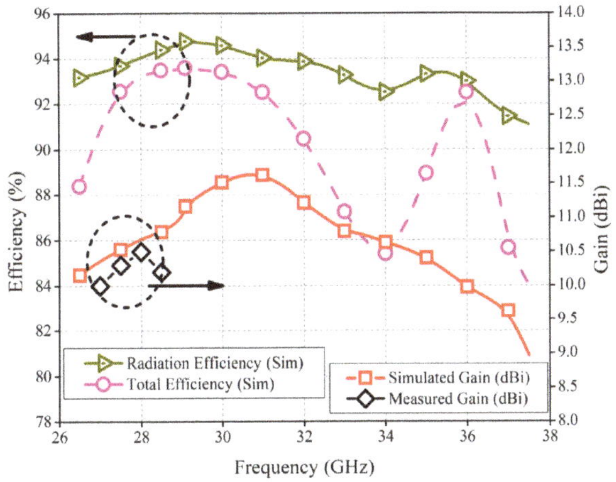

Figure 7. Performance parameters of array systems.

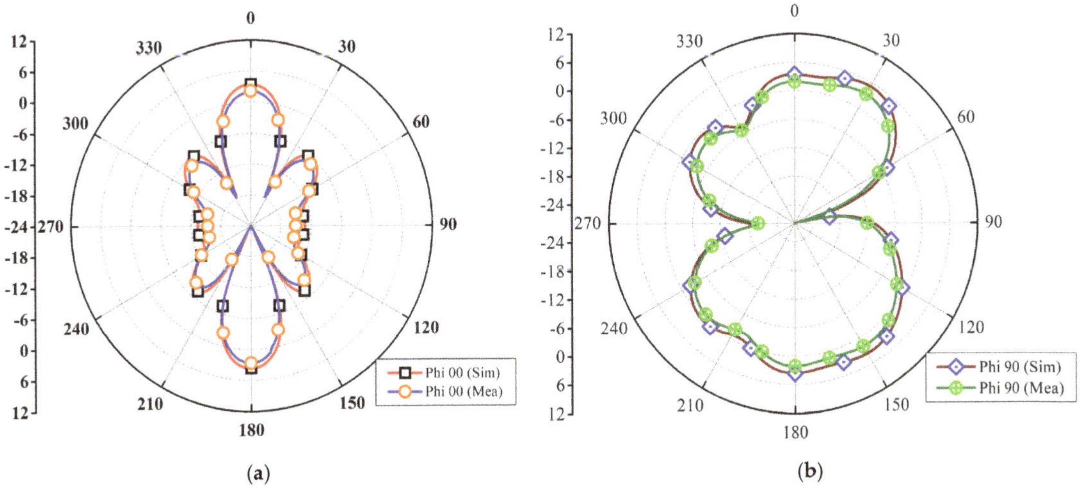

Figure 8. Radiation patterns at 28 GHz (**a**) Phi = 0 (**b**) Phi = 90.

Table 1 summarizes the proposed antenna contribution in the literature. The proposed antenna system is compared with Refs. [4,11,13,14,18,19]. Refs. [3,11,19] are based on the planar monopole antenna array, while Refs. [4,18] are based on the SIW technique. From the table, it is concluded that the proposed antenna array is well designed and delivers high gain with reduced size and wide bandwidth characteristics.

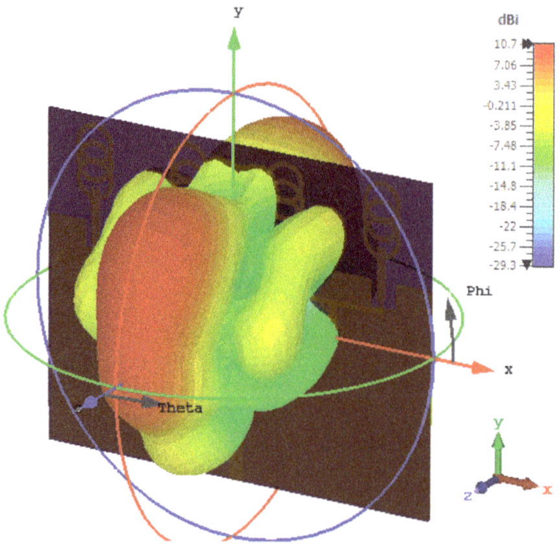

Figure 9. 3D antenna gain.

Table 1. Proposed four-element array comparison with the published literature.

Ref.	Frequency	Antenna Elements	Size	Configuration	Gain	Efficiency
[4]	53–71	2 × 2	23 × 24	SIW	10	82
[11]	26–30	1 × 4	18.5 × 24	Planar	11	94
[13]	25.5–29.5	1 × 4	20 × 22	Planar	10.2	80
[16]	2.45–3.1	1 × 4	155 × 75	Planar	8.39	77
[17]	6.5–100	1	20 × 20	Planar	10	N/A
[18]	25.05–34.92	1 × 4	45 × 20	SIW	12.5	85
[19]	23.34–33.92	1 × 4	37.6 × 14.3	Planar	10.7	90
Proposed	26–38.5	1 × 4	18.5 × 12.5	Planar	11.5	95

4. Conclusions

This paper presented a novel circular ring shape antenna on a thin RO5880 substrate. The antenna exhibited a wideband response from 26.5 to 41 GHz, with gain varying between 3 and 5.4 dBi and radiation efficiency greater than 90% throughout the bandwidth. The proposed antenna was transformed into a four-element linear array system. The size of the array was 18.5 × 12.5, which is quite small, and the bandwidth of the array was observed to be from 26 to 38.5 GHz. The peak gain achieved was 11.5 dBi, while at 28 GHz, it was observed to be 10.7 dBi. The proposed four-element array is fabricated, and the simulated and measured results were found to be in good agreement. Through the performance comparison with the reported literature, the proposed antenna system was found to be a potential candidate for next generation mm-wave communication systems.

Author Contributions: Conceptualization, R.A.A.-A., M.E.M. and S.H.K.; methodology, J.I., M.A.K. and A.G.A.H.; software, M.E.M. and S.H.K.; validation, C.H.S., M.A.K. and A.G.A.H.; formal analysis, N.O.P.; investigation, J.I. and N.O.P.; resources, M.M. and H.M.; data curation, M.M. and H.M.; writing—original draft preparation, M.E.M. and S.H.K.; writing—review and editing, J.I., N.O.P., C.H.S. and S.H.K.; visualization, J.K.; supervision, M.M. and J.K.; project administration, S.H.K.; funding acquisition, M.M., R.A.A.-A. and C.H.S. All authors have read and agreed to the published version of the manuscript.

Funding: Princess Nourah bint Abdulrahman University Researchers Supporting Project number (PNURSP2022R137), Princess Nourah bint Abdulrahman University, Riyadh, Saudi Arabia. The work of Mohamed Marey is supported by Prince Sultan University.

Institutional Review Board Statement: Not applicable.

Informed Consent Statement: Not applicable.

Data Availability Statement: All data have been included in the study.

Acknowledgments: Princess Nourah bint Abdulrahman University Researchers Supporting Project number (PNURSP2022R137), Princess Nourah bint Abdulrahman University, Riyadh, Saudi Arabia. The work of Mohamed Marey is supported by Prince Sultan University.

Conflicts of Interest: The authors declare no conflict of interest.

References

1. Ahmad, A.; Choi, D.-Y.; Ullah, S. A compact two elements MIMO antenna for 5G communication. *Sci. Rep.* **2022**, *12*, 3608. [CrossRef] [PubMed]
2. Farooq, U.; Rather, G. A miniaturised Ka/V dual band millimeter wave antenna for 5G body centric network applications. *Alex. Eng. J.* **2022**, *61*, 8089–8096. [CrossRef]
3. Bilal, M.; Naqvi, S.I.; Hussain, N.; Amin, Y.; Kim, N. High-Isolation MIMO Antenna for 5G Millimeter-Wave Communication Systems. *Electronics* **2022**, *11*, 962. [CrossRef]
4. Luo, Y.; Shen, Y.; Cai, X.; Qian, F.; Xu, S.; Cui, H.; Yang, G. Substrate integrated coaxial line design for mmWave antenna with multilayer configuration. *Int. J. RF Microw. Comput. Eng.* **2022**, *32*, e23090. [CrossRef]
5. Kim, G.; Kim, S. Design and Analysis of Dual Polarized Broadband Microstrip Patch Antenna for 5G mmWave Antenna Module on FR4 Substrate. *IEEE Access* **2021**, *9*, 64306–64316. [CrossRef]
6. Attiah, M.L.; Isa, A.A.M.; Zakaria, Z.; Abdulhameed, M.K.; Mohsen, M.K.; Ali, I. A survey of mmWave user association mechanisms and spectrum sharing approaches: An overview, open issues and challenges, future research trends. *Wirel. Netw.* **2019**, *26*, 2487–2514. [CrossRef]
7. Askari, H.; Hussain, N.; Sufian, A.; Lee, S.M.; Kim, N. A Wideband Circularly Polarized Magnetoelectric Dipole Antenna for 5G Millimeter-Wave Communications. *Sensors* **2022**, *22*, 2338. [CrossRef] [PubMed]
8. Kiani, S.H.; Alharbi, A.G.; Khan, S.; Marey, M.; Mostafa, H.; Khan, M.A. Wideband Three Loop Element Antenna Array for Future 5G mmwave Devices. *IEEE Access* **2022**, *10*, 22472–22479. [CrossRef]
9. Park, S.-J.; Shin, D.-H. Low Side-Lobe Substrate-Integrated-Waveguide Antenna Array Using Broadband Unequal Feeding Network for Millimeter-Wave Handset Device. *IEEE Trans. Antennas Propag.* **2016**, *64*, 923–932. [CrossRef]
10. Li, X.; Xiao, J.; Qi, Z.; Zhu, H. Broadband and High-Gain SIW-Fed Antenna Array for 5G Applications. *IEEE Access* **2018**, *6*, 56282–56289. [CrossRef]
11. Kiani, S.H.; Ren, X.C.; Bashir, A.; Rafiq, A.; Anjum, M.R.; Kamal, M.M.; Din, B.U.; Muhammad, F. Square-Framed T Shape mmwave Antenna Array at 28 GHz for Future 5G Devices. *Int. J. Antennas Propag.* **2021**, *2021*, 2286011. [CrossRef]
12. Kamal, M.; Yang, S.; Kiani, S.; Sehrai, D.; Alibakhshikenari, M.; Abdullah, M.; Falcone, F.; Limiti, E.; Munir, M. A Novel Hook-Shaped Antenna Operating at 28 GHz for Future 5G mmwave Applications. *Electronics* **2021**, *10*, 673. [CrossRef]
13. Raheel, K.; Altaf, A.; Waheed, A.; Kiani, S.; Sehrai, D.; Tubbal, F.; Raad, R. E-Shaped H-Slotted Dual Band mmWave Antenna for 5G Technology. *Electronics* **2021**, *10*, 1019. [CrossRef]
14. Tariq, S.; Naqvi, S.I.; Hussain, N.; Amin, Y. A Metasurface-Based MIMO Antenna for 5G Millimeter-Wave Applications. *IEEE Access* **2021**, *9*, 51805–51817. [CrossRef]
15. Jeong, M.; Hussain, N.; Abbas, A.; Rhee, S.Y.; Lee, S.M.; Gil, S.; Kim, N. Performance improvement of microstrip patch antenna using a novel double-layer concentric rings metaplate for 5G millimeter wave applications. *Int. J. RF Microw. Comput. Eng.* **2020**, *31*, e22509. [CrossRef]
16. Zhou, W.; Arslan, T. A bidirectional planar monopole antenna array for WiFi/Bluetooth and LTE mobile applications. In Proceedings of the 2013 Loughborough Antennas & Propagation Conference (LAPC), Loughborough, UK, 11–12 November 2013; pp. 190–193. [CrossRef]
17. Chu, S.; Hasan, N.; Yan, J.; Chu, C.C. A Planar Super Wideband Annular Ring Monopole Antenna with Time Domain Characterization. In Proceedings of the 2018 Asia-Pacific Microwave Conference (APMC), Kyoto, Japan, 6–9 November 2018. [CrossRef]
18. Ullah, H.; Tahir, F.A. A broadband wire hexagon antenna array for future 5G communications in 28 GHz band. *Microw. Opt. Technol. Lett.* **2019**, *61*, 696–701. [CrossRef]
19. Ullah, H.; Tahir, F.A. A High Gain and Wideband Narrow-Beam Antenna for 5G Millimeter-Wave Applications. *IEEE Access* **2020**, *8*, 29430–29434. [CrossRef]

electronics

Article

A Compact Sub-GHz Wide Tunable Antenna Design for IoT Applications

Rifaqat Hussain [1], Saad I. Alhuwaimel [2], Abdullah M. Algarni [1], Khaled Aljaloud [3] and Niamat Hussain [4,*]

1 Electrical Engineering Department, King Fahd University for Petroleum and Minerals (KFUPM), Dhahran 31261, Saudi Arabia; rifaqat@kfupm.edu.sa (R.H.); algarnia@kfupm.edu.sa (A.M.A.)
2 King Abdulaziz City for Science and Technology, Riyadh 12354, Saudi Arabia; huwaimel@kacst.edu.sa
3 College of Engineering, Muzahimiyah Branch, King Saud University, P.O. Box 2454, Riyadh 11451, Saudi Arabia; kaljaloud@ksu.edu.sa
4 Department of Smart Device Engineering, School of Intelligent Mechatronics Engineering, Sejong University, Seoul 05006, Korea
* Correspondence: niamathussain@sejong.ac.kr

Abstract: This work presents a compact meandered loop slot-line 5G antenna for Internet of Things (IoT) applications. Recently, sub-gigahertz (sub-GHz) IoT technology is widely spreading. It enables long-range communications with low power consumption. The proposed antenna structure is optimized to operate at sub-GHz bands without any additional complex biasing circuitry or antenna structure. A miniaturized design was achieved by a meandered structured loop slot-line that is loaded reactively with a varactor diode. Wideband frequency reconfigurability (FR) was achieved by the use of the varactor diode. The proposed antenna resonates over the frequency band of 758–1034 MHz with a minimum bandwidth of 17 MHz over the entire frequency band. The RO4350 substrate with dimensions of $0.18\lambda_g \times 0.13\lambda_g$ mm^2 is used to design the proposed antenna design. The efficiency and gain values varied from 54–67% and 0.86–1.8 dBi. Compact planar structure, narrow-band operation (suitable for NB-IoT) and simple biasing circuitry, which allows for sub-GHz operation, are unique and attractive features of the design.

Keywords: sub-GHz; meandered loop; frequency agile; IoT antenna; 5G antenna

Citation: Hussain, R.; Alhuwaimel, S.I.; Algarni, A.M.; Aljaloud, K.; Hussain, N. A Compact Sub-GHz Wide Tunable Antenna Design for IoT Applications. *Electronics* **2022**, *11*, 1074. https://doi.org/10.3390/electronics11071074

Academic Editors: Faisel Tubbal, Ladislau Matekovits and Raad Raad

Received: 31 January 2022
Accepted: 27 March 2022
Published: 29 March 2022

1. Introduction

The exponential Internet of things (IoT) growth with the use of small-cell devices results in the massive expansion of the traffic loads. This demand has revolutionized the current technology into 5G, which requires ultra-low latency, a high data rate, and improved capacity. By 2030, multibillion IoT technology users are expected. The growth of sub-GHz low-cost and reliable RF front-end devices is increasing as a requirement of this massive connectivity. Such systems are good candidates as the spectrum below 1 GHz is less crowded and reliable propagation can be ensured in large infrastructures. The Global System for Mobile Communications (GSMA) industry completely described the details of 3GPP standards for both NB-IoT and LTE-M technologies as part of the developing specifications for 5G. NB-IoT and LTE-M technologies are continuously evolving as a part of 5G as per 3GPP. Both technologies are suitable for 5G-enabled NB-IoT devices.

NB-IoT applications include street lighting, electronic appliances, sensors, actuators, and other machines that can be connected to the internet and with each other through device-to-device (D2D) connectivity. NB-IoT networks are suitable for a large number of connected devices with extended battery life. For such applications, long-range and low-power consumption communications are required to work in the sub-GHz bands. As a result, designing compact structure sub-GHz antenna systems that work over multiple frequency bands lower than 1 GHz has become a necessity. This is attainable using frequency reconfigurable (FR) antennas that have omni-directional radiation characteristics, which extend the antenna coverage area [1]. The proposed NB-IoT antenna design is suitable for numerous applications, such as healthcare, pet tracking, kid monitoring, smart metering, parking alarms, event detectors, and home appliances.

With the fast-growing IoT applications, the need for efficient power management and communication over long distances are challenging requirements. In the literature, several IoT antenna designs that address various IoT application challenges at different frequencies have been reported. The reported solutions include compact and low-profile antenna designs for multi-standard IoT antennas with wide-range frequency coverage in sub-6 GHz bands. These various antenna designs include inverted-F antennas, loop antennas, and monopole antennas.

Antenna designs based on monopoles for IoT applications were presented in [2–4]. Such solutions could be appropriate for wideband antenna operation intended to cover many wireless communication systems standards. In [2], a compact (20 mm × 30 mm) multi-band single element antenna is presented. The proposed antenna covers the following frequency bands: 1.79–2.63 GHz, 3.46–3.97 GHz, 4.92–5.85 GHz, and 7.87–8.40 GHz. This is an integrated 4G/5G antenna design to operate at sub-6 GHz and in the mm-wave band. Slots and monopoles were utilized to achieve the desired integrated operation. This is one of the pioneering works in 4G/5G integrated solutions. A compact structure monopole antenna that is resonating over the frequency band from 3 to 12 GHz, the board dimensions are 9.45 mm × 18.5 mm was presented in [3]. In this work, a slot antenna based 4G/5G integrated solution was provided using a dual-function planar connected array. The authors in [4], a multi-band antenna was presented for IoT applications. The main antenna features were its suitability for near-field communications in both microwave frequency bands and ultra-high frequency (UHF) bands. In addition, the antenna is suitable for low-frequency bands operating in IoT applications. This work also reported a 4G/5G integrated solution based on a monopole antenna array.

Several patch antenna designs for IoT applications were investigated in [5,6]. A 2.4 GHz high fractional bandwidth compact patch antenna was presented in [5]. The proposed antenna consists of an inverse S-shaped meander line that is connected to the rectangular box with a slot. This technique is utilized to reduce the antenna dimensions. The board dimensions are 40 mm × 1.6 mm. Both parasitic patch and capacitive loading were used to achieve a gain of −0.256 dBi and an overall efficiency of 79%. The work in [6] presented a miniaturized patch antenna design that operates in sub-GHz. A 95% miniaturization was achieved, compared to a conventional patch antenna, over the frequency band 805 to 835 MHz. The use of high folding, slots, and slits with inductive loading, utilizing vias, helped in achieving this excellent miniaturization. The dimensions of the antenna board with the elevated structure are 30 mm × 31.84 mm × 4.37 mm.

Other antenna types for IoT applications are reported in the literature. For example, the following: monopole antennas loaded with inverted-L shaped stubs in [7], 3D-printed antennas [8], loop antennas [9], and glass frame antennas [10]. The solutions provided in [7,8] are based on monopole antennas and non-planar antenna structures. In [9], a 2.45 GHz transparent conductor antenna for IoT applications was presented. The variations in the realized antenna gain in the presence of metal frames nearby have been discussed in detail. The antenna is compact in size, with dimensions of 36 mm × 36 mm × 0.5 mm. In this work, a dual-band shared-aperture based antenna is presented with high channel isolation. Some other recent work on antenna designs for IoT applications was also reported in [11–14]. All these antennas operate above 1 GHz and hence are not suitable for narrow band IoT applications for long-range communication.

Most of the antenna designs operating in sub-GHz are elevated printed inverted F-antennas (PIFA), monopole, or dipole antennas [15–19]. In [15], a folded miniaturized patch antenna was presented for IoT device operating at 805–835 MHz. In [16], digitally tunable capacitor (DTC) was used to tune the antenna from 600–900 MHz. The digital control and the low power consumption make this antenna a good candidate for IoT applications. In [17], a transparent double folded loop antenna was presented for IoT applications. The antenna was operating at 2.4 GHz with a bandwidth of 500 MHz. In [18], a 3D printed antenna-on-package antenna was presented with isotropic patterns for an IoT application operating at 2.4 GHz. Similarly, in [19], an energy harvesting rectenna was presented for IoT applications. The proposed antenna design would utilize an energy harvesting mechanism using rectenna to mitigate the challenges of battery constraint issues. The antenna was operated at a frequency band of 2.4 GHz.

Also, other frequency reconfigurable antennas closely resembling the proposed antenna are reported in [20–23]. The solutions provided are based on either reactive loading or meandering structures. However, neither of them was able to achieve a compact planar structure with sub-GHz operation. Also, the solutions provided in [20–23] are non-planar structures or non-reconfigurable structures with large antenna dimensions.

Thus, the non-compact and non-planar antennae limited the tuning capabilities, affecting their suitability for small IoT applications and terminal devices. Only a few slot-based antenna designs that operate in sub-GHz bands are reported in the literature, such as [24,25]. The solutions provided in [24,25] used a simple slot structure to tune the antenna in the desired frequency. Most of the reported frequency reconfigurable antenna designs are operating in frequency bands that are above 2 GHz. The proposed design in [24] covers the frequency bands of 0.9, 1.8, 1.9, and 2.4 GHz. The antenna size is 38 mm × 16 mm. In [25], a dual antenna design that covers the 0.860 and 2.45 GHz frequency bands was presented. The antenna is 50 mm × 50 mm in size.

In [26], an energy harvesting antenna design for IoT applications was presented. The proposed antenna consisted of a rectangular patch along with a filter and rectifier circuitry operating at a frequency band of 2.4 GHz. The authors provided an energy harvesting solution without optimizing the antenna. The authors in [27] present a 60 GHz mm-wave antenna for the IoT. The total bandwidth of this antenna is 9.8 GHz having a peak gain of 9.6 dBi. The performance of the antenna was thoroughly evaluated to determine the radiation coverage. The given antenna design is a well-suited option for body-centric IoT applications because of its compact structure.

Very few antennas for sub-GHz were reported, as it is quite challenging to design a compact antenna structure. This is because of the poor input impedance matching with the small antenna size. In [28], a detailed survey based on fractal antennas was presented. The significance of the fractal antennas for IoT application is well studies and provided a thorough literature survey on fractal antennas. Various IoT antennas are reported in the literature [29–31]. In [29], a transparent 2-element 5G MIMO antenna for sub-6 GHz applications is described. Thus, this work provided a guideline on the use of transparent for IoT applications. Similarly, a low-profile single-band and dual-band antenna were presented in [30] for IoT applications. The slot structure was optimized to operation in single as well as dual band. in [31], the authors proposed an artificial magnetic conductor-backed compact wearable antenna was proposed. Such antennas are good candidates for IoT applications. At the expense of complex antenna geometry. Comparing to other works, the proposed antenna design outperformed in terms of various features are comprehensively discussed in this article.

The novelty and distinguishing features of the proposed work are described in detail as follows:

(1) To overcome the challenging requirement for IoT devices to have extended coverage with minimal power consumption, it is highly desirable to design a compact antenna for lower frequency bands. The proposed antenna structure is an attempt to meet the desired performance and standards of narrow-band IoT (NB-IoT) applications for 5G-enabled IoT devices. None of the available literature [2–6,15–23], and [32–36] can be used for NB-IoT in sub-GHz with wide tuning capability;

(2) A compact meandered loop slot-line antenna with frequency agility suitable for NB-IoT applications that operate in sub-GHz bands is proposed. The work cited in [4,6,15–19,21,23] covered sub-GHz bands, but the majority of them were not suitable for NB-IoT operations;

(3) The proposed antenna benefits from the following unique features: simple biasing circuitry, compact planar structure, and it operates over the sub-GHz bands which fit for NB operations, but most of the existing antennas are not suitable for NB operation [2,3,6,16–19,21–23,32–36];
(4) Miniaturization was achieved using a unique combination of meandered loop slot-line along with a reactively loaded slot antenna. A 173% size reduction in area is obtained using this technique. As per the authors' information, none of the works achieved such a high level of miniaturization;
(5) Resonating bands' smooth variation is noticed over a very wide band from 758 to 1034 MHz with antenna dimensions of 60 × 27 mm^2;
(6) Moreover, a step-by-step antenna design procedure and analysis are provided to give general guidelines to understanding how to scale its design for any desired frequency band;
(7) As per 3GPP Release 13 and 15, the proposed antenna design is suitable to be utilized in the following NB-IoT bands: B-5, B-8, B-13, B-14, B-17, B-18, B-19, B-20, and B-26. None of the reported literature can cover such a large number of sub-1 GHz bands.
(8) The proposed NB-IoT antenna is best suited for 5G-enabled NB-IoT devices.

Section 2 of this paper discusses the proposed antenna design details and the theoretical analysis. The simulated and measured scattering parameters results and the radiation characteristics of the proposed antenna design are presented in Section 3. Finally, Section 4 provides a set of conclusions about the presented work.

2. Design Details

The major challenges for IoT antenna designs are the requirements for compact antenna structures and sub-GHz operations for better power management along with long-distance communication. In this section, the detailed antenna geometry of the proposed design is presented, equivalent circuit diagram and the physics behind the antenna's operation, followed by a step-by-step antenna design procedure.

2.1. Antenna Geometry

Figure 1 shows the proposed a meandered loop slot-line antenna design. The antenna was designed and fabricated using an RO4350 substrate with dimensions of 27 mm × 60 mm × 0.76 mm. The relative permittivity (ε_r) of the substrate is 3.48 and its loss tangent (tanδ) is 0.0036. The Rogers substrate can be replaced with FR-4 to obtain good resonance sweeps but with slightly lower radiation efficiency values. For the Rogers-4350 board, the temperature coefficient of the dielectric constant is among the lowest of any circuit board material. This makes it an ideal substrate for broadband applications. After optimizing the antenna design, the design is mainly composed of the following: an etched slot-line structure from the ground plane and a meandered-shaped rectangular slot-line used to increase the radiating structure's effective electrical length. The proposed antenna design dimensions are the following: l_1 = 38, l_2 = 36, l_3 = 19, l_4 = 11, l_5 = 7.5, w_1 = 5.5, w_2 = 3.5, w_3 = 6, w_4 = 4.

The proposed antenna top layer detailed view is shown in Figure 1a. It is composed of a matched 50 Ω input impedance (Z_{in}), a microstrip feedline, and the biasing circuitry of the varactor diode. The varactor diode biasing circuit (Dv) consisted of a series of combinations of RF chokes (L_1 and L_2) and current limiting resistors (R_1 and R_2). Sorting posts (SP) were used to connect the bottom layer varactor diode to the biasing circuitry in the top layer. RF chokes were used to isolate the radiating structure from the power supply. The reverse-biased varactor diode acted as a DC blocking capacitor. Hence, the DC biasing part and the RF radiating structure are well isolated and have a very minimal effect on the antenna's performance. SMV 1233 varactor diode is used [37]. The varactor diode (SMV1233) maximum reverse bias current is 20 nA. From the maximum current specifications, the maximum biasing circuit power loss is −87.7 dBm.

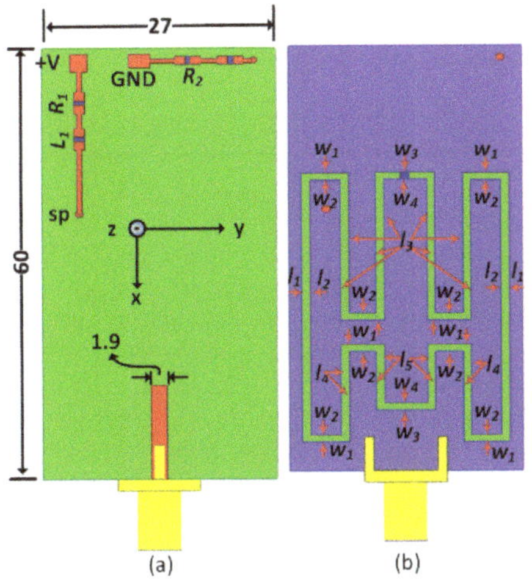

Figure 1. Proposed antenna (**a**) Top view (**b**) Bottom view (All dimensions are in millimeters, mm).

The ground plane detailed view is shown in Figure 1b. After optimization, the rectangular slot operates at below 1 GHz frequency band. After several parametric sweeps of the structure's length and width, the resultant structure is shown. The 1 mm wide uniform slot-line is used throughout the meandered structure. The final optimized antenna dimensions are shown in Figure 1b. The varactor diodes' exact positions and placements are critical as they help in achieving wide frequency band tuning. The LPKF S103 machine [38] was used to fabricate the antenna. The fabricated model top and bottom layers are shown in Figure 2a,b, respectively. The other antenna design dimensions are shown in Figure 1. An increase in the slot length will increase the electrical length of the antenna, and hence the antenna will resonate at a lower frequency band. On the other hand, any change in the slot width will result in poor matching. The given optimal slot width gave us the best Zin matching.

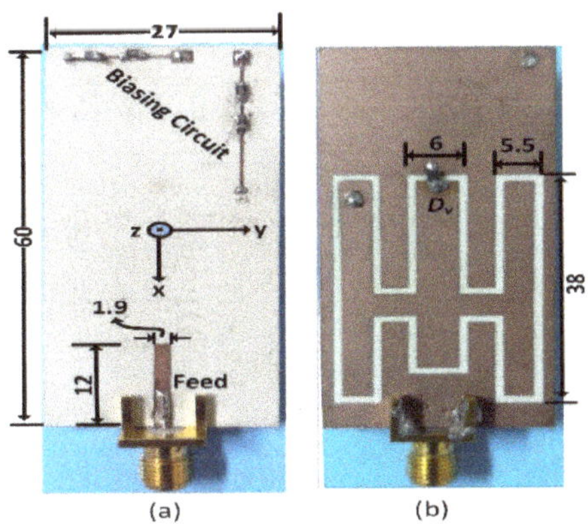

Figure 2. Fabricated antenna (**a**) Top view (**b**) Bottom view (dimensions are in millimeters, mm).

2.2. Antenna Operation

Slot-based FR antennas gain more attraction as they are low-profile planar structures, easy to integrate, and able to operate in wide frequency bands. In the literature, open and closed-ended slot designs were investigated and reported. The closed ends or short-circuited slot antenna can be modeled as $\lambda/2$ transmission line which corresponds to its fundamental frequency of resonance [39]. These antennas may be loaded with capacitive reactance. The antenna resonance can be optimized over a wide frequency band. The rectangular slot antenna fundamental resonance frequency is given by the following [40]:

$$f_r = \frac{c}{(l_m + w_m)} \times \frac{c}{4\, f_{res} \sqrt{\frac{\varepsilon_r + 1}{1.5\varepsilon_r}}} \tag{1}$$

$l_m = l_1 + l_2 + 4l_3 + 2l_4 + 2l_5$, $w_m = 4w_1 + 4w_2 + w_3 + w_4$, c is the speed of light in free space, εr is the relative permittivity of the substrate, f_r is the modified rectangular antenna's fundamental resonance frequency, and the term (l_m + w_m) is the mean circumference of the meandered rectangular ring slot antenna. In the proposed design, the effective mean circumference can be an estimate of several rectangular slots that determine the overall fundamental resonance frequency.

The meandered loop-slot structure was reactively loaded using a varactor diode (D_v), as shown in Figures 1 and 2. The varactor diode was utilized to vary the slot-line capacitance at a certain point. As a result, it helped in bringing the resonance frequency down to the lower band. The reactive loading is a non-uniform operation, and it can be determined using the location of the varactor diode (L_{v1}), its capacitance value C_v and the impedance (Z_o) of the slot line structure. The resonance frequency of the reactive loaded slot can be determined using the transmission line equivalent circuit model as given in the following [41]:

$$\tan(\beta\, L_t) + \tan(\beta\, (L_t - L_v)) - \omega C_v Z_{in} \tan(\beta\, L_1) \tan\beta\, (L_t - L_v) = 0 \tag{2}$$

where L_t is the total length of the slot (L_t = l_m + w_m), β is the propagation constant. This constant depends on the frequency of operation L_v is the distance of diode placement from the feedline, C_v is the varactor diode capacitance, and ω is the angular frequency. The reactively loaded slot antenna resonance frequency can be numerically determined by solving Equation (2). The resonance frequencies obtained for the optimized design using HFSS were also compared with theoretical values based on Equation (2). The two results (theoretical, HFSS-based) obtained are (750 MHz and 764 MHz), (826 MHz, 836 MHz), (901 MHz, 885 MHz), (981 MHz, 964 MHz), (1052 MHz, 1029 MHz). A small variation in values is observed. Thus, the mathematical expression derived is useful to obtain an insight into the effective reactive loading of the slot antenna structure.

Figure 3 shows the equivalent circuit model of the proposed meandered loop-slot antenna. The series combination of a microstrip feedline (series $L_f C_f$ circuit) and an RLC resonating circuit that represents the loop-slot radiating structure is shown in Figure 3a.

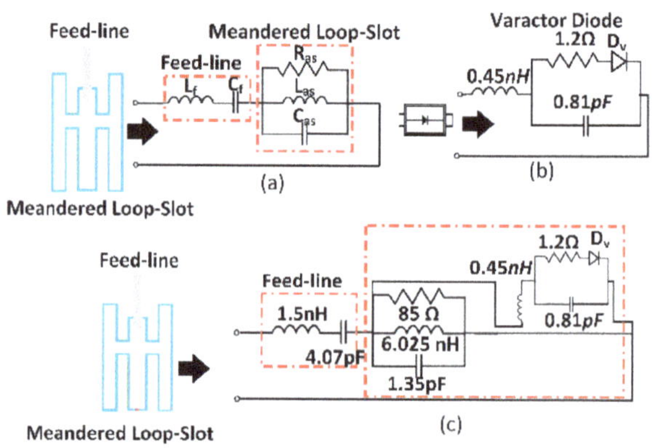

Figure 3. Equivalent circuit model (**a**) Meandered loop-slot with feedline (**b**) Varactor diode (**c**) Complete circuit model.

The varactor diode equivalent circuit and the complete antenna design equivalent circuit model are shown in Figure 3b,c, respectively [42]. The antenna equivalent circuit model analysis assists in gaining a better understanding of the proposed antenna design multi-band operation, the slot reactive loading, and FR antenna operation. ADS can be used to obtain the antenna circuit element values and the antenna reactive loading parts can be extracted by using the S-parameters [43].

2.3. Antenna Design Procedure

This section provides a step-by-step antenna design procedure and optimization steps to obtain the final design. The first step to designing the proposed antenna was designing a rectangular slot-line antenna structure with dimensions 24 mm × 15 mm, fed with 50 Ω microstrip line. The antenna was resonating above the 2 GHz bands. The dimensions of the slot were optimized to make it resonate above 1.5 GHz by increasing the electrical length of the radiating slot. To further reduce the resonance frequency to lower bands, the rectangular structure was turned into a meandered structure to achieve a resonance frequency that is below 2 GHz, as shown in Figure 4a. For the given meandered structure, the slot structure was resonating at 1.75 GHz. The antenna was further optimized by increasing its electrical length, as shown in Figure 4b. The given structure was working at 1.305 GHz. The width of each slot as well as the distance between different meandered slots were optimized to tune the antenna to be effectively loaded with varactor capacitance to cover the maximum sub-GHz bands of the proposed antenna structure. For the same antenna structure as shown in Figure 4b, the theoretical resonance value calculated based on Equation 1 was 1.287 GHz. Hence, a very close agreement between the theoretical and simulated values was observed.

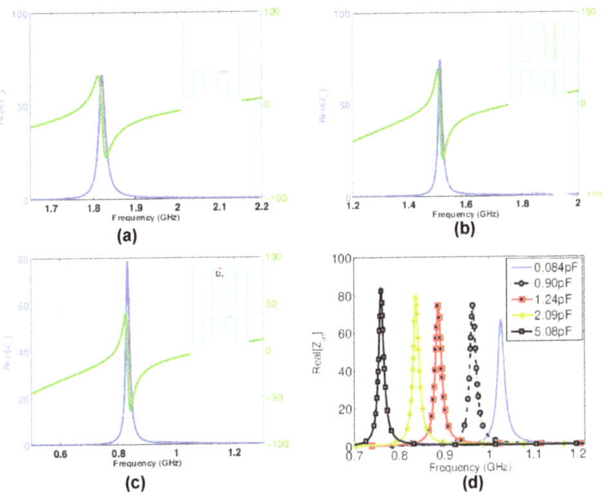

Figure 4. Zin curves (**a**) Stage-1 Z_{in} curves (**b**) Z_{in} curves of the optimized design (**c**) Z_{in} curves with varactor loading (**d**) Re{Z_{in}} for various capacitance values.

A single varactor diode was utilized to bring the resonance frequency down to sub-GHz bands, as well as a continuous sweep of frequencies was obtained over a wide band. Figure 4a–c show the real(R_e) and imaginary(I_m) parts of input impedance (Z_{in}). It can be seen that Re{Z_{in}} is around 50Ω while the Im{Z_{in}} is crossing the zero value at the resonating bands. The placement of the diode was very critical as the input impedance (Z_{in}) matching was dependent on it. The varactor diode's various locations resulted in different reactive loading effects on the radiating structure. To obtain an optimum Zin matching, several parametric analyses were performed, including its placement on the slot structure.

The antenna optimized dimensions and the placement of diodes resulted in a continuous frequency sweep from 758∼1034 MHz, which can thus support several narrow-band IoT applications below 1 GHz. Figure 4d shows the Re{Z_{in}} for different capacitance values of 0.84, 0.90, 1.24, 2.09, and 5.08 pF. It has been observed that the capacitive loading helped in matching the Z_{in} at different resonating bands for different values of reverse bias voltages across the varactor diode.

3. Results of Simulation and Measurements

For the proposed meandered loop-slot antenna, the simulation and modeling were performed using HFSS^{TM}. The various dimensions of antenna design have been optimized to achieve the sub-1 GHz band. This included optimal varactor diode placement, maximum effective reactive loading, as well as antenna placement on the PCB board. The S-parameters were measured using the Agilent N9918A VNA, while the antenna efficiencies and gain patterns were computed using the SATIMO Star lab anechoic chamber.

3.1. Reflection Coefficient Curve

The reflection curves characterize the proposed antenna design. The simulated and measured results of the reflection curves are shown in Figure 5. Figure 5a shows the proposed antenna design simulated reflection coefficient curves (S_{11}). The resulted curves are for the design with varactor diode capacitance values from 0.84–5.08 pF and the corresponding reverse bias voltage values ranges from 15 to 0 V. A smooth variation in the resonating bands was observed from 753–1040 MHz with a −10 dB bandwidth of 17 MHz. The wide frequency sweeping with narrowband operations is the proposed antenna design's key characteristic. The proposed slot radiating structure can easily cover several IoT bands in the sub-GHz range.

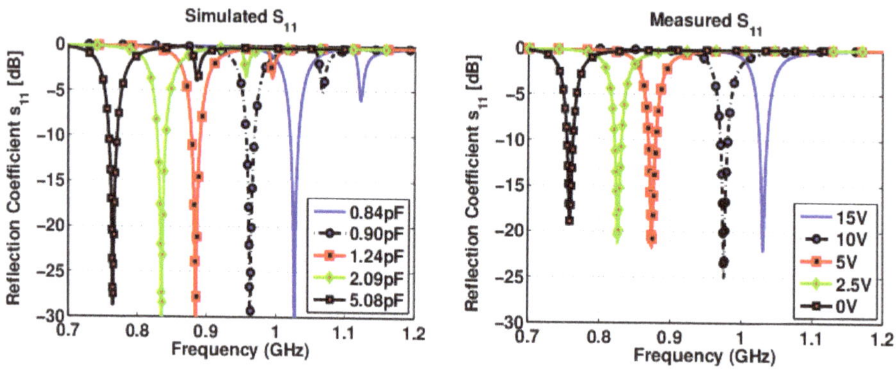

Figure 5. Reflection coefficient curves of the antenna (**a**) Simulated S_{11} (**b**) Measured S_{11}.

The proposed antenna's measured S-parameters are shown in Figure 5b. It is clear from the figure that the measured results are in good agreement with the simulation results. In the event of any small variation between the two results, it can be compensated for as continuous frequency sweeps have been obtained for both. This demonstrates the main advantage of using varactor diode capacitive loading for these types of operations.

The slight variations in both simulated and measured S_{11} are mainly due to the substrate properties, fabrication tolerances, and varactor diode modeling flexibility in HFSS [44,45]. The HFSS does not consider diode losses and packaging effects. Also, manual soldering of the SMA connector might affect the values of input impedance Z_{in}. The additional harmonics are the higher-order resonances of the slot antenna. In an ideal simulating software environment, higher-order modes are poorly matched with the input impedance Z_{in}, and hence are visible. However, the measured Z_{in} is only matched with its fundamental mode of operation within the given frequency bands, and the poorly matched bands did not appear for the measured values. Hence, no harmonics are observed for the measured S_{11} results.

3.2. Current Density Analysis

For single antenna designs, current density analysis is usually performed to understand the radiation characteristics, determine the active parts of the radiating structure, and find the effective electrical length of the antenna. For the proposed antenna design, the current density distribution is plotted to understand the antenna's behavior at the resonating frequency bands. It has been observed that current density distribution varies over different operating bands. It has been observed that a large portion of the antenna is radiating at lower frequency bands and vice versa. Such analysis can be utilized to optimize the antenna's dimensions by eliminating non-radiating antenna parts. The surface current distributions of the proposed antenna design were analyzed at various bands of resonance. Figure 6a,b shows the current distribution for non-optimized design for frequency bands at 1.95 GHz and 1.75 GHz, respectively. For the optimized design, the surface current densities at frequency bands of 1029 MHz and 764 MHz are shown in Figure 6c,d, respectively. From the given figures, it can be seen that the current density has a different distribution for different frequency bands. Figure 6d shows the current density distribution at 764 MHz. It can be seen that the maximum current variation is along the outermost part of the loop slot structure. The effective electrical length can be mapped to the corresponding resonating band. The slot length variations affect the first and other resonating bands for different values of capacitive loading. Thus, the analysis helped in understanding the antenna's operation and gave useful insights about operating at different frequency bands.

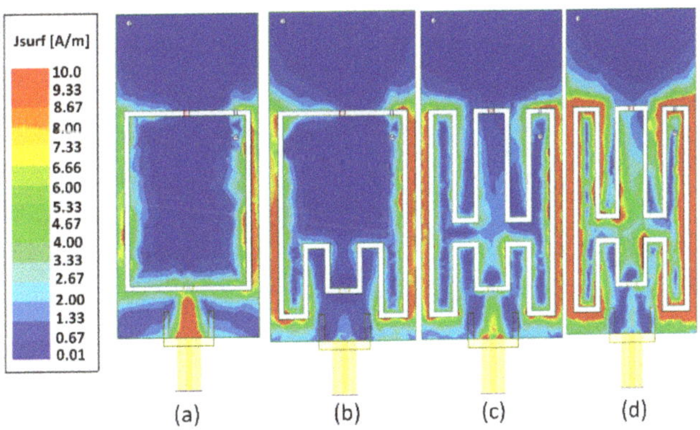

Figure 6. Surface current density at (**a**) 1950 MHz, (**b**) 1750 MHz, (**c**) 964 MHz, and (**d**) 764 MHz.

3.3. Radiation Patterns

The radiation characteristics of the proposed antenna design at different frequency bands characterize the antenna. The proposed antenna simulated 3-D gain patterns at 885 MHz and 1029 MHz are shown in Figure 7a,b, respectively. The gain patterns of the proposed antenna design, omnidirectional behavior was observed at the resonating bands. The gain patterns that result from simulation and measurements and antenna efficiency ($\%\eta$) values of the antenna are listed in Table 1 and are shown in Figure 8. From the resulting values, it can be concluded that the proposed antenna will have good performance when operating in the sub-GHz bands.

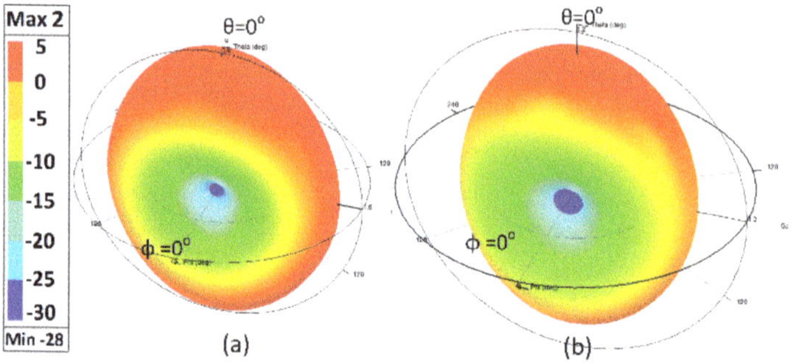

Figure 7. Simulated 3-D antenna gain patterns at (**a**) 885 MHz (**b**) 1029 MHz.

Table 1. Peak gain (PG) and efficiency (%η) values of the proposed antenna.

Simulated Results			Measured Results		
f_s (MH)	PG (dBi)	%η	f_m (MHz)	PG (dBi)	%η
764	0.86	55	758	-	-
836	0.98	60	827	0.86	54
885	1.2	63	875	0.96	59
964	1.4	67	976	1.13	64
1029	1.8	70	1033	1.42	67

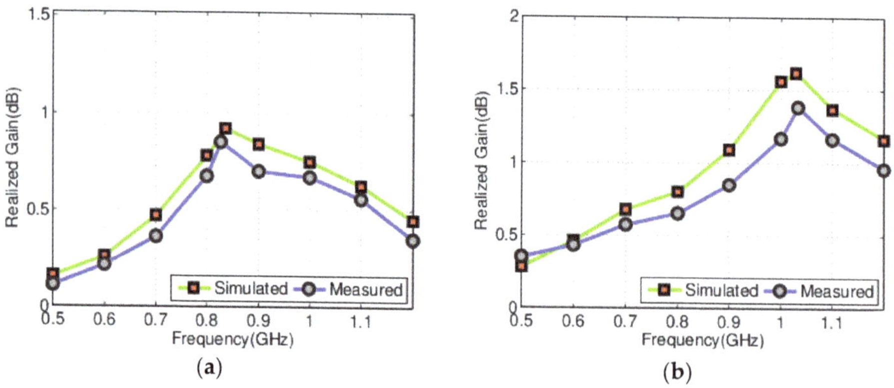

Figure 8. Simulated and measured realized gains (**a**) 2.09 pF, 2.5V (**b**) 0.84 pF, 15V.

The proposed antenna design radiation characteristics were validated by performing measurements with the setup shown in Figure 9. The 2D simulated and measured gain patterns at 885 MHz and 1029 MHz are shown in Figure 10 for $\varphi = 0°$ and $\varphi = 90°$ cut. It can be observed that in the given mode of operation at each resonating band, the antenna has omnidirectional behavior. Figure 10a,b show the simulated and measured radiation patterns for $\varphi = 0°$ cut while Figs. Figure 10c,d show the simulated and measured radiation patterns for $\theta = 90°$. The proposed work is intended to be utilized in 5G-enabled IoT devices that mostly operate at lower power. Therefore, with typical signal power densities, this antenna would work well.

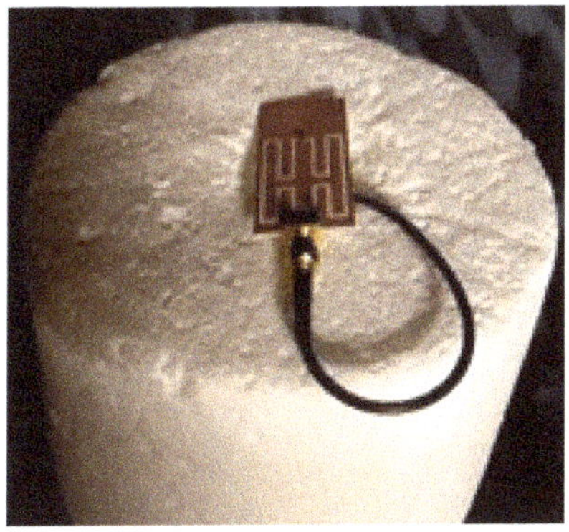

Figure 9. Antenna's farfield measurement setup.

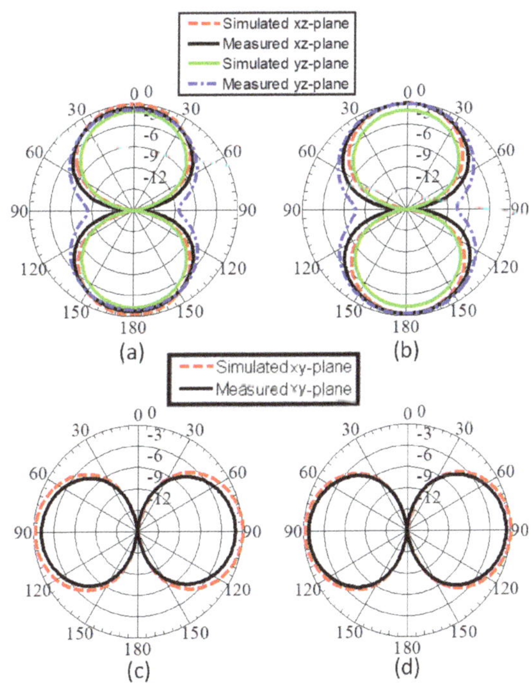

Figure 10. Simulated and measured total gain patterns (**a**) φ = 0° at 885 MHz (**b**) φ = 0° at 1029 MHz (**c**) θ = 90° at 885 MHz (**d**) θ = 90° at 1029 MHz.

Table 2 presents a detailed comparison that compares the distinguishing features of the most relevant IoT antennas available in the literature. The various features included antenna type, size, bands covered, suitability for NB-IoT operation, %η, planar structure, percentage miniaturization obtained, continuous frequency sweep, and the number of sub-GHz NB-IoT bands covered. Most of the IoT antennas reported are either wide-band monopole (MP) or non-planar PIFA designs. For such antennas, it is quite challenging to obtain a sub-GHz continuous frequency sweep over a wide band. Although, the proposed antenna design is competitive in terms of its compactness with some of the available designs, it outperformed in sub-GHz bands with wide tunability and NB-IoT operation for better power management, long battery life, and enabled devices for long-distance communications, etc.

Table 2. Proposed IoT antenna versus related works.

Ref.	Ant. Type	Ant. Size mm^2/mm^3	Bands GHz	NB IoT?	%η	Planar?	sub-GHz Bands?	Gain (dBi)
[2]	monopole	$0.34\lambda_g \times 0.23\lambda_g$	2.2, 3.5, 5.5, 8	No	54–86.3	Yes	0	6.9
[3]	monopole	$0.34\lambda_g \times 0.23\lambda_g$	3.1–8.5 (wide-band)	No	-	Yes	0	−0.78
[4]	monopole	$0.18\lambda_g \times 0.36\lambda_g$	0.915, 2.45	Yes	93, 95	Yes	1	1.9, 3.85
[5]	monopole	$0.41\lambda_g \times 0.58\lambda_g$	2.4	Yes	79	Yes	0	−0.256
[6]	patch	$0.15\lambda_g \times 0.61\lambda_g$	0.82	No	-	No	1	−2.5
[15]	monopole	$0.503\lambda_g \times 0.19\lambda_g$	0.9, 1.8, 2.3, 2.6	Yes	80–90	Yes	1	3–4
[16]	monopole	$0.15\lambda_g \times 0.16\lambda_g$	0.85, 2.1	No	40–67.2	Yes	1	1–3.2
[17]	monopole	$0.14\lambda_g \times 0.1\lambda_g \times 0.28$	0.85, 2.4	No	70, 90	Yes	1	-
[18]	PIFA	$0.23\lambda_g \times 0.05\lambda_g$	0.829, 1.95	No	60–73	No	1	-
[19]	PIFA	$0.044\lambda_g \times 0.16\lambda_g$	0.836, 2.2, 3.6	No	48–83	No	1	0.5–5.2
[20]	slot	$0.035\lambda_g \times 0.035\lambda_g$	1.32–1.49, 1.75–5.2	Yes	45–81	Yes	0	0.5–4.5
[21]	PIFA	$0.15\lambda_g \times 0.12\lambda_g$	0.22, 0.80, 4.7, 4.96	No	60–85	No	2	-
[22]	patch	$0.75\lambda_g \times 0.75\lambda_g$	3.01, 3.83, 4.83, 5.76	No	42–60	Yes	0	1.43–3.06
[23]	PIFA	$1.55\lambda_g \times 1.55\lambda_g$	0.41–0.91, 2.1–3.5	No	10–50	No	1	−6.1, 4.9
[32]	slot	$1.58\lambda_g \times 1.94\lambda_g$	1.82, 1.93, 2.1	No	85–89	No	0	3.2–6.3
[33]	slot	$1.2\lambda_g \times 1.33\lambda_g$	5∼6	No	-	Yes	0	1.36–1.87
[34]	slot	$0.65\lambda_g \times 0.36\lambda_g$	3.0	No	-	Yes	0	−0.44–2.97
[35]	slot	$0.55\lambda_g \times 0.16\lambda_g$	5.73–5.97	No	60	No	0	1.28
[36]	slot	$0.014\lambda_g \times 0.0425\lambda_g$	2.45–3	No	-	No	0	−25
Prop.	**slot**	$\mathbf{0.18\lambda_g \times 0.13\lambda_g}$	**0.758–1.034**	**Yes**	**54–67**	**Yes**	**9**	**0.86–1.8**

In Table 2, the proposed is compact and better than [2–6,15–23], and [32–36] in terms of antenna size. Although, some work as cited in [4,6,15–19,21,23] covered single sub-GHz bands but without tuning capability to switch between different bands. Some of the designs as presented in [4,6,15–19,21,23] are good for covering the sub-GHz band, but the majority of them are not suitable for NB-IoT operations. Both planar and non-planar IoT antennas, [2–6,15–23,32–36], are reported in Table 2. None of the antennas as reported in [2–6,15–23,32–36] have a continuous frequency sweep in the sub-GHz band. The proposed antenna is the only candidate that can cover sub-GHz bands over a wide frequency range.

4. Conclusions

A miniaturized meandered loop slot-line antenna that is suitable for IoT applications was proposed in this work. The presented antenna was optimized using bending, meandering, and reactively loading the slot methods to operate in the sub-GHz bands. Recently, sub-GHz IoT technology is becoming more popular in order to enable devices to achieve long-range communications with low power consumption. A very wideband tuning, 758 to 1034 MHz, was achieved. The antenna's reconfigurability was also investigated in this work. A good understanding of the general antenna design guidelines for this antenna system type was developed. The proposed antenna design was fabricated on a RO4350 substrate with dimensions of 60 mm × 27 mm. The compact planar structure of the antenna, its simple biasing circuitry, its ability to operate over a wide band of sub-GHz, and its narrow-band operation are unique features of this design.

Author Contributions: Conceptualization, R.H.; software, R.H., A.M.A. and S.I.A.; validation, R.H. and A.M.A.; writing—original draft preparation, R.H.; writing—review and editing, R.H., N.H., S.I.A. and K.A. All authors have read and agreed to the published version of the manuscript.

Funding: This research received no external funding.

Acknowledgments: The authors would like to acknowledge the support provided by Researchers Supporting Project number (RSP2022R474), King Saud University, Riyadh, Saudi Arabia.

Conflicts of Interest: The authors declare no conflict of interest.

References

1. Sabban, A. Wearable Circular Polarized Antennas for Health Care, 5G, Energy Harvesting, and IoT Systems. *Electronics* **2022**, *11*, 427. [CrossRef]
2. Dong, J.; Yu, X.; Hu, G. Design of a compact quad-band slot antenna for integrated mobile devices. *Int. J. Antennas Propag.* **2016**, *2016*, 3717681. [CrossRef]
3. Bekasiewicz, A.; Koziel, S. Compact UWB monopole antenna for internet of things applications. *Electron. Lett.* **2016**, *52*, 492–494. [CrossRef]
4. Romputtal, A.; Phongcharoenpanich, C. Iot-linked integrated NFC and dual band UHF/2.45 GHz RFID reader antenna scheme. *IEEE Access* **2019**, *7*, 177832–177843. [CrossRef]
5. Islam, M.S.; Islam, M.T.; Ullah, M.A.; Beng, G.K.; Amin, N.; Misran, N. A modified meander line microstrip patch antenna with enhanced bandwidth for 2.4 GHz ISM-band Internet of Things (IoT) applications. *IEEE Access* **2019**, *7*, 127850–127861. [CrossRef]
6. Bichara, R.M.; Asadallah, F.A.; Costantine, J.; Awad, M. A folded miniaturized antenna for IoT devices. In Proceedings of the 2018 IEEE Conference on Antenna Measurements & Applications (CAMA), Västerås, Sweden, 3–6 September 2018; pp. 1–3.
7. Hussain, N.; Abbas, A.; Park, S.M.; Park, S.G.; Kim, N. A compact tri-band Antenna based on inverted-L stubs for smart devices. *Comput. Mater. Contin.* **2022**, *70*, 3321–3331. [CrossRef]
8. Su, Z.; Klionovski, K.; Bilal, R.M.; Shamim, A. 3D antenna- on-package with near-isotropic radiation pattern for IoT (WiFi based) applications. In Proceedings of the 2018 IEEE International Symposium on Antennas and Propagation & USNC/URSI National Radio Science Meeting, Boston, MA, USA, 8–13 July 2018; pp. 1431–1432.
9. Koga, Y.; Kai, M. A transparent double folded loop antenna for IoT applications. In Proceedings of the 2018 IEEE-APS Topical Conference on Antennas and Propagation in Wireless Communications (APWC), Cartagena, Colombia, 10–14 September 2018; pp. 762–765.
10. Shafique, K.; Khawaja, B.; Khurram, M.D.; Sibtain, S.M.; Siddiqui, Y.; Mustaqim, M.; Chattha, H.T.; Yang, X. Energy harvesting using a low-cost rectenna for Internet of Things (IoT) applications. *IEEE Access* **2018**, *6*, 30932–30941. [CrossRef]
11. Kumar, P.; Urooj, S.; Malibari, A. Design and implementation of quad-element super-wideband MIMO antenna for IoT applications. *IEEE Access* **2020**, *8*, 226697–226704. [CrossRef]
12. Houret, T.; Lizzi, L.; Ferrero, F.; Danchesi, C.; Boudaud, S. DTC-enabled frequency-tunable inverted-F antenna for IoT applications. *IEEE Antennas Wirel. Propag. Lett.* **2019**, *19*, 307–311. [CrossRef]
13. Oh, J.I.; Jo, H.W.; Kim, K.S.; Cho, H.; Yu, J.W. A Compact Cavity-Backed Slot Antenna Using Dual Mode for IoT Applications. *IEEE Antennas Wirel. Propag. Lett.* **2021**, *20*, 317–321. [CrossRef]
14. Chattha, H.T.; Ishfaq, M.K.; Khawaja, B.A.; Sharif, A.; Sheriff, N. Compact multiport MIMO antenna system for 5g IoT and cellular hand- held applications. *IEEE Antennas Wirel. Propag. Lett.* **2021**, *20*, 2136–2140. [CrossRef]
15. Yang, Y.; Chu, Q.; Mao, C. Multiband MIMO antenna for GSM, DCS, and LTE indoor applications. *IEEE Antennas Wirel. Propag. Lett.* **2016**, *15*, 1573–1576. [CrossRef]
16. Dong, J.; Yu, X.; Deng, L. A decoupled multiband dual-antenna system for WWAN/LTE smartphone applications. *IEEE Antennas Wirel. Propag. Lett.* **2017**, *16*, 1528–1532. [CrossRef]
17. Zhang, S.; Zhao, K.; Ying, Z.; He, S. Adaptive quad-element multi- wideband antenna array for user-effective LTE MIMO mobile terminals. *IEEE Trans. Antennas Propag.* **2013**, *61*, 4275–4283. [CrossRef]
18. Zheng, M.; Wang, H.; Hao, Y. Internal hexa-band folded monopole/dipole/loop antenna with four resonances for mobile device. *IEEE Trans. Antennas Propag.* **2012**, *60*, 2880–2885. [CrossRef]
19. Wong, K.L.; Liao, Z.G. Passive reconfigurable triple-wideband antenna for LTE tablet computer. *IEEE Trans. Antennas Propag.* **2014**, *63*, 901–908. [CrossRef]
20. Hussain, R.; Sharawi, M.S.; Shamim, A. 4-element concentric pentagonal slot-line-based ultra-wide tuning frequency reconfigurable MIMO antenna system. *IEEE Trans. Antennas Propag.* **2018**, *66*, 4282–4287. [CrossRef]
21. Mopidevi, H.; Damgaci, Y.; Rodrigo, D.; Jofre, L.; Cetiner, B.A. A quad-band antenna for public safety applications. *IEEE Antennas Wirel. Propag. Lett.* **2014**, *13*, 1231–1234. [CrossRef]
22. Boukarkar, A.; Lin, X.Q.; Jiang, Y.; Yu, Y.Q. Miniaturized single-feed multiband patch antennas. *IEEE Trans. Antennas Propag.* **2016**, *65*, 850–854. [CrossRef]
23. Young, M.W.; Yong, S.; Bernhard, J.T. A miniaturized frequency reconfigurable antenna with single bias, dual varactor tuning. *IEEE Trans. Antennas Propag.* **2015**, *63*, 946–951. [CrossRef]
24. Yuan, B.; Cao, Y.; Wang, G. A miniaturized printed slot antenna for six-band operation of mobile handsets. *IEEE Antennas Wirel. Propag. Lett.* **2011**, *10*, 854–857. [CrossRef]
25. Calabrese, C.; Marrocco, G. Meandered-slot antennas for sensor-RFID tags. *IEEE Antennas Wirel. Propag. Lett.* **2008**, *7*, 5–8. [CrossRef]
26. Wang, Y.; Zhang, J.; Peng, F.; Wu, S. A glasses frame antenna for the applications in Internet of Things. *IEEE Internet Things J.* **2019**, *6*, 8911–8918. [CrossRef]
27. Ur-Rehman, M.; Kalsoom, T.; Malik, N.A.; Safdar, G.A.; Chatha, H.T.; Ramzan, N.; Abbasi, Q.H. A wearable antenna for mmWave IoT applications. In Proceedings of the 2018 IEEE International Symposium on Antennas and Propagation & USNC/URSI National Radio Science Meeting, Boston, MA, USA, 8–13 July 2018; pp. 1211–1212.
28. Vignesh, L.B.; Kavitha, K. A survey on fractal antenna design. *Int. J. Pure Appl. Math.* **2018**, *120*, 10941–10959.

29. Desai, A.; Palandoken, M.; Elfergani, I.; Akdag, I.; Zebiri, C.; Bastos, J.; Rodriguez, J.; Abd-Alhameed, R.A. Transparent 2-Element 5G MIMO Antenna for Sub-6 GHz Applications. *Electronics* **2022**, *11*, 251. [CrossRef]
30. Abdulkawi, W.M.; Sheta, A.F.A.; Elshafiey, I.; Alkanhal, M.A. Design of Low-Profile Single- and Dual-Band Antennas for IoT Applications. *Electronics* **2021**, *10*, 2766. [CrossRef]
31. Shahzad, M.A.; Paracha, K.N.; Naseer, S.; Ahmad, S.; Malik, M.; Farhan, M.; Ghaffar, A.; Hussien, M.; Sharif, A.B. An Artificial Magnetic Conductor-Backed Compact Wearable Antenna for Smart Watch IoT Applications. *Electronics* **2021**, *10*, 2908. [CrossRef]
32. Majid, H.A.; Rahim, M.K.; Hamid, M.R.; Ismail, M.F. Frequency and pattern reconfigurable slot antenna. *IEEE Trans. Antennas Propag.* **2014**, *62*, 5339–5343. [CrossRef]
33. Erfani, E.; Nourinia, J.; Ghobadi, C.; Niroo-Jazi, M.; Denidni, T.A. Design and implementation of an integrated UWB/reconfigurable-slot antenna for cognitive radio applications. *IEEE Antennas Wirel. Propag. Lett.* **2012**, *11*, 77–80. [CrossRef]
34. Scardelletti, M.C.; Ponchak, G.E.; Jordan, J.L.; Jastram, N.; Mahaffey, V. Tunable reduced size planar folded slot antenna utilizing varactor diodes. In Proceedings of the 2010 IEEE Radio and Wireless Symposium (RWS), New Orleans, LA, USA, 10–14 January 2010; pp. 547–550.
35. Choi, D.H.; Park, S.O. Varactor-tuned active integrated antenna using slot antenna. In Proceedings of the 2005 IEEE Antennas and Propagation Society International Symposium, Washington, DC, USA, 3–8 July 2005; pp. 221–224.
36. Ray, A.; De, A.; Bhattacharyya, T.K. A miniaturized and reconfigurable on-chip slot antenna for RFID applications. In Proceedings of the 2019 IEEE International Symposium on Antennas and Propagation and USNC-URSI Radio Science Meeting, Atlanta, GA, USA, 7–12 July 2019; pp. 2203–2204.
37. Smv1233 Datasheet. Available online: https://datasheetspdf.com/datasheet/SMV1233.html (accessed on 20 January 2022).
38. Lpkf Protomat s103. Available online: https://www.lpkfusa.com (accessed on 5 January 2022).
39. Balanis, C.A. *Antenna Theory: Analysis and Design*; John Wiley & Sons: Hoboken, NJ, USA, 2016.
40. Wong, K.L.; Huang, C.C.; Chen, W.S. Printed ring slot antenna for circular polarization. *IEEE Trans. Antennas Propag.* **2002**, *50*, 75–77. [CrossRef]
41. Behdad, N.; Sarabandi, K. Dual-band reconfigurable antenna with a very wide tunability range. *IEEE Trans. Antennas Propag.* **2006**, 54409–54416. [CrossRef]
42. Marqués, R.; Mesa, F.; Martel, J.; Medina, F. Comparative analysis of edge-and broadside-coupled split ring resonators for metamaterial design-theory and experiments. *IEEE Trans. Antennas Propag.* **2003**, *51*, 2572–2581. [CrossRef]
43. Bouezzeddine, M.; Schroeder, W.L. Design of a wideband, tunable four-port MIMO antenna system with high isolation based on the theory of characteristic modes. *IEEE Trans. Antennas Propag.* **2016**, *64*, 2679–2688. [CrossRef]
44. Ghaffar, A.; Li, X.J.; Awan, W.A.; Naqvi, A.H.; Hussain, N.; Alibakhshikenari, M.; Limiti, E. A Flexible and Pattern Reconfigurable Antenna with Small Dimensions and Simple Layout for Wireless Communication Systems Operating over 1.65–2.51 GHz. *Electronics* **2021**, *10*, 601. [CrossRef]
45. Ojaroudi Parchin, N.; Jahanbakhsh Basherlou, H.; Al-Yasir, Y.I.A.; Abdulkhaleq, A.M.; Abd-Alhameed, R.A. Reconfigurable Antennas: Switching Techniques—A Survey. *Electronics* **2020**, *9*, 336. [CrossRef]

electronics MDPI

Article

A Novel Meander Bowtie-Shaped Antenna with Multi-Resonant and Rejection Bands for Modern 5G Communications

Yanal S. Faouri [1], Sarosh Ahmad [2,3,*], Naser Ojaroudi Parchin [4], Chan Hwang See [4] and Raed Abd-Alhameed [5]

1 Department of Electrical Engineering, The University of Jordan, Amman 11942, Jordan; y.faouri@ju.edu.jo
2 Department of Signal Theory and Communication, Universidad Carlos III de Madrid, Leganes, 28911 Madrid, Spain
3 Department of Electrical Engineering and Technology, Government College University Faisalabad, Faisalabad 38000, Pakistan
4 School of Engineering and the Built Environment, Edinburgh Napier University, Edinburgh EH10 5DT, UK; n.ojaroudiparchin@napier.ac.uk (N.O.P.); c.see@napier.ac.uk (C.H.S.)
5 Faculty of Engineering and Informatics, University of Bradford, Bradford BD7 1DP, UK; r.a.a.abd@bradford.ac.uk
* Correspondence: saroshahmad@ieee.org

Abstract: To support various fifth generation (5G) wireless applications, a small, printed bowtie-shaped microstrip antenna with meandered arms is reported in this article. Because it spans the broad legal range, the developed antenna can serve or reject a variety of applications such as wireless fidelity (Wi-Fi), sub-6 GHz, and ultra-wideband (UWB) 5G communications due to its multiband characterization and optimized rejection bands. The antenna is built on an FR-4 substrate and powered via a 50-Ω microstrip feed line linked to the right bowtie's side. The bowtie's left side is coupled via a shorting pin to a partial ground at the antenna's back side. A gradually increasing meandering microstrip line is connected to both sides of the bowtie to enhance the rejection and operating bands. The designed antenna has seven operating frequency bands of (2.43–3.03) GHz, (3.71–4.23) GHz, (4.76–5.38) GHz, (5.83–6.54) GHz, (6.85–7.44) GHz, (7.56–8.01) GHz, and (9.27–13.88) GHz. The simulated scattering parameter S_{11} reveals six rejection bands with percentage bandwidths of 33.87%, 15.73%, 11.71, 7.63%, 6.99%, and 12.22%, respectively. The maximum gain of the proposed antenna is 4.46 dB. The suggested antenna has been built, and the simulation and measurement results are very similar. The reported antenna is expanded to a four-element design to investigate its MIMO characteristics.

Keywords: multi-band; UWB; 5G communications; sub-6 GHz; notches; bowtie-shaped; multiband; MIMO; time-domain analysis

Citation: Faouri, Y.S.; Ahmad, S.; Parchin, N.O.; See, C.H.; Abd-Alhameed, R. A Novel Meander Bowtie-Shaped Antenna with Multi-Resonant and Rejection Bands for Modern 5G Communications. *Electronics* **2022**, *11*, 821. https://doi.org/10.3390/electronics11050821

Academic Editors: Faisel Tubbal, Ladislau Matekovits and Raad Raad

Received: 7 February 2022
Accepted: 4 March 2022
Published: 6 March 2022

1. Introduction

Modern wireless communication devices, which have evolved fast over the last four decades, are required to support a variety of applications, including real-time voice communication, text messaging, Wi-Fi, Bluetooth, Global Positioning System (GPS), video apps, and among others. All these applications operate in distinct frequency bands, necessitating the use of frequency reconfigurable antennas or multiband antennas to handle several applications with a single antenna. Frequency diversity can be configured electronically by utilizing varactors [1–3], micro-electromechanical systems (MEMS) [4], PIN diodes [5,6], or liquid metal [7]. On the other hand, designing an antenna in which its reflection coefficient spans the UWB range has grown in popularity, for its lowliness, inexpensive production costs, small power consumption, simplicity of production, and large bandwidth; so, the federal communication commissions (FCC) allowed UWB to use the unlicensed operating band from 3.1 to 10.6 GHz in 2002 [8]. Several UWB antennas utilize this band completely, as in [9], or obtained the other definition of the UWB by covering more than 500 MHz of bandwidth, as in [10].

Numerous distinct geometries of multiband and or UWB antennas with one or more notches have been documented in the literature. A lanky leaf antenna was reported in [11] to produce two rejection bands within the UWB range of 2.8–10 GHz. A small monopole antenna was adjusted in [12] to produce triple rejection bands within the UWB of 2.285–19.35 GHz with 5.88 dB gain and utilizes three PIN diodes. Four nested hexagonal fractal antennas were designed in [13] to operate in multiple frequency bands. For one of their designs, 'Antenna III' has five resonant frequencies in the UWB of 1.92–13.45 GHz with only one notch at a level around −6 dB and a peak average gain of 2.96 dB. A hexagonal shape microstrip patch with its edges being replaced by round curves operating in the frequency range 3–27.57 GHz [14] was adjusted by adding an inverting stub and other two slots of triangular shape to create triple frequency notches with bandwidths of 1.63 GHz, 1.09 GHz, and 0.76 GHz in an SWB with a 179.4% percentage bandwidth (PBW) [15]. In [16], a monopole antenna based on the split-ring resonator technique is well-designed to produce three operating bands. The antenna has a resonance at 2.45 GHz in addition to dual bands within the UWB range 3.4–11.8 GHz. A unit-cell metamaterial of dimensions 10×10 mm^2 was extended into a 2×1 and 2×2 MIMO in [17]. These models provide two distinct bands to operate in the s- and x-band with resonances at 4.27 GHz, 5.42 GHz, and 12.4 GHz. A single-sided bowtie or a monopole bowtie in one layer with meander arms on the other layer was investigated in [18] to operate at three resonating bands. An isosceles triangular microstrip antenna coupled electromagnetically to unequal an arms V-shaped parasitic has been proposed in [19] to provide triple spectrums. This configuration can support six resonating bands at resonant frequencies of 2.88 GHz, 3.64 GHz, 3.95 GHz, 4.38 GHz, 4.81 GHz, and 5.6 GHz with five rejection bands between them where two of these rejection bands have a reflection coefficient level larger than 5 dB. Four spectrums for sub-6 GHz and mm-wave applications that have been radiated from a slotted patch of conical shape connected to a small triangular patch were proposed in [20]; the design resonant frequencies were 2.4 GHz, 5.2 GHz, 5.8 GHz, and 27.5 GHz. An antenna element of the 'F' shape placed above a truncated ground plane has been proposed in [21] to operate in four reconfigurable frequency modes.

The design and exploration of a UWB/multi-band 5G antenna with Hexa-frequency stopping bands and seven passing bands are presented in this work, incorporating a meander line structure that can be also integrated into a designing coupler [22] or power divider [23]. This design incorporates both multiple resonant modes and frequency rejection strategies to provide seven operational bands. The antenna was designed in the style of a bowtie, with each side made up of a meander line. The antenna exhibits a directional pattern with a major lobe and back lobe at lower frequencies as an alternative to the normal omnidirectional radiation pattern that would be formed from a similar limited ground arrangement. The proposed antenna has been further investigated compared to [24] by addressing an equivalent circuit model, and the circuit components' value effect is parametrized in addition to the time-domain analysis of the output signal. The suggested antenna has been manufactured, and the similarity between simulation and measured results were observed. Finally, an expansion of the designed antenna has been extended to form a four-port MIMO antenna to investigate the mutual coupling along with other MIMO characteristics at all resonating bands.

2. Proposed Antenna Design

A full parametric study is conducted on all variables to acquire the dimensions reported in Table 1. The proposed antenna with allocated parameters is illustrated in Figure 1. The antenna was built on a double-sided FR-4 substrate with a dielectric constant of 4.4 and a loss tangent of 0.02. The antenna's overall dimensions are $30 \times 30 \times 1.6$ mm^3. The right side of the bowtie is linked to the feed line, while the other portion of the bow tie is grounded using a shorting via to enhance the resonances since without the shorting pin; only two resonances are developed. Both bowtie-shaped resonators are located at the top layer as shown in Figure 1a where each side of the bowtie has a gradually increasing meander line arms. The arm's length, width, and spacing have been optimized to produce several resonances with an acceptable bandwidth for the utilized services and several notches with an acceptable rejection level to filter out other services. Figure 1b shows the modified ground plane that has been adjusted to sweep over the UWB scope. Figure 1c displays a side view of the board to indicate the shorting pin placement.

Table 1. Suggested Antenna Lengths in mm.

Parameter	Value	Parameter	Value
WS	30	LS	30
LG	11	W1	1.5
W2	2.5	W3	2.8
W4	0.6	W5	0.4
W6	0.5	W7	1
W8	13.2	W9	15.3
L1	18.75	L2	13.25
L3	9.25	L4	1.75
L5	4.25	L6	6.75
L7	0.25	L8	1.25
L9	5.25	L10	23.25

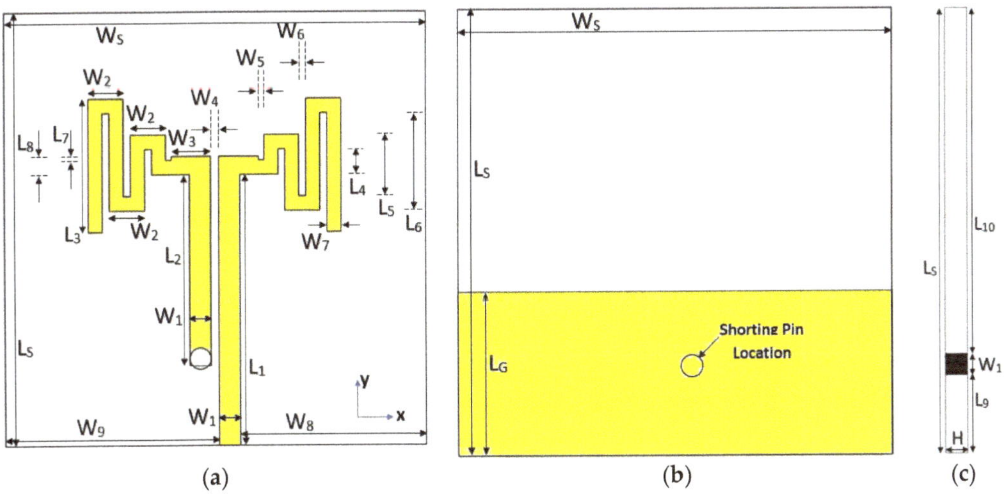

Figure 1. The suggested antenna; (**a**) front view, (**b**) back view, and (**c**) side view.

To reach the proposed design, it first passes through mainly two steps as shown in Figure 2a. First, a meander bowtie antenna is considered on a full ground plane (ANT I) which produces three frequency bands that require to be enhanced as depicted in Figure 2b. Then in (ANT II), a partial ground plane is utilized for better antenna characteristics that result in three operating bands in addition to other bands that required impedance matching. To optimize the matching impedance, the meandered arm's length, width, and spacing are considered as shown in (ANT III), which produces the proposed seven resonating bands with optimized rejection bands.

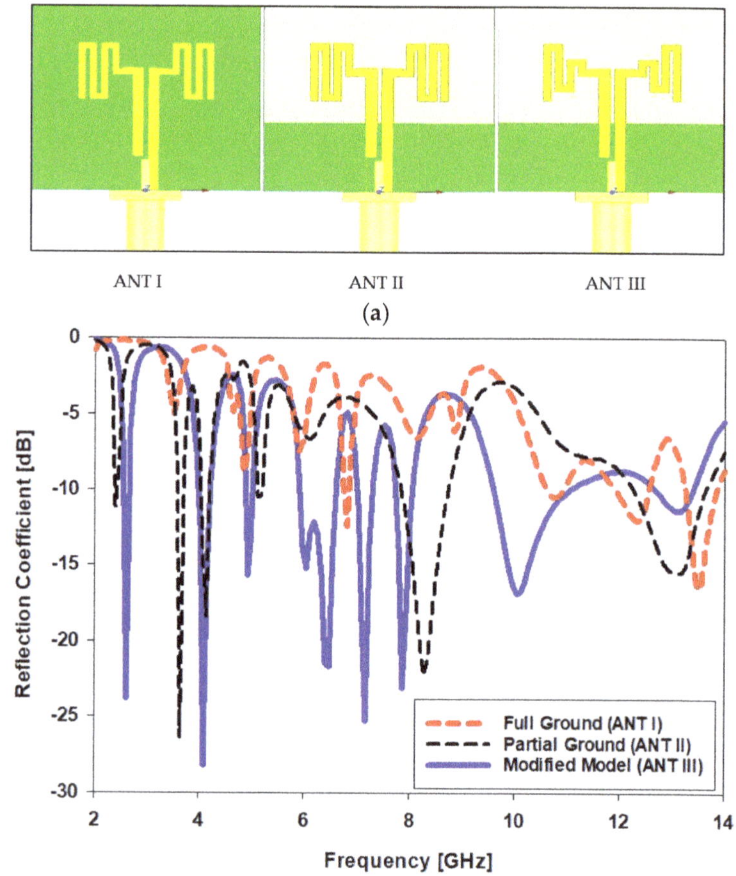

Figure 2. (**a**) Proposed antenna design steps; (**b**) reflection coefficient comparison for the design steps.

3. Results and Discussion

The designed antenna is built as shown in Figure 3a to validate the simulation results that are obtained through high-frequency structure simulator (HFSS) software, which were fine-tuned using an intense parametric optimization procedure to generate the reflection coefficient displayed in Figure 3b, where the simulated and measured S_{11} are in good agreement. The antenna 6-dB bandwidth ranges from 2.52 GHz to 13.83 GHz, and its 10-dB bandwidth spans over 2.55–10.94 GHz, which satisfies UWB requirements. Due to the multi-band behavior of the proposed antenna; the 6-dB bandwidth can be considered to provide seven resonating bands. These bands have bandwidths of (2.52–2.68) GHz, (3.9–4.4) GHz, (4.9–5.2) GHz, (5.8–6.7) GHz, (6.8–7.4) GHz, (7.6–8.2) GHz, and (9.3–13.8) GHz, respectively, in addition to an acceptable 10-dB impedance bandwidth in both simulation and measurements if they were considered. Besides these resonating bands, the suggested antenna was also acquired with six frequency notches with a notch level ≥ -5 dB. The rejecting bands' centered frequencies 'f_c' are 3.24 GHz, 4.71 GHz, 5.48 GHz, 6.81 GHz, 7.51 GHz, and 8.7 GHz and the percentage bandwidths (PBW) are 41.36%, 14.65%, 15.15%, 5.29%, 5.99%, and 18.04%, respectively. Together, several resonances modes and notch-band techniques have been used in this design to pass or suppress different communication services.

(a)

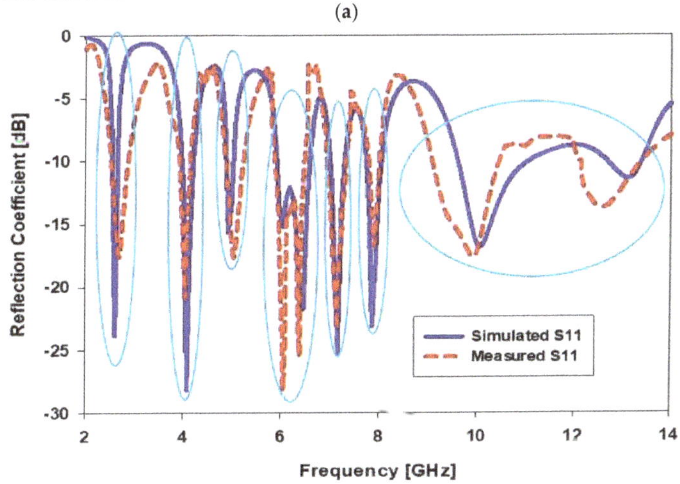

(b)

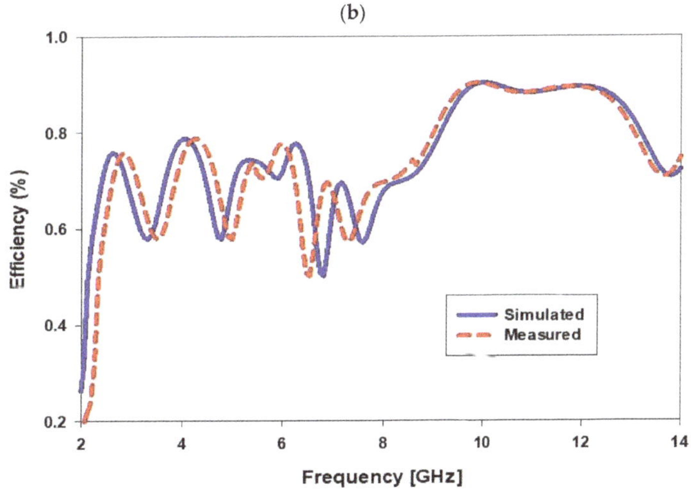

(c)

Figure 3. *Cont.*

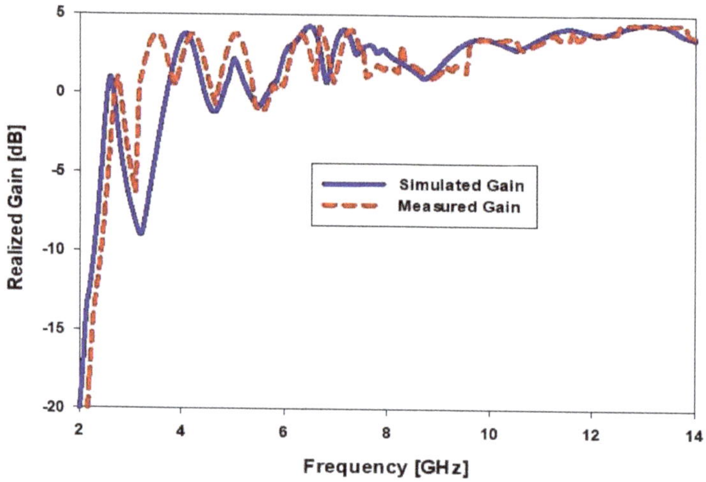

(d)

Figure 3. (**a**) The top and bottom layers of the fabricated proposed antenna, (**b**) simulated and measured antenna reflection coefficient, (**c**) proposed antenna design steps, and (**d**) reflection coefficient comparison for the design steps.

The simulated and measured peak realized gain graph is depicted in Figure 3c. A good agreement is noticed between the two traces where the gain tends to increase as the frequency increases with a sharp reduction at the rejecting band's center frequencies. The simulated results of the gain at the six notches' center frequencies are −8.2515 dB, −0.6843 dB, −0.2025 dB, 1.6513 dB, 2.5284 dB, and 0.78 dB, respectively. Table 2 summarizes the measured values of different aspects of the proposed antenna at these rejecting bands and the seven resonant frequencies. The proposed antenna peak realized gains within each resonance are tabulated in Table 2, where the maximum gain reaches 4.46 dB at 13.74 GHz. The graphs of the simulated and measured radiation efficiency are depicted in Figure 3d, and the results well agreed with the efficiencies attained at the operating band and a reduction at the rejecting bands. The maximum efficiency is found to be 90.3% at 9.82 GHz, while the lowest radiation efficiency reaches 50.14% at 6.53 GHz, which belongs to the fourth notch.

The fabricated antenna is tested in its characteristics by mounting the antenna in a suitable anechoic chamber to conduct the required measurements as shown in Figure 4a. The radiation pattern at all resonances-centered frequencies is measured and compared with the simulation. The pattern at eight selected frequencies in which the S_{11} curve has deepest locations is plotted in Figure 4b–i for both the E-plane, which represent the YZ-plane ($\phi = 90^\circ$), and the H-plane, which represents the XZ-plane ($\phi = 0^\circ$). In the H-plane, the antenna has shown an omnidirectional pattern at the first two resonances and starts to exhibit directional radiation pattern towards single or multiple orientations. The E-plane demonstrates an omnidirectional radiation pattern at the first resonant while the beam starts to acquire different directional patterns at the other resonances, and this is due to the partial ground and other modes being involved as frequency was increased. A good agreement between the simulated and measured patterns are noticed for the shown patterns. The radiation pattern at the rejection band-centered frequencies is also monitored, and it has a similar behavior but with much less power compared to the resonating band, and the pattern at the first notch-centered frequency is plotted in Figure 4j.

Table 2. Measured characteristics of the proposed antenna at the notches.

Notch No.	10−dB BW (GHz)	f_c	PBW %	Gain at f_c
1	2.88–3.93	3.1	33.87	−6.35
2	4.15–4.88	4.64	15.73	−0.69
3	5.23–5.88	5.55	11.71	−1.20
4	6.44–6.97	6.95	7.63	0.70
5	7.29–7.82	7.58	6.99	1.07
6	8–9.08	8.84	12.22	0.98
Measured Characteristics of the Proposed Antenna at the Resonances				
Band No.	**6−dB BW (GHz)**	f_r	**PBW %**	**Gain at f_r**
1	2.43–3.03	2.7	22.22	1.02
2	3.71–4.23	4.05	12.84	3.52
3	4.76–5.38	5.05	12.28	3.77
4	5.83–6.54	6.04	11.75	3.82
5	6.85–7.44	7.15	8.25	4.06
6	7.56–8.01	7.90	5.78	3.5
7	9.27–13.88	11.55	39.83	4.46

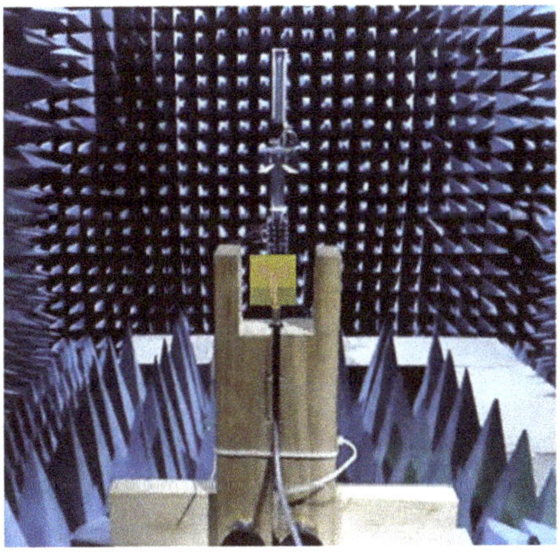

(a)

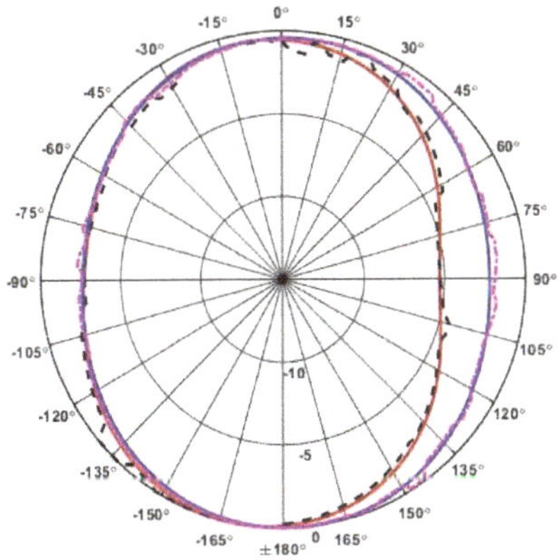

(b)

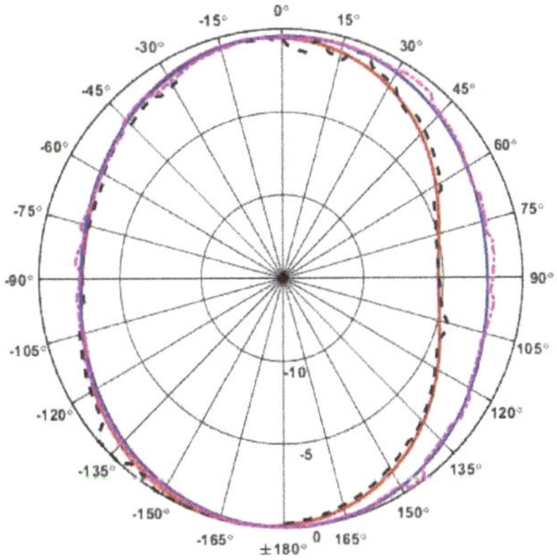

(c)

Figure 4. *Cont.*

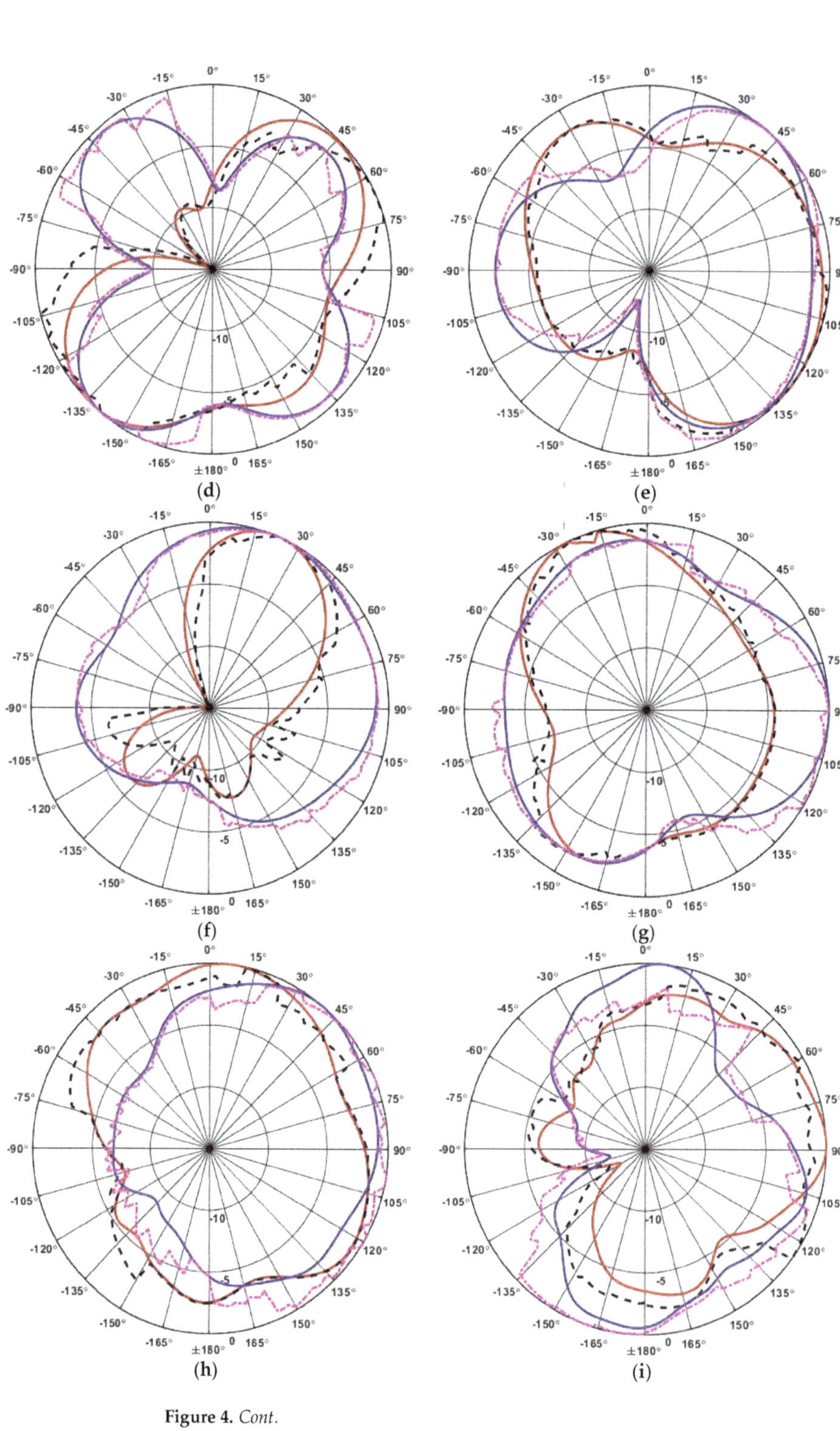

Figure 4. *Cont.*

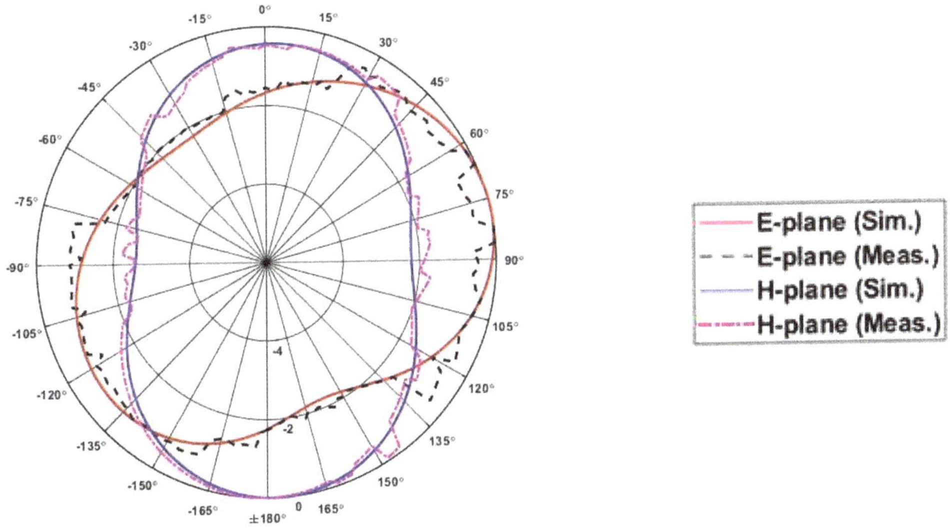

(j)

Figure 4. (**a**) Proposed antenna in the anechoic chamber. Simulated and measured antenna co-polarized radiation pattern for E- and H-planes at (**b**) 2.7 GHz, (**c**) 4.05 GHz, (**d**) 5.05 GHz, (**e**) 6.04 GHz, (**f**) 7.15 GHz, (**g**) 7.90 GHz, (**h**) 11.55 GHz, (**i**) 13.11 GHz, and (**j**) 3.1 GHz.

4. Time-Domain Characteristics

A significant aspect of UWB systems is the computation of the dispersion that happens when the antenna radiates and receives a pulse signal. The pulse-based UWB systems that utilize delivering very tiny pulses in time have a unique set of design requirements for antennas. As a result, more research and examination into the time domain behavior of a UWB antenna is needed. The pulse distortion and fidelity factor of the emitted pulse are the most important time-domain characteristics. They determine the amount of pulse distortion caused by the antenna.

In the proposed design, a normalized 5th order Gaussian derivative is set as an input signal. The pulse duration is 370 ps, as shown in Figure 5a, and it will have the wideband spectrum in the frequency domain as demonstrated in Figure 5b. The nth Gaussian pulse is represented in the time domain by Equation (2) where $H_n(t)$ is the nth Hermit polynomial, and its fifth-order polynomial is given in Equation (3) [25]:

$$G(t) = Ae^{-\frac{t^2}{2\sigma^2}} \tag{1}$$

$$G^n(t) = \frac{d^nG}{dt^n} = (-1)^n \frac{1}{\left(\sqrt{2}\sigma\right)^2} \cdot H_n\left(\frac{t}{\sqrt{2}\sigma}\right) \cdot G(t) \tag{2}$$

$$H_5(t) = 32t^5 - 160t^3 + 120t \tag{3}$$

The Fourier transform of the input signal is depicted in Figure 5c and resembles the reflection coefficient generated in the frequency domain and described in Figure 3b. The output normalized signal is shown in Figure 5d, and it has a pulse width of 3.74 ns, in which the signal distortion is around 50% in the first 1 ps; then, it reduces rapidly after that.

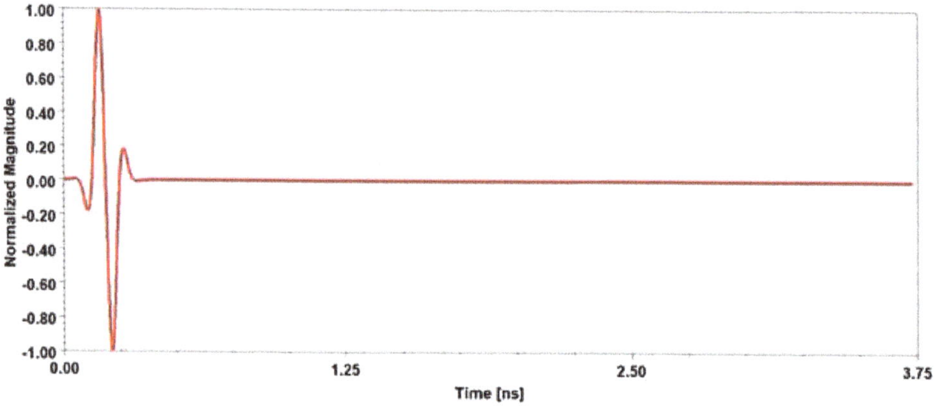

(a)

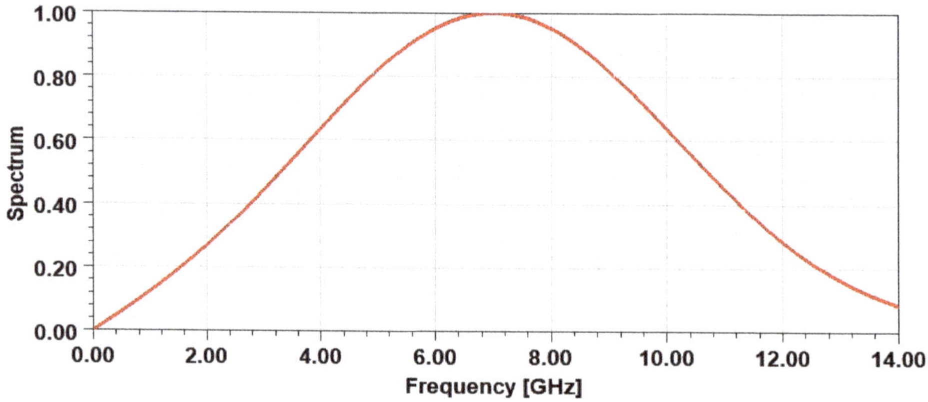

(b)

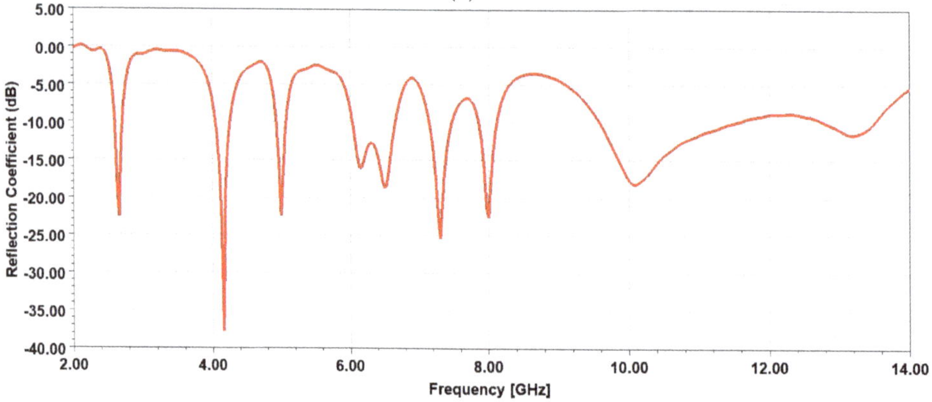

(c)

Figure 5. *Cont.*

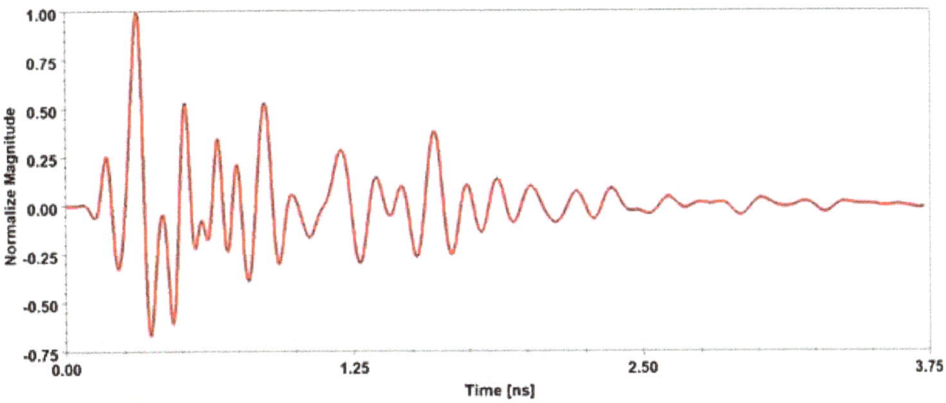

(d)

Figure 5. Time-domain results of the proposed antenna showing (**a**) input pulse, (**b**) input spectrum, (**c**) Fourier transform of the input signal, and (**d**) output signals.

5. Equivalent Circuit Analysis

Creating an equivalent circuit of lumped components for such a design analyzed in this paper is not an easy process. First, the antenna real and imaginary input impedance must be monitored to predict the best resonator to be used and whether they have common behavior or not. As can be noticed from Figure 6, the first six resonances have an impedance close to 50 Ω near the resonating frequency, while for the last wideband, the impedance is widely spanned near the 50 Ω line. An assumed circuit model is displayed in Figure 7, where the inductance (Lf) represents the input feedline and the first parallel RLC model to match the shorting pin used in the design; then, several series resonators are employed to create the first six bands, which are representing the meandered bowtie arms. Finally, the partial ground plane, which has a major effect in the wideband of the seventh band, is modelled by the rest of the circuit.

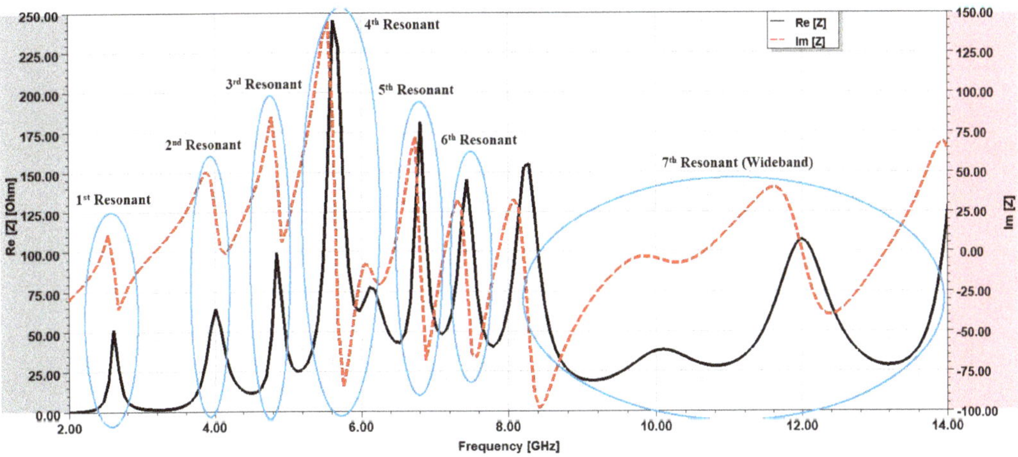

Figure 6. Real and imaginary impedance for the designed model.

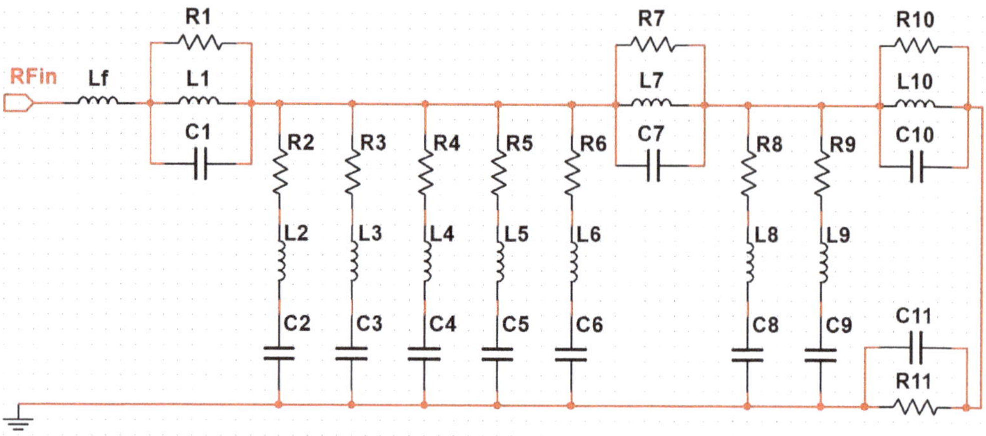

Figure 7. Equivalent circuit model.

The proposed circuit model is optimized using an ADS environment and has an approximated S-parameter result matched favorably with the simulated and measured results discussed before, and it is shown in Figure 8. The optimized values of the lumped circuit components are summarized in Table 3.

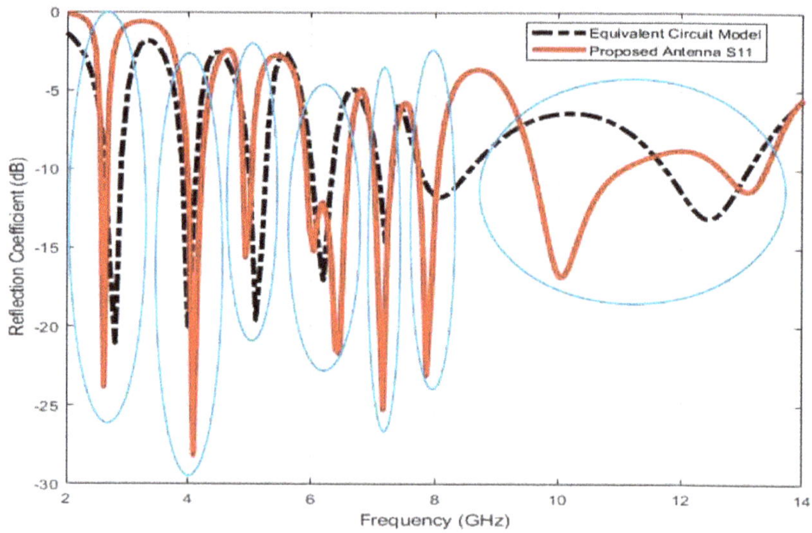

Figure 8. Comparison between simulated and calculated reflection coefficient.

Table 3. Comparison between the simulated and calculated Q and *ROC* of the proposed antenna.

	Notch 1	Notch 2	Notch 3	Notch 4	Notch 5	Notch 6
Q Simulated	3	9.47	7.51	21.28	19.76	7.37
Q Calculated	4.53	6.92	9.33	11.17	18.5	3.55
ROC Simulated	7.68	11.32	8.09	12.43	9.71	4.11
ROC Calculated	8.1	8.23	10.45	7.25	8.57	1.12

Two elements, the quality factor (Q) that represents the ratio of the stored to the dissipated energy in each cycle, and the roll-off criteria (ROC) which represent the variation in the reflected signal versus frequency, are completely linked to the selectivity and sharpness of the rejection mechanism. For the frequency rejection bands, the quality factor of the RLC series and parallel resonators can be calculated as [26].

$$Q = \frac{f_c}{BW_{-3dB}} \tag{4}$$

$$ROC = \frac{\Delta S_{11}(dB)}{\Delta f(GHz)} \tag{5}$$

where BW_{-3dB} represents the notch bandwidth at a level of 3 dB below its maximum, and ΔS_{11}(dB) is the variation in the reflection coefficient within the frequency band represented by Δf(GHz). The calculated quality factor and ROC of the six rejection bands based on Equations (4) and (5) are summarized in Table 4.

Table 4. Lumped component values of the equivalent circuit model (R in Ω, C in pF and L in nH).

R1 = 28.81	R2 = 37.43	R3 = 55.52	R4 = 120.53	R5 = 27.31	R6 = 114.96
L1 = 3.80	L2 = 31.96	L3 = 47.75	L4 = 5.52	L5 = 40.02	L6 = 147.98
C1 = 0.14	C2 = 0.095	C3 = 0.031	C4 = 0.052	C5 = 0.022	C6 = 0.033
R7 = 3.38 k	R8 = 70.24	R9 = 81.69 k	R10 = 9.21 k	Lf = 1.84	
L7 = 902.61	L8 = 34.62	L9 = 26.96	L10 = 1.09 m	R11 = 0.35	
C7 = 1.98	C8 = 0.018	C9 = 1.98 μ	C10 = 0.18	C11 = 0.58	

6. Quad Port MIMO Antenna Analysis

The multiband single element antenna proposed is expanded to a four-port MIMO antenna place such that the adjacent elements are orthogonal to each other. The substrate size is 70 × 70 mm^2 and 10 mm distance between orthogonal antenna elements, as shown in Figure 9. The current distribution at the center frequency of the six rejection bands is plotted and shown in Figure 10a–f to determine the antenna portion that is responsible for each notch. The first notch, at a center frequency of 3.24 GHz, is caused by the first two arms of the right bowtie portion, whereas the left feed line and the left bowtie portion have a dominant effect for creating the second notch at 4.64 GHz. At 5.48 GHz, which is the center of the third notch, the right feed line and the last arm of the bowtie are dominant. For the last three notches at 6.81 GHz, 7.51 GHz, and 8.7 GHz, the responsible potions are the lower portion of the feed line, the two feedlines, and the middle portion of the feedlines, respectively.

The total active reflection coefficient (TARC) of the four-port MIMO antenna is depicted in Figure 11. TARC relates the total incident power to the total radiated power, and it is noticed that the MIMO antenna has a TARC similar to the reflection coefficient of the single element with a similar number of resonating bands and center frequencies. The mutual coupling between all four ports is plotted in Figure 12. The average mutual coupling between the elements at the operating bands center frequency is below −17 dB across the seven resonating bands, while it is a little higher (−13 dB) at the first resonant between the adjacent elements, and it is much lower between diagonal elements.

The Envelope Correlation Coefficient (ECC) and diversity gain (DG) are useful metrics for assessing the diversity of MIMO antenna elements [4]. The ECC (much lower than 0.5) and DG (≅10 dB) of the proposed MIMO are displayed in Figure 13, and they are calculated from the scattering parameters between the two ports of high mutual coupling as given by Equations (6) and (7), respectively. Other MIMO characteristics such as mean effective gain (MEG) and channel capacity loss (CCL) for the seven resonating bands are tabulated in Table 5, and the results are satisfying the determined limit for these quantities, such as the CCL, is below 0.4 bit/s/Hz; the MEG ratios between the investigated ports are close or equal to unity.

$$ECC = \frac{|S_{11} * S_{12} + S_{21} * S_{22}|^2}{\left(1 - |S_{11}|^2 - |S_{21}|^2\right)\left(1 - |S_{22}|^2 - |S_{12}|^2\right)} \tag{6}$$

$$DG = 10\sqrt{1 - |ECC|^2} \tag{7}$$

Table 5. Proposed MIMO antenna characteristics at the resonant frequencies.

	MEG 1 (dB)	MEG 2 (dB)	MEG 1/MEG 2	CCL (bit/s/Hz)
Resonant 1	−3.57	−3.52	0.99	0.38
Resonant 2	−3.09	−3.12	1.01	0.06
Resonant 3	−3.22	−3.21	1	0.14
Resonant 4	−3.27	−3.27	1	0.18
Resonant 5	−3.07	−3.03	0.99	0.03
Resonant 6	−3.05	−3.05	1	0.03
Resonant 7	−3.15	−3.16	1	0.09

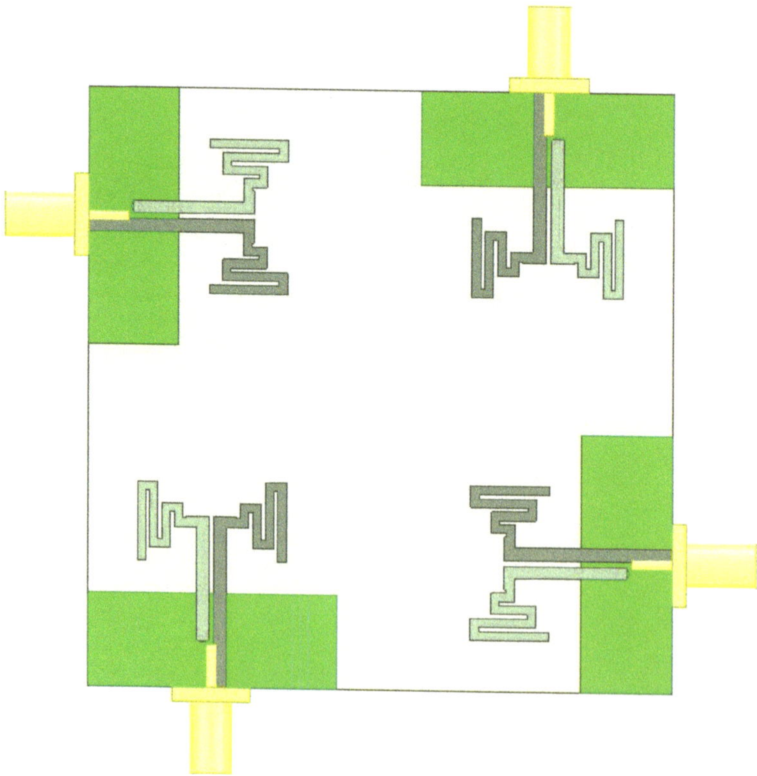

Figure 9. Four-element MIMO antenna configuration.

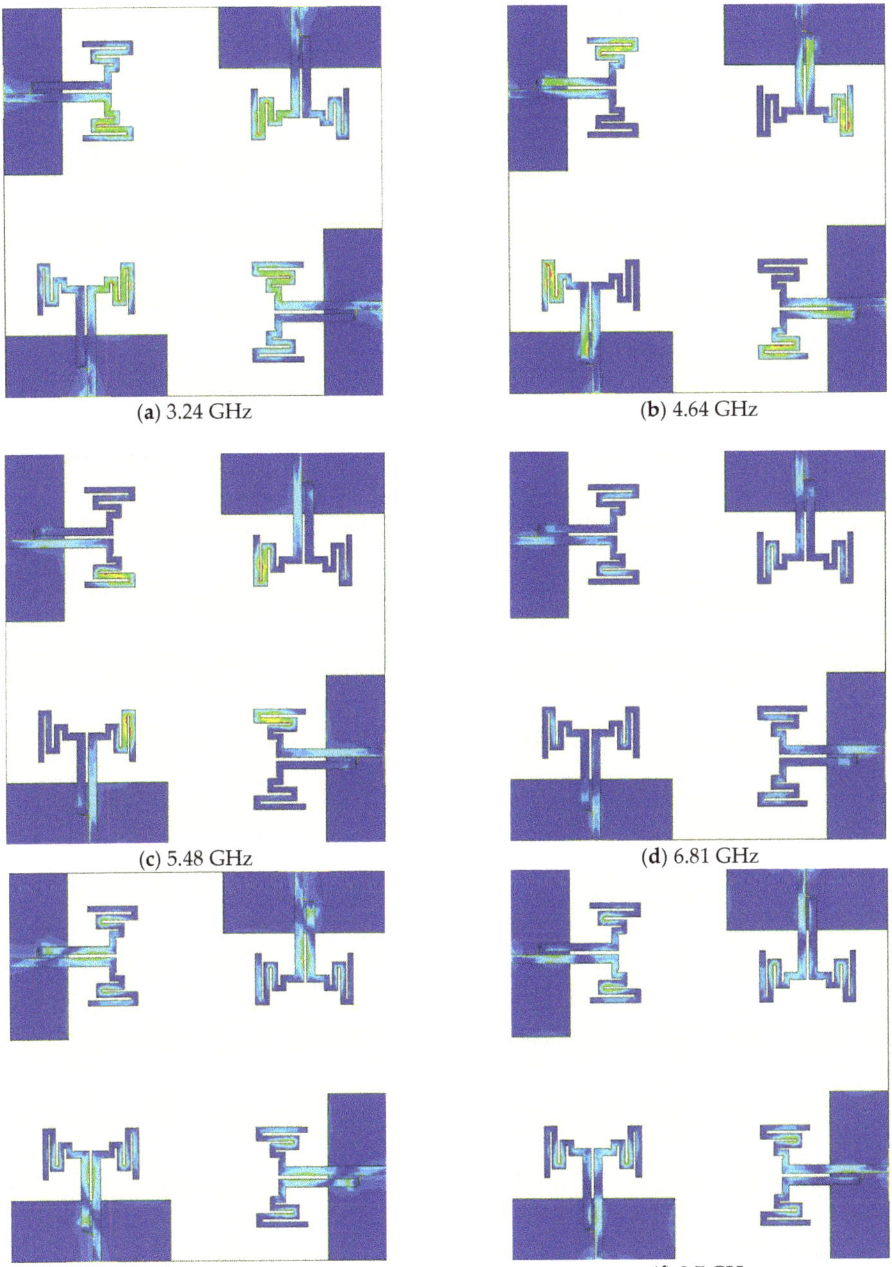

(**a**) 3.24 GHz (**b**) 4.64 GHz (**c**) 5.48 GHz (**d**) 6.81 GHz (**e**) 7.51 GHz (**f**) 8.7 GHz

Figure 10. Current distribution at the rejection bands center frequency; (**a**) at 3.24 GHz, (**b**) at 4.64 GHz, (**c**) at 5.48 GHz, (**d**) at 6.81 GHz, (**e**) at 7.51 GHz, (**f**) at 8.7 GHz.

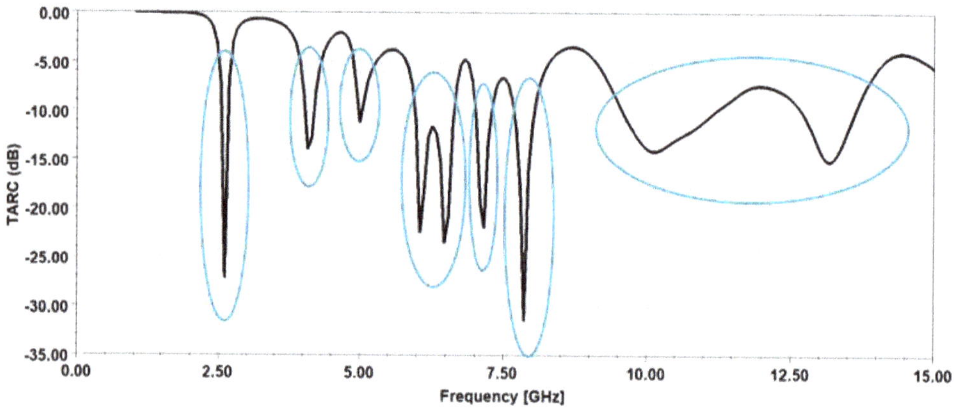

Figure 11. The total active reflection coefficients of the MIMO antenna.

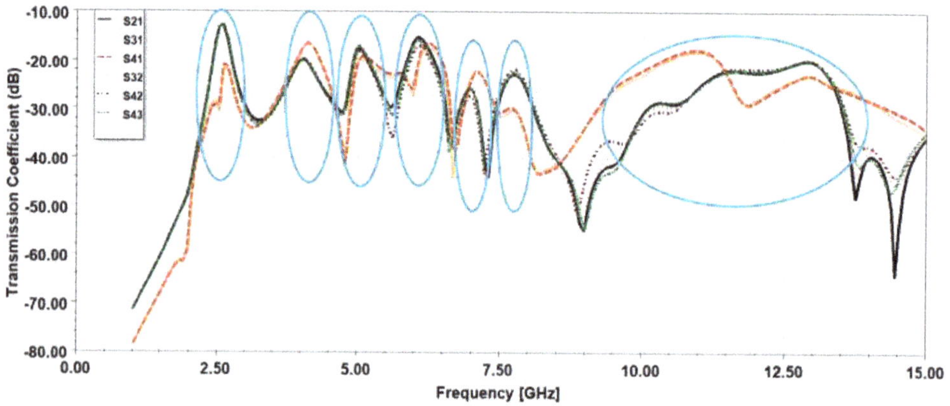

Figure 12. The transmission coefficients between all four ports.

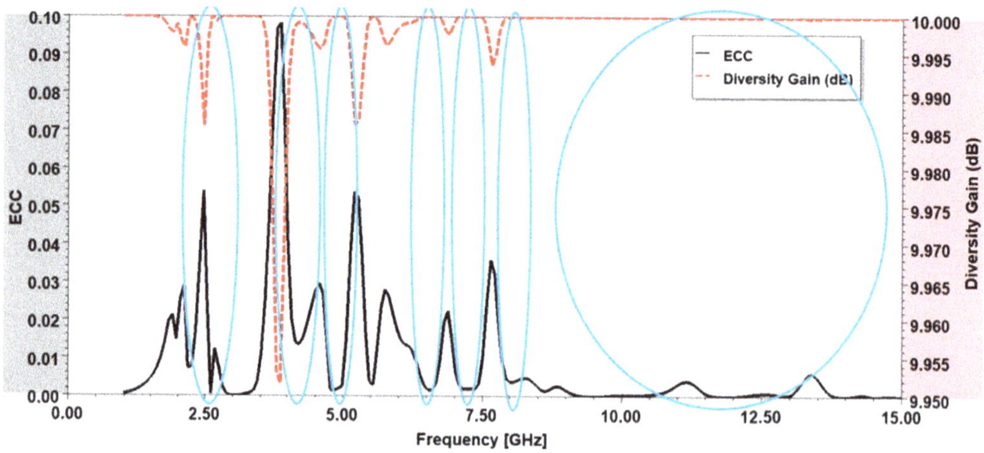

Figure 13. ECC and DG for the four-port MIMO Design.

7. Comparison with Related Works and Served Applications

The proposed antenna results are compared with the related works available in the literature for different aspects and the comparisons are tabulated in Table 6. The proposed antenna has a relatively smaller size and can pass low frequencies due to the meander arm structure. Additionally, the proposed antenna can operate at seven distinguished bands by satisfying the UWB license range. The maximum gain is acceptable for such a monopole antenna, and it is lower than some works in the literature due to the multiband mode of operations.

Table 6. Comparison of this work with other multiband antennas.

Ref.	Radiator Shape	Resonances (GHz)	Gain (dB)	Size (mm^2)	No. of Spectrums	Covered Bands
[16]	Split Ring	2.45, 5.2, 9.7	7	40 × 30	3	S, C, X
[17]	Split Ring metamaterial	4.27, 5.42, 12.4	-	20 × 20	3	C, X
[18]	Bowtie and meandered lines	1.62, 4.22, 7.13	8.2	20 × 10.4	3	L, C
[19]	Triangle and V-parasitic	2.88, 3.64, 3.95, 4.38, 4.81, 5.6	10.5	70 × 50	6	S, C
[20]	Slotted conical patch	2.4, 5.2, 5.8, 27.5	5.85	30 × 30	4	S, C, Ka
[21]	F-shape element on a truncated ground plane	2.1, 2.4, 3.35, 3.5, 5.28, 5.97	3.88	40 × 35	6	S, C
Proposed Antenna	Meandered Bowtie	2.7, 4.05, 5.05, 6.04, 7.15, 7.9, 11.55	4.46	30 × 30	7	S, C, X

8. Conclusions

The proposed design can be used for broadband radio applications within the first resonance, long-distance radio telecommunications; sub-6 GHz and other C-Band applications can be served by utilizing the second to fifth spectrums; X-Band radar applications can be supported by the sixth spectrum (7.6–8.2) GHz, and space communications, terrestrial broadband, satellite communications, amateur radio systems aided by the last vast spectrum (9.27–13.83) GHz. In addition, it can support the fifth generation (5G) FR1 band that is intended to cover up to 7.125 GHz. The achieved notches are intended to avoid interfering with WiMAX through the first notch, aeronautical radio navigation, which falls within the second and the sixth notches, the 5 GHz Wi-Fi band within the third rejecting band, and fixed and mobile satellite transmission within the fourth and fifth stopbands.

A printed microstrip meandered bowtie antenna is investigated and fabricated on an FR-4 substrate. The upper layer contains the two portions of the bowtie, where the right segment of the bowtie is coupled to the feedline, and the left portion is linked to the ground in the back layer through a shorting pin. Although the structure seems small and simple, the antenna exhibits seven resonating bands and six rejecting bands by optimizing the meandered arm's length, width, number, and spacing, making the antenna suitable for sub-6 GHz and UWB 5G communications. The antenna has a maximum gain and radiation efficiency of 4.46 dB and 90.3%, respectively. The MIMO configuration of the proposed antenna shows an acceptable value for ECC, DG, MEG, and CCL. The antenna is best suited for an end-user customer to accept and reject different communication services by just utilizing the corresponding band.

Author Contributions: Conceptualization, Y.S.F. and S.A.; methodology, Y.S.F. and S.A.; software, Y.S.F.; validation, Y.S.F., N.O.P., C.H.S., R.A.-A. and S.A.; formal analysis, N.O.P., C.H.S., R.A.-A. and Y.S.F.; investigation, Y.S.F. and S.A.; resources, Y.S.F. and S.A.; data curation, Y.S.F. and R.A.-A. writing—original draft preparation, Y.S.F.; writing—review and editing, Y.S.F., C.H.S. and S.A.; visualization, Y.S.F., N.O.P., C.H.S. and R.A.-A.; supervision, Y.S.F.; project administration, Y.S.F., funding; N.O.P., C.H.S., R.A.-A. All authors have read and agreed to the published version of the manuscript.

Funding: This paper is also partially funded by British Council "2019 UK-China-BRI Countries Partnership Initiative" program, with project titled "Adapting to Industry 4.0 oriented International Education and Research Collaboration.

Data Availability Statement: The data is available with the author upon request.

Acknowledgments: The authors would like to thank the Deanship of Scientific Research at the University of Jordan for providing the facilities for conducting this research for the year 2019–2022. The authors wish to express their thanks to the support provided by British Council "2019 UK-China-BRI Countries Partnership Initiative" programme with project titled "Adapting to Industry 4.0 oriented International Education and Research Collaboration".

Conflicts of Interest: The authors declare no conflict of interest.

References

1. Wang, M.; Tang, M.; Zhang, H.C.; Xu, J.; Cui, T.J.; Mao, J. Miniaturization of Frequency-Reconfigurable Antenna Using Periodic Slow-Wave Structure. *IEEE Trans. Antennas Propag.* **2021**, *69*, 7889–7894. [CrossRef]
2. Qwakneh, M.M.; AlHammami, K.A.; Omar, A.R.S.; Faouri, Y.S. Slot Antenna for Multi-Band Frequency Diversity. In Proceedings of the 2020 International Conference on Radar, Antenna, Microwave, Electronics, and Telecommunications (ICRAMET), Tangerang, Indonesia, 18–20 November 2020; pp. 287–290.
3. Faouri, Y.S.; Sharif, H.; Smadi, L.; Jamleh, H.O. Low Pass and Quad Band Pass Tunable Filter Based on Stub Resonators Technique. *Jordanian J. Comput. Inf. Technol.* **2019**, *5*, 124–136. [CrossRef]
4. Hassan, M.M.; Zahid, Z.; Khan, A.A.; Rashid, I.; Rauf, A.; Maqsood, M.; Bhatti, F.A. Two element MIMO antenna with frequency reconfigurable characteristics utilizing RF MEMS for 5G applications. *J. Electromagn. Waves Appl.* **2020**, *34*, 1210–1224. [CrossRef]
5. Al-Hammami, K.A.; Omar, A.R.S.; Qwakneh, M.M.; Faouri, Y.S. Hexagonal Patch Shaped MIMO Antenna for Frequency Agility. In Proceedings of the 2020 IEEE International Conference on Communication, Networks and Satellite (Comnetsat), Batam, Indonesia, 17–18 December 2020; pp. 301–305.
6. Omar, A.R.S.; Al-Hammami, K.A.; Qwakneh, M.M.; Faouri, Y.S. Frequency Reconfigurable Antenna Inspired by Tri-SRR Metamaterial. In Proceedings of the 2020 IEEE International Conference on Communication, Networks and Satellite (Comnetsat), Batam, Indonesia, 17–18 December 2020; pp. 296–300.
7. Song, L.; Gao, W.; Chui, C.O.; Rahmat-Samii, Y. Wideband frequency reconfigurable patch antenna with switchable slots based on liquid metal and 3-D printed microfluidics. *IEEE Trans. Antennas Propag.* **2019**, *67*, 2886–2895. [CrossRef]
8. Ahmad, S.; Ijaz, U.; Naseer, S.; Ghaffar, A.; Qasim, M.A.; Abrar, F.; Parchin, N.O.; See, C.H.; Abd-Alhameed, R. A Jug-Shaped CPW-Fed Ultra-Wideband Printed Monopole Antenna for Wireless Communications Networks. *Appl. Sci.* **2022**, *12*, 821. [CrossRef]
9. Al-Gburi, A.J.A.; Ibrahim, I.B.M.; Zakaria, Z.; Ahmad, B.H.; Bin Shairi, N.A.; Zeain, M.Y. High Gain of UWB Planar Antenna Utilising FSS Reflector for UWB Applications. *Comput. Mater. Contin.* **2022**, *70*, 1419–1436. [CrossRef]
10. Garbaruk, M. A Planar Four-Element UWB Antenna Array with Stripline Feeding Network. *Electronics* **2022**, *11*, 469. [CrossRef]
11. Li, X.P.; Xu, G.; Ma, M.R.; Duan, C.J. UWB dual-band-notched lanky-leaf-shaped antenna with loaded half-square-like slots for communication system. *Electronics* **2021**, *10*, 1991. [CrossRef]
12. Sharma, M.; Awasthi, Y.K.; Singh, H. Compact multiband planar monopole antenna for Bluetooth, LTE, and reconfigurable UWB applications including X-band and Ku-band wireless communications. *Int. J. RF Microw. Comput.-Aided Eng.* **2019**, *29*, e21668. [CrossRef]
13. Sharma, N.; Bhatia, S.S. Performance enhancement of nested hexagonal ring-shaped compact multiband integrated wideband fractal antennas for wireless applications. *Int. J. RF Microw. Comput. -Aided Eng.* **2020**, *30*, e22079. [CrossRef]
14. Al-Faouri, Y.S.; Awad, N.M.; Abdelazeez, M.K. Enhanced Ultra-Wide Band Hexagonal Patch Antenna. *Jordanian J. Comput. Inf. Technol.* **2018**, *4*, 150–158.
15. Faouri, Y.S.; Awad, N.M.; Abdelazeez, M.K. Hexagonal patch antenna with triple band rejections. In Proceedings of the 2019 IEEE Jordan International Joint Conference on Electrical Engineering and Information Technology (JEEIT), Amman, Jordan, 9–11 April 2019; pp. 446–448.
16. Yeboah-Akowuah, B.; Tchao, E.T.; Ur-Rehman, M.; Khan, M.M.; Ahmad, S. Study of a printed split-ring monopole for dual-spectrum communications. *Heliyon* **2021**, *7*, e07928. [CrossRef] [PubMed]
17. Mansoul, A.; Seddiki, M.L. Multiband reconfigurable Bowtie slot antenna using switchable slot extensions for WiFi, WiMAX, and WLAN applications. *Microw. Opt. Technol. Lett.* **2018**, *60*, 413–418. [CrossRef]
18. Qi, J.; Ren, J.; Qiu, J. A triple-band antenna based on bowtie and meander-line structures. In Proceedings of the 2017 IEEE International Symposium on Antennas and Propagation & USNC/URSI National Radio Science Meeting, San Diego, CA, USA, 9–14 July 2017; pp. 2479–2480.
19. Elsadek, H.; Nashaat, D.M. Multiband and UWB V-shaped antenna configuration for wireless communications applications. *IEEE Antennas Wirel. Propag. Lett.* **2008**, *7*, 89–91. [CrossRef]

20. Khan, Z.; Memon, M.H.; Rahman, S.U.; Sajjad, M.; Lin, F.; Sun, L. A single-fed multiband antenna for WLAN and 5G applications. *Sensors* **2020**, *20*, 6332. [CrossRef] [PubMed]
21. Ullah, S.; Ahmad, S.; Khan, B.A.; Flint, J.A. A multi-band switchable antenna for Wi-Fi, 3G Advanced, WiMAX, and WLAN wireless applications. *Int. J. Microw. Wirel. Technol.* **2018**, *10*, 991–997. [CrossRef]
22. Roshani, S.; Roshani, S. A compact coupler design using meandered line compact microstrip resonant cell (MLCMRC) and bended lines. *Wirel. Netw.* **2021**, *27*, 677–684. [CrossRef]
23. Roshani, S. A Wilkinson power divider with harmonics suppression and size reduction using meandered compact microstrip resonating cells. *Frequenz* **2017**, *71*, 517–522. [CrossRef]
24. Salim, A.; Baqaleb, S.; Faouri, Y. Multiband Meander UWB Bowtie Antenna with Six Rejection Bands. In Proceedings of the 11th International Conference on Information and Communication Systems (ICICS), Irbid, Jordan, 7–9 April 2020.
25. Mighani, M.; Akbari, M. New UWB monopole planer antenna with dual band notched. *Prog. Electromagn. Res. C* **2014**, *52*, 153–162. [CrossRef]
26. Sanyal, R.; Sarkar, P.P.; Sarkar, S. Octagonal nut shaped monopole UWB antenna with sextuple band notched characteristics. *AEU-Int. J. Electron. Commun.* **2019**, *110*, 152833. [CrossRef]

electronics

MDPI

Article

A Wide-Angle Pattern Diversity Antenna System for mmWave 5G Mobile Terminals

Karthikeya Gulur Sadananda [1], Issa Elfergani [2,3,*], Chemseddine Zebiri [4], Jonathan Rodriguez [2], Shiban Kishen Koul [5] and Raed A. Abd-Alhameed [3,6]

1 Centre for Antennas and Radio Frequency Systems, Department of Electronics and Telecommunication Engineering, Ramaiah Institute of Technology, Bengaluru 560054, India; gsk@msrit.edu
2 Instituto de Telecomunicações, Campus Universitário de Santiago, 3810-193 Aveiro, Portugal; jonathan@av.it.pt
3 School of Engineering and Informatics, University of Bradford, Bradford BD7 1DP, UK; r.a.a.abd@bradford.ac.uk
4 Laboratoired'Electronique de Puissance et Commande Industrielle (LEPCI), Department of Electronics, University of Ferhat Abbas, Sétif -1-, Sétif 19000, Algeria; czebiri@univ-setif.dz
5 Centre for Applied Research in Electronics, IIT Delhi, New Delhi 110016, India; shiban_koul@hotmail.com
6 Information and Communication Engineering Department, Basrah University College of Science and Technology, Basrah 24001, Iraq
* Correspondence: i.t.e.elfergani@av.it.pt or i.elfergani@bradford.ac.uk

Abstract: A shared ground shared radiator with wide angular coverage for mmWave 5G smartphones is proposed in this paper. A four-element corporate-fed array with conventional impedance matched power divider is designed. Stepped impedance transformers are integrated with the corner most elements to achieve pattern diversity with wide angular coverage without significant compromise in gain. The proposed three-port shared radiator conformal commercial antenna could be easily integrated with commercial mmWave 5G smartphones. All the three ports' excitations operate in the 28 GHz band. Radiation pattern bandwidth of the multi-port system is high. The gain variation is from 6 to 11 dBi amongst the ports and across the operating spectrum. The highest mutual coupling is 10 dB, in spite of the electrically connected structure. The proposed shared radiator element has a wide angular coverage of 100°, maintaining high front-to-back ratio when the respective port is excited. Simulation and measurement results for the proposed structure are illustrated in detail.

Keywords: shared radiator; conformal; mmWave 5G; 28 GHz band

Citation: Sadananda, K.G.; Elfergani, I.; Zebiri, C.; Rodriguez, J.; Koul, S.K.; Abd-Alhameed, R.A. A Wide-Angle Pattern Diversity Antenna System for mmWave 5G Mobile Terminals. *Electronics* **2022**, *11*, 571. https://doi.org/10.3390/electronics11040571

Academic Editor: Sotirios K. Goudos

Received: 22 January 2022
Accepted: 10 February 2022
Published: 14 February 2022

1. Introduction

Due to the tremendous growth in the smartphone data traffic, it is expected that millimeter waves would be useful for decongesting the current communication channels in the microwave domain [1–3]. Experimental characterizations have proved that higher frequency carriers could be well suited for cellular communication links [4,5]. Thus, exploration of hardware design centered at 28 GHz would be an important design topic. As per the Friis transmission formula [6], millimeter wave frequencies would experience additional losses due to the nature of the transmission link. It must also be noted that the real-world 28 GHz signal would suffer from additional losses due to the building materials and multipath effects [4]. Hence, high-gain antennas on the mobile phone are the only sensible choice to realize a mmWave link, given the constraints of the receiver sensitivity in the Ka-band.

Numerous articles and conference proceeding papers have been reported in the past decade. A high-gain corporate fed array is reported in [7], the electrical footprint is high and hence might not be suitable for direct integration with commercial 5G devices. Phased array presented in [8,9] is complex to fabricate and has low-gain yield for the electrical size occupied. Printed dipole of [10] would occupy a large footprint within the smartphone limits, proving to be unsuitable for the application at hand. The design reported in [11] lacks gain switch ability and is incompatible with the modern commercial smartphones.

Gain switch ability is required in mmWave 5G systems to cater to broadcast and data link modes. In the broadcast mode, a low-gain or wide-beam antenna is required and in the data link mode, a high gain or narrow beam is required. The same could be realized using phased arrays by selectively firing up the designated ports [10,11]. The CPW (coplanar waveguide)-fed antenna design of [12] works in the microwave band, the authors have demonstrated an orthogonal polarization system, but the same concept might not be applicable in the mmWave regime. Similarly, the design concept of [13] would lead to bidirectional radiation patterns, when the antenna is scaled up to millimeter wave frequencies. The corporate-fed array illustrated in [14] would occupy a larger physical footprint, when designed for the 28 GHz band. The MIMO concept illustrated in [15] might not be scaled down to Ka band. However, the design and deployment of phased arrays would be expensive. Therefore, an electrically compact three-port-wide angular coverage shared ground antenna system is proposed in this paper. The novel features of the proposed antenna design are listed below:

- Shared radiator design with minimal physical footprint.
- Shared ground design which is compatible with the system ground plane of the typical commercial device.
- Conformal shared radiator design which could be panel mountable with commercially available smartphones.
- Wide angular coverage with three ports without the use of phase shifters.
- Achievement of high gain for the respective port for a minimal occupied physical footprint.

2. Proposed Pattern Diversity Antenna

The multiport electrically connected antenna system is displayed in Figure 1a. It is a corporate-fed, inset-fed patch antenna array [16–19] with multiple ports. The dielectric used to realize this microstrip-fed design is Rogers 5880, which has a dielectric constant of 2.2 ± 0.02 with a corresponding loss tangent of 0.0009. Lower dielectric constant is necessary to facilitate radiation, especially in the higher frequency of operation. It must also be noted that the forward gain loss due to higher dielectric loss tangent could be avoided by this substrate. Rogers substrate is not very flexible as the dielectric composition is rigid, but with a thickness of 20 mil, flexibility could be expected. However, this substrate cannot be bent and planarized multiple times as this would fracture the substrate and hamper the dielectric and metal integrity of the design.

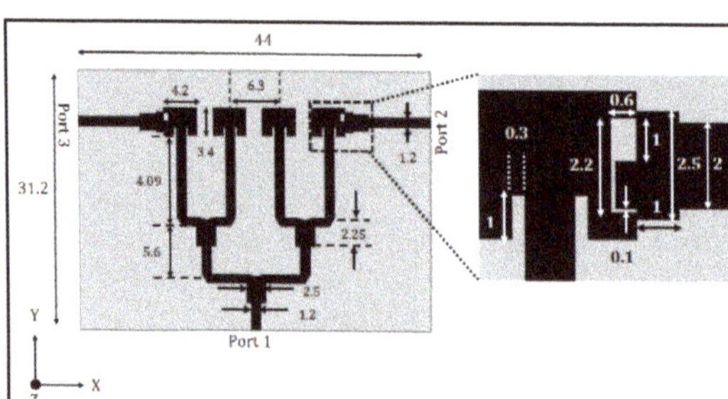

(a)

Figure 1. *Cont.*

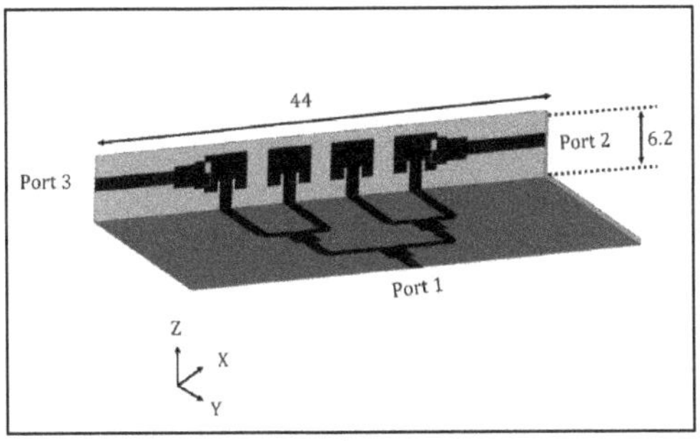

(b)

Figure 1. (**a**) Proposed three-port electrically connected antenna, (**b**) Corner bent antenna (All dimensions are in mm).

The proposed design is an electrically connected multiport radiator with an electrically large ground with a conventional microstrip feeding technique. A 1.2 mm feed line has been incorporated to closely match the characteristic impedance SMA (sub-miniature A) connector. The entire structure is based on the corporate-fed array concept. Here, a four-element array is designed to achieve wide angular coverage as well as reasonably high gain for the available physical aperture. The feed line connects to a two-way non-Wilkinson-based power divider. A Wilkinson-based power divider would have enhanced the isolation between the terminated loads of the corresponding lines, the inset-fed patch antennas in this case. However, implementing a Wilkinson-based power divider at higher frequencies would be difficult due to the size of the miniature resistors and the requirement of a thin-tip solder. The notch at the T-junction of the power divider facilitates in impedance matching of the patch radiators and the feeding network. The primary radiators are spaced at an approximately half-wavelength distance on the substrate.

The outermost elements of the corporate-fed array are also connected to the ports, as observed in Figure 1a. These ports are connected to the edge along the Y axis of the corner most elements. The 50 Ω lines originating from ports 2 and 3 are impedance matched to the shorter edge of the radiator through two low impedance-stepped transformers. The 2.2 mm × 0.6 mm slot, displayed in the inset of Figure 1a, does not contribute to radiation due to the electrically small size of the slot. When port 2 is excited, the right corner element contributes to radiation and hence a low gain and wide beam is observed. The same phenomenon is observed when the connected structure is excited with port 3. As the corner elements are electrically away from the phase center of the corporate-fed array, beam tilt is expected for the corresponding port excitation.

The current design of Figure 1a would not be suitable for direct integration with the mobile terminal, as this would be a broadside radiator and the signal strength would drastically reduce when it is blocked by a human torso. Hence, the topology of placement of the proposed shared radiator must be such a way as to reorient radiation away from the user, when the antenna module is mounted on the mobile device. Hence, corner bent topology is investigated as illustrated in Figure 1b. Here, the height of the corner bent radiator is 6.2 mm, which is compliant with the industry standard height of 7 mm. It must also be noted that the proposed design has an electrically massive system ground plane; hence, it would be favorable for integration with commercial smartphones. The feeding network of the proposed topology is in the orthogonal plane as that of the radiating aperture. Hence, the geometry itself acts as an isolating network between the feed and the active radiating aperture. The design could be further miniaturized by bending the portion of the feeding structure of ports 2 and 3. A photograph of the fabricated prototype is illustrated in Figure 2.

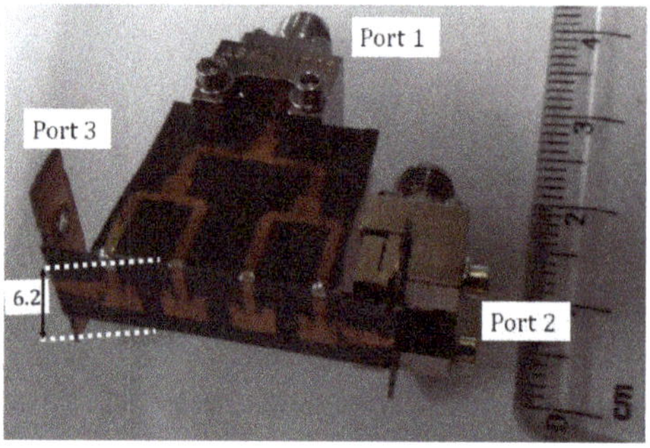

Figure 2. Photograph of the electrically connected antenna.

The input impedance bandwidth corresponding to port 1 is 11.4%, centered at 28.1 GHz, as seen in Figure 3. The wideband characteristics are realized due to the topology of the feeding network and the impedance transformers at ports 2 and 3.

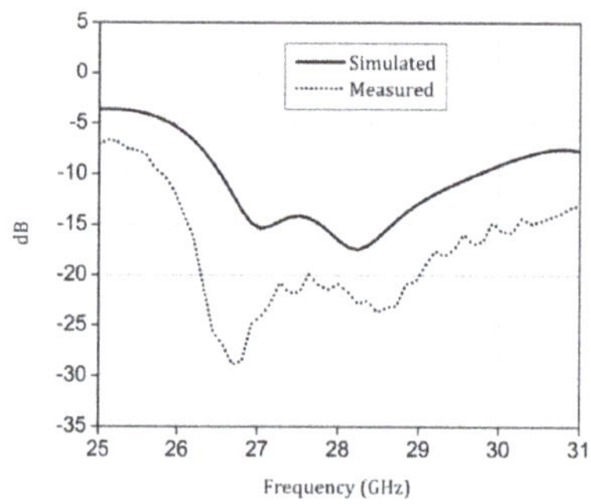

Figure 3. Input reflection coefficient of Port 1.

$|S_{11}|$ for ports 2 and 3 is graphed in Figure 4 and both are operational in the 28 GHz band. All the S-parameters measurements were performed using Agilent E8364C. It must be observed that the S-parameters' measurements are difficult in the present context, as the connectors used are electrically large and are in close proximity with the antenna structure. The clearance for the connector cables is also very minimal; hence, the deviation between simulated and measured results are pretty evident. Additionally, the fabrication errors due to bending has resulted in the deviation between simulated and measured results.

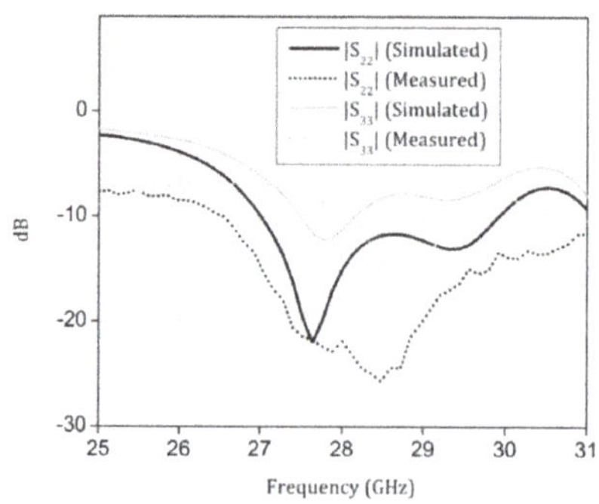

Figure 4. Input reflection coefficients of port 2 and 3.

The composite radiation patterns when the respective ports are activated are demonstrated in Figure 5. When port 1 is activated, the high-gain, low-beamwidth mode is activated. All four elements have in-phase E-fields which aid in beamforming leading to narrow beams. However, ports 2 and 3 have a wider beam, as the radiators on either edge are the primary contributors to the wide beam. The higher back lobe specifically for excitation of port 3 is due to the following reasons: Dual bending of the substrate which creates cracks within the copper trace of the feeding network, hence leading to radiation leakages from the discontinuities. Faulty assembly of the electrically large connector with the feeding line corresponding to port 3, leading to poor transition from the trace pin of the connector to the feed line. The overall angular coverage is 100° indicating a pretty wide angular coverage. The front-to-back ratio is greater than 10 dB and is maintained throughout the operating spectrum indicating minimal radiation towards the user.

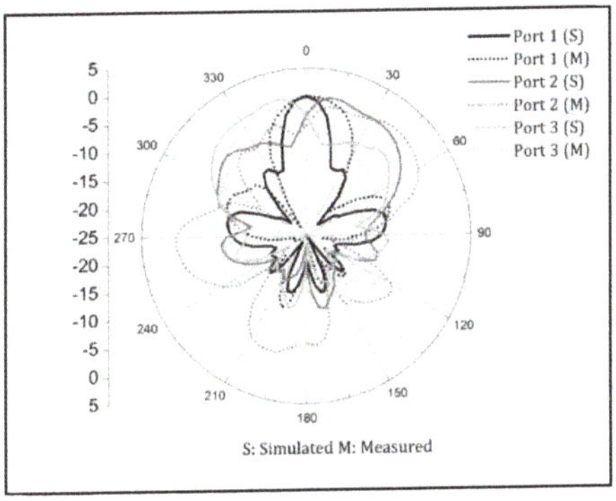

Figure 5. Radiation patterns of the proposed shared radiator.

The measured mutual coupling is less than 15 dB, as observed in Figure 6, in spite of the electrically connected geometry. The low mutual coupling is primarily due to the design of impedance transformers at ports 2 and 3. The forward gains of the designed antenna system is illustrated in Figure 7a, the radiation efficiency is shown in Figure 7b. The wider beams for ports 2 and 3 result in lower gain compared to its counterpart of port 1. An example placement of the proposed design with a commercial smartphone is demonstrated in Figure 8.

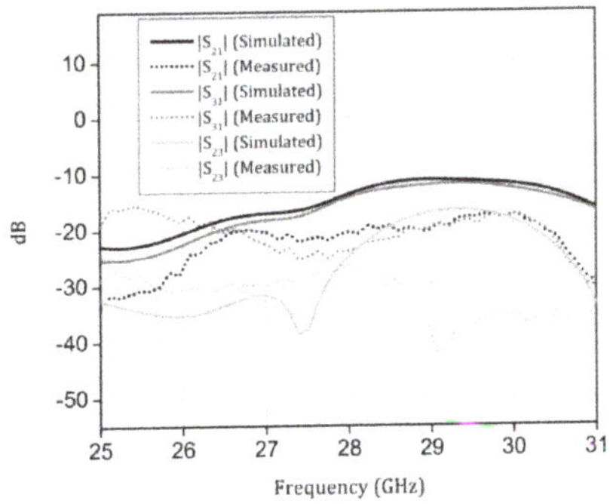

Figure 6. Mutual coupling between the ports.

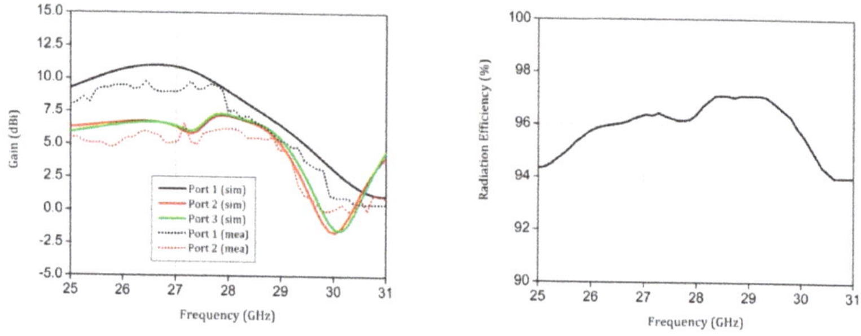

(a) (b)

Figure 7. (**a**) Gains of various ports of the shared radiator. (**b**) Radiation efficiency.

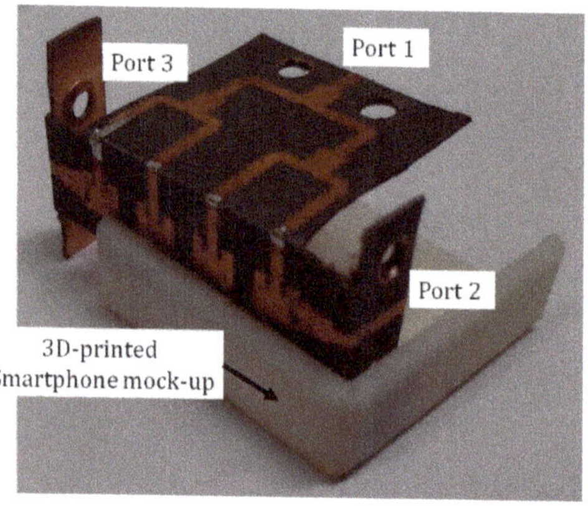

Figure 8. Example placement of the proposed design with a commercial smartphone.

Table 1 illustrates various figures of merit of the proposed dual conformal antenna system in comparison to previously published designs. It is clear that the proposed system has wide angular coverage with a panel height less than 7 mm.

Table 1. Comparison of the proposed antenna with previously reported articles.

Ref	Frq	AS	MC	AE	AC	ERV	GS	MI	Con	SG
[17]	28	15 × 12	NA	NA	70	0.138	Yes	No	No	Yes
[10]	28	45 × 15	<10	NA	90	0.207	Yes	No	No	Yes
[20]	28	20 × 20	<15	NA	80	0.05	No	No	No	Yes
[21]	28	42 × 12	<10	97	90	0.006	No	No	No	Yes
[22]	28	60 × 70	<7	90	NA	0.027	No	No	No	Yes
[23]	28	5 × 5	NA	68	90	0.005	Yes	Yes	No	Yes
[24]	28	14 × 12	NA	78	NA	0.026	No	No	No	Yes
[25]	26	22 × 11	<15	NA	NA	0.192	No	No	No	-NA-
PRW	**28**	**24 × 6.2**	**<10**	**97**	**100**	**0.07**	**Yes**	**Yes**	**Yes**	**Yes**

Ref = Reference, Frq = Frequency (GHz), AS = Antenna size (in mm × mm), MC = Mutual coupling (dB), ERV = Effective Radiating Volume (λ_0^3), AE = Antenna Efficiency (%), AC = Angular coverage (°), G = Gain (dBi), GS = Gain Switchability, MI = Mobile Integration, Con = Conformal, SG = Shared Ground, PRW = Proposed Work.

3. Conclusions

A three-port conformal antenna system operating in the 28 GHz is proposed, wherein a standard 50 Ω line is fed to the two-way power divider, which in turn is connected to the four-way power divider. The four-way power divider is loaded with four inset-fed patch antennas operating in the 28 GHz band. The corner elements of the corporate-fed array are in turn connected to the two ports through impedance transformers. Port 1 excitation would lead to a high-gain–narrow-beam mode preferable for a data link and ports 2 or 3 excitation would lead to a low-gain–wide-beam mode preferable for broadcast application. A wide angular coverage of 100° is achieved for various excitations of the ports within the shared radiator antenna system. Mutual coupling is less than 10 dB across the spectrum and across the ports, in spite of the electrically connected structure. The proposed antenna could be a potential candidate for future 5G applications.

Author Contributions: Conceptualization, K.G.S., S.K.K. and I.E.; methodology, K.G.S., I.E. and C.Z.; software, K.G.S. and C.Z.; validation, J.R., K.G.S. and I.E.; investigation, J.R. and K.G.S.; writing—original draft preparation, K.G.S. and S.K.K.; writing—review and editing, R.A.A.-A., I.E. and C.Z.; supervision R.A.A.-A. and I.E.; project administration, I.E. and J.R.; funding acquisition, J.R. All authors have read and agreed to the published version of the manuscript.

Funding: This work is supported by the Moore4Medical project, funded within ECSEL JU in collaboration with the EU H2020 Framework Programme (H2020/2014-2020) under grant agreement H2020-ECSEL-2019-IA-876190, and Fundação para a Ciência e Tecnologia (ECSEL/0006/2019).

Data Availability Statement: All data are included within manuscript.

Acknowledgments: This work is funded by the FCT/MEC through national funds and when applicable co-financed by the ERDF, under the PT2020 Partnership Agreement under the UID/EEA/50008/2020 project. This work is also partially funded by ECSEL JU in collaboration with the EU H2020 Framework Programme H2020-MSCA-RISE-2019-2023-872878.

Conflicts of Interest: The authors declare no conflict of interest.

References

1. Pi, Z.; Khan, F. An introduction to millimeter-wave mobile broadband systems. *IEEE Commun. Mag.* **2011**, *49*, 101–107. [CrossRef]
2. Wang, H.; Zhang, P.; Li, J.; You, X. Radio propagation and wireless coverage of LSAA-based 5G millimeter-wave mobile communication systems. *China Commun.* **2019**, *16*, 1–18. [CrossRef]
3. Sagazio, P.; Callender, S.; Shin, W.; Orhan, O.; Pellerano, S.; Hull, C. Architecture and Circuit Choices for 5G Millimeter-Wave Beamforming Transceivers. *IEEE Commun. Mag.* **2018**, *56*, 186–192. [CrossRef]
4. Rappaport, T.S.; Sun, S.; Mayzus, R.; Zhao, H.; Azar, Y.; Wang, K.; Wong, G.; Schulz, J.K.; Samimi, M.; Gutierrez, F. Millimeter Wave Mobile Communications for 5G Cellular: It Will Work! *IEEE Access* **2013**, *1*, 335–349. [CrossRef]
5. Roh, W.; Seol, J.Y.; Park, J.; Lee, B.; Lee, J.; Kim, Y.; Cho, J.; Cheun, K.; Aryanfar, F. Millimeter wave beamforming as an enabling technology for 5G cellular communications: Theoretical feasi-bility and prototype results. *IEEE Commun. Mag.* **2014**, *52*, 106–113. [CrossRef]
6. Friis, H.T. A Note on a Simple Transmission Formula. *Proc. IRE* **1946**, *34*, 254–256. [CrossRef]
7. Haraz, O.M.; Elboushi, A.; Alshebeili, S.A.; Sebak, A.-R. Dense Dielectric Patch Array Antenna with Improved Radiation Characteristics Using EBG Ground Structure and Dielectric Superstrate for Future 5G Cellular Networks. *IEEE Access* **2014**, *2*, 909–913. [CrossRef]
8. Bondarik, A.; Sjöberg, D. Gridded parasitic patch stacked microstrip array antenna for 60 GHz band. *IET Microw. Antennas Propag.* **2020**, *14*, 712–717. [CrossRef]
9. Zhang, S.; Chen, X.; Syrytsin, I.; Pedersen, G.F. Pedersen A planar switchable 3-D-coverage phased array antenna and its user effects for 28-GHz mobile terminal applications. *IEEE Trans. Antennas Propag.* **2017**, *65*, 6413–6421. [CrossRef]
10. Ta, S.X.; Choo, H.; Park, I. Broadband Printed-Dipole Antenna and Its Arrays for 5G Applications. *IEEE Antennas Wirel. Propag. Lett.* **2017**, *16*, 2183–2186. [CrossRef]
11. Moreno, R.M.; Kurvinen, J.; Ala-Laurinaho, J.; Khripkov, A.; Ilvonen, J.; van Wonterghem, J.; Viikari, V. Dual-Polarized mm-Wave End-Fire Chain-Slot Antenna for Mobile Devices. *IEEE Trans. Antennas Propag.* **2020**, *69*, 25–34. [CrossRef]
12. Agarwal, S.; Rafique, U.; Ullah, R.; Ullah, S.; Khan, S.; Donelli, M. Double Overt-Leaf Shaped CPW-Fed Four Port UWB MIMO Antenna. *Electronics* **2021**, *10*, 3140. [CrossRef]
13. Al-Gburi, A.J.A.; Ibrahim, I.B.M.; Zakaria, Z.; Ahmad, B.H.; Bin Shairi, N.A.; Zeain, M.Y. High Gain of UWB Planar Antenna Utilising FSS Reflector for UWB Applications. *Comput. Mater. Contin.* **2022**, *70*, 1419–1436. [CrossRef]
14. Khan, J.; Ullah, S.; Tahir, F.A.; Tubbal, F.; Raad, R. A Sub-6 GHz MIMO Antenna Array for 5G Wireless Terminals. *Electronics* **2021**, *10*, 3062. [CrossRef]
15. Arumugam, S.; Manoharan, S.; Palaniswamy, S.K.; Kumar, S. Design and Performance Analysis of a Compact Quad-Element UWB MIMO Antenna for Automotive Communications. *Electronics* **2021**, *10*, 2184. [CrossRef]
16. Stutzman, W.L.; Thiele, G.A. *Antenna Theory and Design*; John Wiley & Sons: Hoboken, NJ, USA, 2012.

17. Yang, B.; Yu, Z.; Dong, Y.; Zhou, J.; Hong, W. Compact Tapered Slot Antenna Array for 5G Millimeter-Wave Massive MIMO Systems. *IEEE Trans. Antennas Propag.* **2017**, *65*, 6721–6727. [CrossRef]
18. Koul, S.K.; Poddar, A.K.; Sadananda, K.G.; Rohde, U.L. Conformal Antenna Module With 3D-Printed Radome. U.S. Patent Application 17/221,965, 9 December 2021.
19. Koul, S.K. *Antenna Architectures for Future Wireless Devices*; Springer: Singapore, 2021. [CrossRef]
20. Wani, Z.; Abegaonkar, M.P.; Koul, S.K. Millimeter-wave antenna with wide-scan angle radiation characteristics for MIMO ap-plications. *Int. J. RF Microw. Comput.-Aided Eng.* **2018**, *29*, e21564. [CrossRef]
21. Shim, J.-Y.; Go, J.-G.; Chung, J.-Y. A 1-D Tightly Coupled Dipole Array for Broadband mmWave Communication. *IEEE Access* **2019**, *7*, 8258–8265. [CrossRef]
22. Ikram, M.; Al Abbas, E.; Nguyen-Trong, N.; Sayidmarie, K.H.; Abbosh, A. Integrated Frequency-Reconfigurable Slot An-tenna and Connected Slot Antenna Array for 4G and 5G Mobile Handsets. *IEEE Trans. Antennas Propag.* **2019**, *67*, 7225–7233. [CrossRef]
23. Hwang, I.-J.; Ahn, B.; Chae, S.-C.; Yu, J.-W.; Lee, W.-W. Quasi-Yagi Antenna Array with Modified Folded Dipole Driver for mmWave 5G Cellular Devices. *IEEE Antennas Wirel. Propag. Lett.* **2019**, *18*, 971–975. [CrossRef]
24. Hasan, M.N.; Bashir, S.; Chu, S. Dual band omnidirectional millimeter wave antenna for 5G communi-cations. *J. Electromagn. Waves Appl.* **2019**, *33*, 1581–1590. [CrossRef]
25. Pan, Y.M.; Qin, X.; Sun, Y.X.; Zheng, S.Y. A Simple Decoupling Method for 5G Millimeter-Wave MIMO Dielectric Resonator Antennas. *IEEE Trans. Antennas Propag.* **2019**, *67*, 2224–2234. [CrossRef]

electronics

Article

A Multi-Slot Two-Antenna MIMO with High Isolation for Sub-6 GHz 5G/IEEE802.11ac/ax/C-Band/X-Band Wireless and Satellite Applications

Abdullah G. Alharbi [1], Jayshri Kulkarni [2,*], Arpan Desai [3], Chow-Yen-Desmond Sim [4] and Ajay Poddar [5]

1 Department of Electrical Engineering, Faculty of Engineering, Jouf University, Sakaka 42421, Saudi Arabia; a.g.alharbi@ieee.org
2 Department of Electronics and Telecommunication Engineering, Vishwakarma Institute of Information Technology, Kondhwa, Pune 411048, India
3 Department of Electronics and Communication Engineering, CSPIT, Charotar University of Science and Technology (CHARUSAT), Changa 388421, India; arpandesai.ec@charusat.ac.in
4 Department of Electrical Engineering, Feng Chia University, Taichung 407, Taiwan; cysim@fcu.edu.tw
5 Synergy Microwave Corporation, 201 McLean Boulevard, Paterson, NJ 07504, USA; poddar_ajay@yahoo.com
* Correspondence: jayah2113@gmail.com; Tel.: +91-9552466969

Abstract: A tapered symmetrical coplanar waveguide (S-CPW) fed monopole antenna is initially studied. To achieve multiband characteristics, the radiating element of this monopole antenna is loaded with multiple narrow slots and multiple slotted stubs (MSS). The designed slot-loading monopole is further transformed into a two-antenna MIMO type with a gap distance of only 0.12λ (at 5 GHz), and thus it has a small overall size of 32 × 20 × 0.8 mm^3. By deploying five concentric ring elements between the two adjacent antenna elements, the desirable isolation of better than 20 dB is yielded. As the low band and high band operation of the proposed two-antenna MIMO is 81.08% (3.3–7.8 GHz) and 40% (8.0–12.0 GHz), respectively, it can therefore satisfy the Sub-6 GHz 5G New Radio (NR) n77/78/79, IEEE 802.11ac/ax, X-band/C-band wireless and satellite applications. Furthermore, it has shown a desirable gain of above 3 dBi and a radiation efficiency greater than 69% throughout the two bands of interest.

Keywords: monopole antenna; two-antenna MIMO; slot loaded; Sub-6 GHz; IEEE 802.11ac/ac; X-band; C-band

Citation: Alharbi, A.G.; Kulkarni, J.; Desai, A.; Sim, C.-Y.-D.; Poddar, A. A Multi-Slot Two-Antenna MIMO with High Isolation for Sub-6 GHz 5G/IEEE802.11ac/ax/C-Band/X-Band Wireless and Satellite Applications. *Electronics* **2022**, *11*, 473. https://doi.org/10.3390/electronics11030473

Academic Editors: Faisel Tubbal, Ladislau Matekovits and Raad Raad

Received: 15 January 2022
Accepted: 4 February 2022
Published: 5 February 2022

1. Introduction

The main challenging part of designing a compact size MIMO antenna is to obtain a very high isolation of >20 dB between the two adjacent antenna elements without affecting the scattering, radiation, and diversity performances. Therefore, several decoupling techniques have been reported in recent years for compact size planar MIMO antennas with multi-band or broadband operations [1–21]. In [1], a two-antenna MIMO with frequency reconfigurable characteristics is designed using radio frequency micro electro mechanical systems (RF-MEMS) switches. This antenna has a designed footprint of 32 × 98 × 1 mm^3, and it can switch among the 0.6, 1.8, 2.4, 3.5, and 5.5 GHz bands with isolation >15 dB. However, the use of the RF-MEMS switch makes the antenna design more complex. To reduce the complexity, Ref. [2] has proposed two symmetrically located spider-shaped antennas (of size 37 × 56 × 1.6 mm^3) that can operate in the Wi-Fi/WiMAX/Bluetooth and C-band applications. However, it does not cover the entire 5 GHz band. A dual band, a Two-port antenna functioning in a 2.4 GHz and 5 GHz band with a dimension of 46 × 20 × 1.6 mm^3, is proposed in [3], but the isolation is only around 12 dB. Therefore, a Wang shaped triple-band MIMO antenna with a high isolation of 31 dB is reported in [4], but this antenna has occupied a large area of 70 × 52 mm^2. To achieve smaller dimensions for the MIMO antenna, Ref. [5] has designed a two-antenna MIMO with a planar size of only 24 × 20 mm^2.

Here, a single complementary split-ring resonator (S-CSRR) is loaded into the radiating element to enhance the performance, as well as operating in the X- and Ku-band. In [6], a two-port MIMO monopole antenna with a small size of 27 × 16 × 0.8 mm^3 is reported. This antenna is a single band operation (4.85–7.32 GHz), and it has shown good isolation of >15 dB [6]. To achieve dualband operation with higher isolation, a dualband MIMO antenna (size 27 × 21 × 1.6 mm^3) using orthogonal polarisation has been reported [7]. Even though it can yield very high isolation of >22.5 dB, the two operating bands are rather narrow at 4.15% (5.19–5.41 GHz) and 4.81% (7.30–7.66 GHz). To further achieve triple-band operation, a MIMO antenna with a compact size of 20 × 14.75 mm^2 has been reported in [8]. The antenna element is composed of an asymmetric coplanar strip (ACS) feedline, along with an inverted L-shaped slot and a meander line to operate in the WLAN (2.4/5.2/5.8 GHz) and WiMAX (2.5/3.5/5.5 GHz) applications. Notably, all the MIMO antennas reported in [1–8] do not apply any decoupling structure between adjacent antenna elements. Hence, they will suffer from low isolation if a wider operating bandwidth is implemented.

The easiest way to reduce the mutual coupling of any two adjacent antenna elements is to apply the spatial diversity method. The advantage of this method is that the MIMO antenna does not require any decoupling structure, such as the one reported in [9] that has a gap distance of 36 mm (>1/4 wavelength) between the two adjacent antennas, and good isolation of >30 dB is achieved. However, applying the spatial diversity method will result in occupying more space and, hence, an increased dimension. To reduce the mutual coupling without occupying much gap distance between the antenna elements, several MIMO antenna designs with different decoupling structures, such as the U-shaped slots [10], T-shape stub [11], comb-shaped structures [12], modified T-shape stub at ground plane [13], and defected ground structures [14–17], have been reported. However, the volume size of [10–17] is between (20 × 35 × 0.8 mm^3, 560 mm^3) and (50 × 50 × 7 mm^3, 17,500 mm^3), and none of these MIMO designs exhibit a very wide operational bandwidth (>80%) and high isolation >20 dB for modern wireless applications with a physical volume size of below or near 500 mm^3.

To achieve high isolation without increasing the overall volume of the MIMO antenna, recent decoupling techniques such as the loading of the EBG structure [18,19] and metamaterials [20,21] have been widely used. In [18], the fractal EBG technique is applied to enhance the isolation to 24.67 dB, but the two-antenna MIMO has a volume size of 38.2 × 95.94 × 1.6 mm^3, and it covers only a narrow single ISM band from 2.39 to 2.48 GHz. Even though the volume size of the antenna in [19] (55 × 28 × 1.6 mm^2) is much less than that of [18], and it has exhibited a wider operating bandwidth of 64.42% (2.01–3.92 GHz), one can see that it has exhibited isolation of only >15 dB after applying the EBG structure. In [20], the textile-based antenna has applied a unique metamaterial inspired decoupling network to enhance the isolation between adjacent antenna elements. The two bands of interest, in this case, are 56% (1.34–3.92 GHz) and 37.4% (4.34–6.34 GHz), and isolation of >18 dB was measured. However, this antenna has a very large dimension of 100 × 60 × 1 mm^3. To achieve a small volume size of 47.5 × 40 × 1.6 mm^3, [21] has proposed a single band, two-antenna MIMO monopole that has applied a metamaterial split ring resonator unit to improve the isolation. However, the antenna is a single band operation with a 10-dB impedance bandwidth of 12.3% (3.35–3.78 GHz) and the isolation is only 15 dB.

From the study of literature [1–21], it is observed that the reported MIMO antennas can yield high isolation, but with a narrow operational band or vice-versa. Furthermore, many of them have required a very large volume size. Therefore, in this paper, a compact volume size (32 × 20 × 0.8 mm^3, 512 mm^3) two-antenna monopole MIMO with a dual wideband operation (3.3–7.8 GHz and 8.0–12.0 GHz), as well as high isolation (>20 dB) for the wireless and satellite applications is proposed. The proposed two-antenna MIMO functions in the Sub-6 GHz 5G NR n77 (3.3–4.2 GHz)/n78 (3.3–3.8 GHz)/n79 (4.4–5.0 GHz), standard C-band uplink (3.7–4.2 GHz)/downlink (5.92–6.425 GHz), extended C-band uplink (3.4–3.7 GHz)/downlink (6.425–6.725 GHz), 802.11ac (5.15–5.85 GHz)/802.11ax

(5.92–7.125 GHz), and X-band radar (8.0–12.0 GHz). Therefore, an integration of the Sub-6 GHz 5G, IEEE 802.11ac/ax, X-band, and C-band into a single antenna along with MIMO configuration would be a good candidate for future wireless and satellite applications. This manuscript is organized as: Sections 2 and 3 explain the geometry and evolution of the monopole element and its characteristics, Section 4 analyses the geometry layout of the two-antenna MIMO along with its working principles. Sections 5 and 6 discuss the results and diversity performances of the proposed two-antenna MIMO. Section 7 compares the proposed two-antenna MIMO with the pioneering state of arts and finally, Section 8 concludes the paper.

2. Design Layout of Single Monopole Antenna Element

The design layout and geometry of the single monopole antenna element that will be further applied for realising the proposed two-antenna monopole MIMO are depicted in Figure 1. Here, the radiating section (rectangular radiator) of the monopole antenna is loaded with two narrow L-shaped slots, an inverted-U narrow slot, and two symmetrical MSS (deployed on the right and left side of the rectangular radiator). To achieve better impedance matching, a two-steps co-planar waveguide (CPW) feeding line (of width 1.5 mm and 2 mm) is applied, and the tapered structure that is loaded at the bottom of the rectangular radiator (linked to the feeding line of the CPW) is for yielding a wider operating bandwidth. As seen from Figure 1, the two co-planar ground planes have an area of 4 × 7 mm^2 each, and it is slightly truncated, with an air gap of 0.5 mm due to the two-steps CPW feeding line. The antenna structure is engraved on a 0.8 mm FR-4 substrate having a relative permittivity (εr) of 4.3 and a loss tangent ($tan\delta$) of 0.025. The overall planar size of this single monopole antenna element is 20 × 13 mm^2.

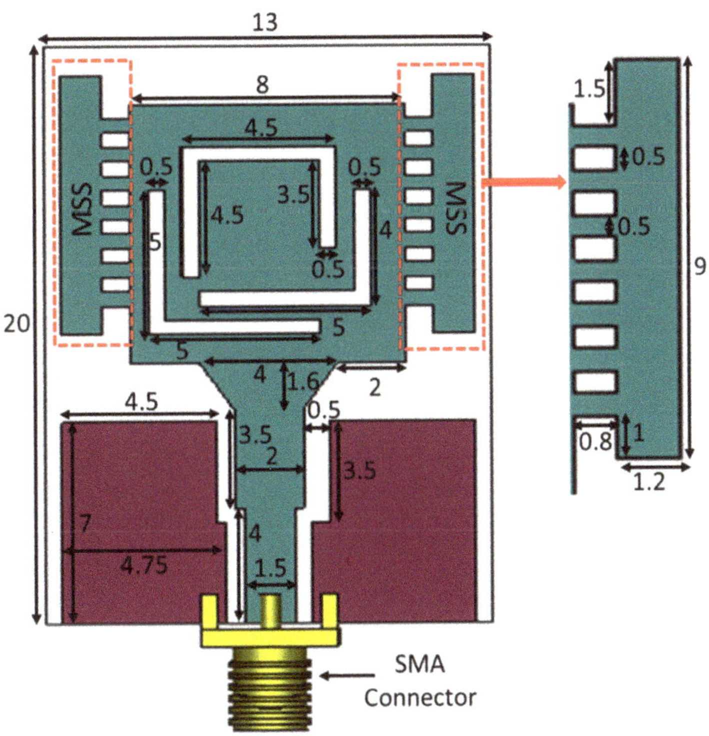

Figure 1. Layout and geometry of single monopole antenna element (all units in mm).

3. Antenna Evolution Mechanism

To comprehend the excitation of the two wide operational bandwidths from the single monopole antenna element, this section explains the antenna evolution mechanism including the step-wise design and its associated reflection coefficient curve.

3.1. Step-1: Design of Rectangular Radiator

The antenna design begins with a two-step CPW-fed rectangular radiator with a tapered structure at the bottom, as illustrated in Figure 2a, and it is denoted as ANT#1. Here, ANT#1 was analysed and numerically simulated using the CST Microwave Studio® (CST MWS) software. The width and length of the rectangular radiator is optimised using the CST MWS.

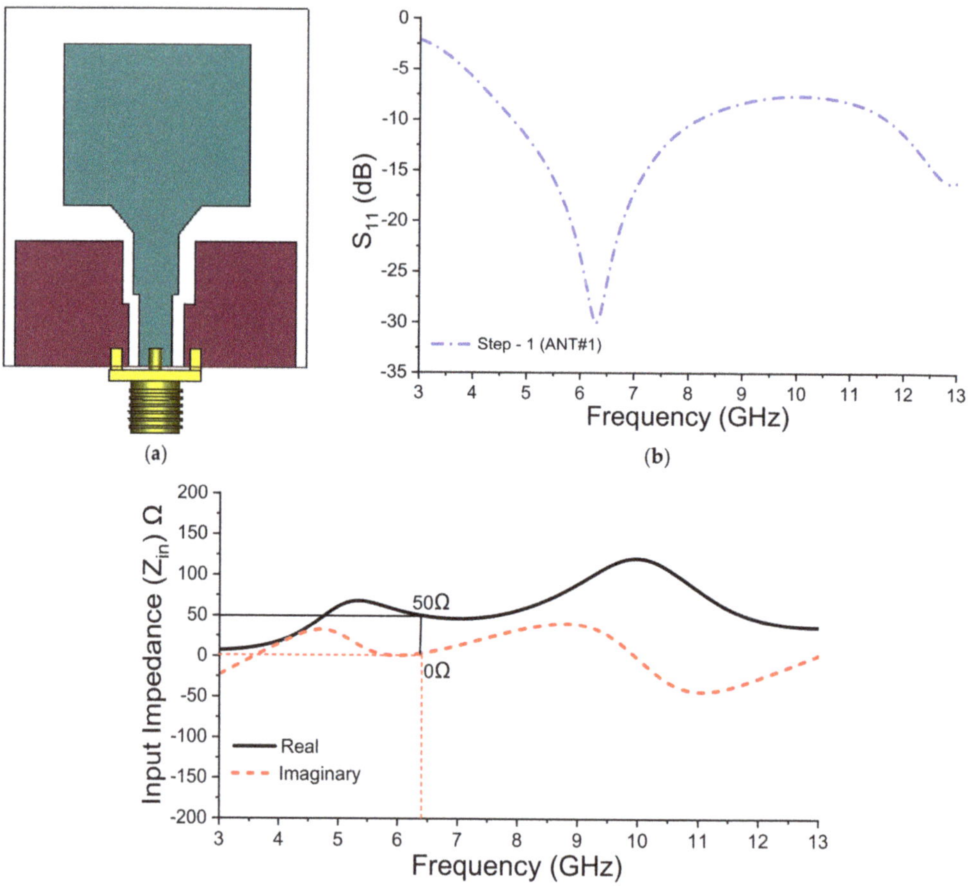

(a) (b) (c)

Figure 2. Step-1 of single monopole antenna element, (**a**) ANT#1 structure, (**b**) S_{11}, (**c**) input impedance.

From Figure 2b, it is clearly observed that ANT#1 has successfully induced a wideband operation with a 10-impedance bandwidth of 4.74–8.10 GHz (centred at 6.3 GHz). Thus, it can meet the wideband demand for X- and C-band applications. Figure 2c validates that the incorporation of a rectangular patch, truncated ground planes and a tapered structure offers equal amounts of inductive and capacitive reactance at a resonating frequency of 6.3 GHz. Hence, achieving good impedance matching throughout the operating band (4.74–8.10 GHz).

3.2. Step-2: Deployment of MSS on Right and Left Side of ANT#1

As can be seen in Figure 2c, the frequency band generated by ANT#1 is not enough to cover the standard and extended C-band. Therefore, to shift the frequency band towards the lower spectrum and to match the load impedance of 100 Ω to the 50 Ω feed line, a series open circuit MSS is embedded into the left and right side of the ANT#1 (which forms the ANT#2), without disturbing the feeding arrangement of ANT#1, as shown in Figure 3a. Notably, there are six small slots (each has a size of 0.8×0.5 mm^2) in each MSS (see Figure 1), and these slots are acting as capacitors and behave like an open circuit to block the high inductive reactance across the operating frequency range.

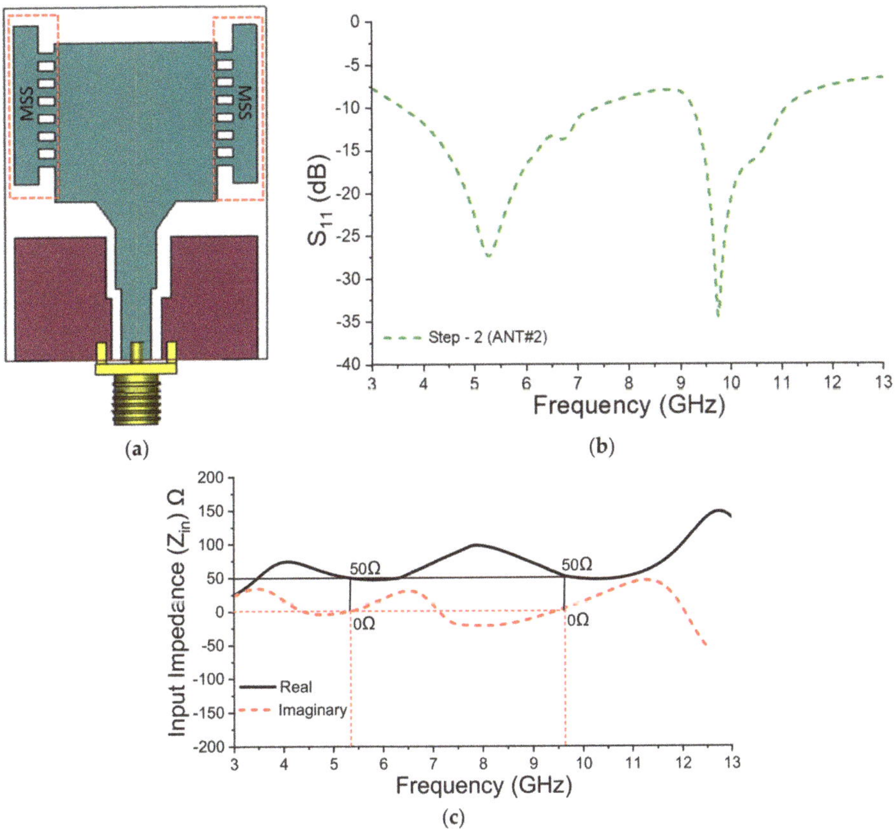

Figure 3. Step-2 of single monopole antenna element, (**a**) ANT#2 structure, (**b**) S_{11}, (**c**) input impedance.

From the reflection coefficient curve of Figure 3b, it is visualised that the incorporation of the two symmetrical MSS into ANT#1 (that forms the ANT#2) can shift the previous frequency mode (6.3 GHz) towards the lower spectrum at approximately 5.3 GHz, and a good 10-dB impedance bandwidth of approximately 3.4–7.0 GHz (low band) is achieved. Meanwhile, ANT#2 can also generate another high band operation with 10-dB impedance bandwidth of 9.4–11.1 GHz. Therefore, ANT#2 is able to operate in the IEEE 802.11ac, C-band and partial X-band applications.

To comprehend the contributions of the MSS that achieves good impedance matching across the two bands of interest, Figure 3c depicts the input impedance Zin (Ω) diagram of ANT#2. Here, one can clearly see that good impedance matching has been achieved throughout the two operating bands of 3.4–7.0 GHz and 9.4–11.1 GHz, as their resistive impedances are very much closer to 50 Ω and their corresponding reactive impedances are near 0 Ω. This validates that the two symmetrical MSS can aid in achieving good impedance matching across the low band and exciting a new high band.

3.3. Step-3: The Loading of L-Shaped Slots and Inverted U-Shaped Slot

To further integrate the Sub-6 GHz NR 5G bands along with the IEEE 802.11 ac/ax, C-band, and X-band into a single antenna simultaneously, ANT#3 (as shown in Figure 4a) is developed by further loading 3 narrow slots, namely, two L-shaped slots and an inverted U-shaped slot. When these slots are loaded into the rectangular radiator, they get interlocked with each other and aid in reducing the capacitive reactance. Moreover, these slots also help to widen the previous frequency band by forcing the current distribution on the surface of the radiator to flow for a longer time period, as well as diverting the current paths to flow in various directions with various velocities. This results in the merging of all the currents coming from various directions with different velocities and offers a wide bandwidth, which can be seen in Figure 4b that plots the current distributions at 5 GHz for ANT#1 and ANT#3. To illustrate the wideband characteristics of ANT#3, its corresponding reflection coefficient is plotted in Figure 4c. It is clearly seen that ANT# 3 is able to yield a dual bandwidth operation spanning in the range of 81.08% (3.3–7.8 GHz) and 40% (8.0–12.0 GHz) at the resonant frequency of 5.0 GHz and 10.5 GHz, respectively. Figure 4d shows the input impedance diagram of ANT#3. Here, one can see excellent impedance matching throughout the two bands of interest, in which the resistive and reactive impedances are around 50 Ω and near 0 Ω across the two bands, respectively. Therefore, ANT#3 is considered for further analysis and will be applied for MIMO applications. The step-wise configuration of ANT#1 to ANT#3 and their associated operating bands are mentioned in Table 1.

Table 1. Step-wise configuration and operating bands of ANT#1 to ANT#3.

Step	Sub-6 GHz 5G NR	IEEE 802.11ac	IEEE 802.11ax	C-Band	X-Band	Impedance Matching
ANT#1	-	Yes	Yes	-	-	Good
ANT#2	-	Yes	-	Yes	-	Good
ANT#3	Yes	Yes	Yes	Yes	Yes	Good

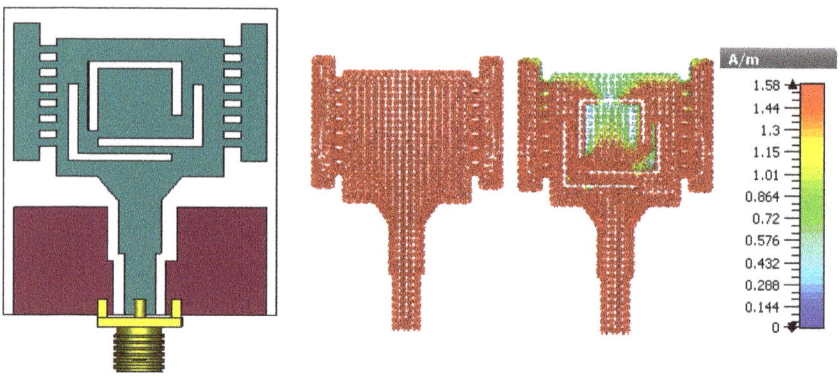

(a) (b)

Figure 4. *Cont.*

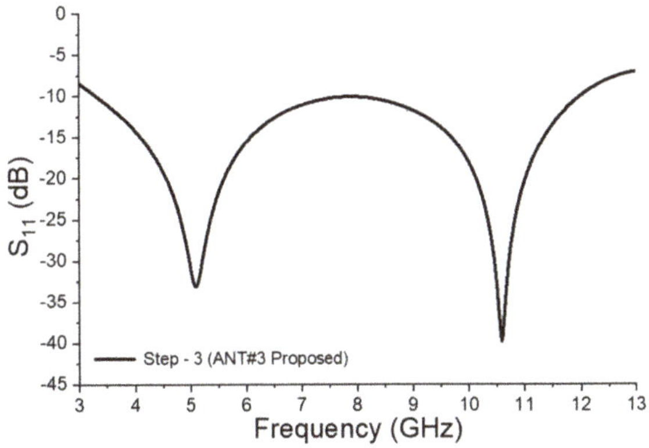

(c)

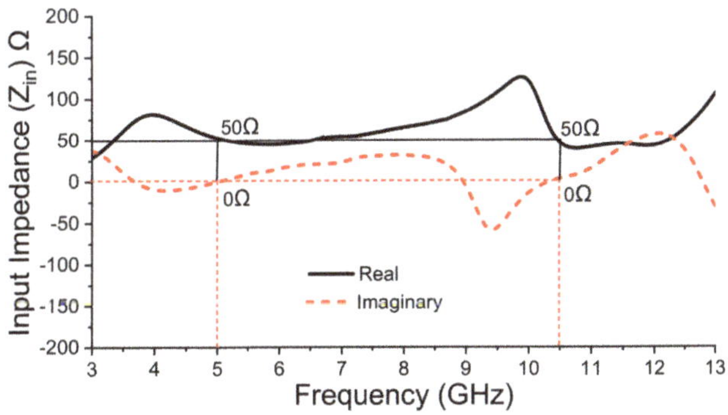

(d)

Figure 4. Step-3 of single monopole antenna element, (**a**) proposed monopole antenna element, (**b**) current distributions at 5 GHz, (**c**) S_{11}, (**d**) input impedance.

4. Geometry, Design and Analysis of the Proposed Two-Antenna MIMO

Figure 5 depicts the geometry of the proposed two-antenna MIMO without and with loading the decoupling structure. As seen from Figure 5a, two identical monopole antenna elements (ANT#3) are closely deployed side by side of each other, with a gap distance of only 0.12λ (λ is the free space wavelength at 5 GHz), and it is much narrower than the one reported in [9], with a gap distance of >1/4λ. Nevertheless, this gap distance can still provide enough space for the deployment of a decoupling structure (5 concentric ring elements), as shown in Figure 5b. Furthermore, this narrow gap distance can also ensure that the desired antenna and MIMO diversity performances remain unaffected in a rich multipath fading environment.

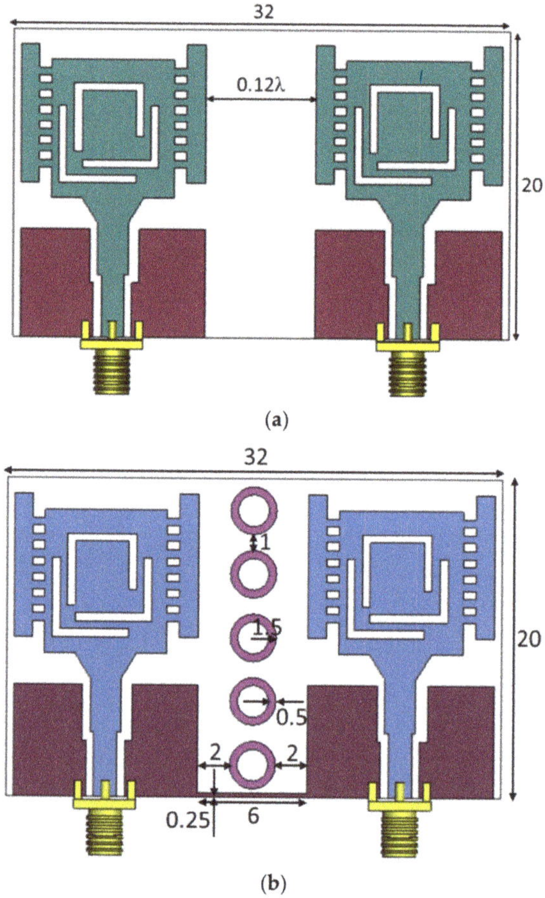

Figure 5. Geometry of the proposed two-antenna MIMO, (**a**) without decoupling structure, (**b**) with decoupling structure.

4.1. Analysis of the Two-Antenna MIMO without Decoupling Structure

Figure 6 plots the simulated reflection coefficient (S_{11} and S_{22}) and isolation (S_{12} and S_{21}) curves of the proposed two-antenna MIMO without loading the decoupling structure. Here, both the S_{11} and S_{22} are almost identical, and they have shown wide 10-dB impedance bandwidths of 81.08% (3.3–7.8 GHz) and 40% (8.0–12.0 GHz). However, the isolation level (S_{12} or S_{21}) between the two antenna elements at around 4 GHz is approximately 10 dB, which is undesirable because, as per the requirement of industry and IEEE standards, the minimum isolation between two adjacent antenna elements should be greater than 15 dB so that each antenna element will produce independent communication paths, resulting in a higher data rate as well as uninterrupted internet access, wireless and satellite services.

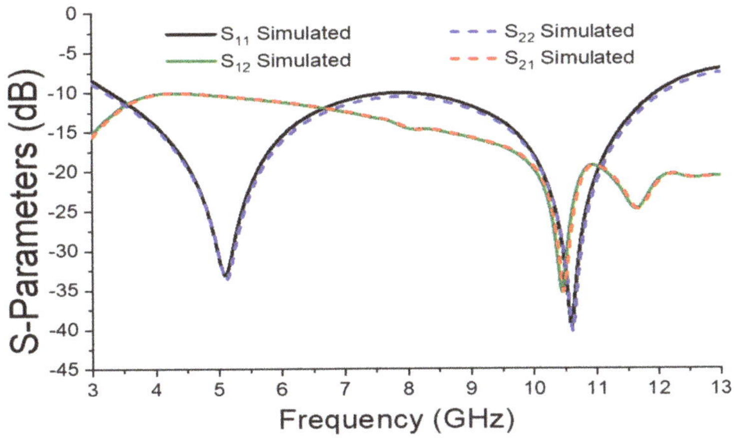

Figure 6. Simulated S-parameters of the proposed two-antenna MIMO without loading the decoupling structure.

Figure 7 shows the electric field intensity (V/m) distribution across the proposed two-antenna MIMO at a resonant frequency of 5GHz. As shown in Figure 7a, when antenna element#1 is excited (while antenna element#2 is terminated by a 50 Ω load impedance), one can see that the antenna element#2 is highly influenced (coupled) by the strong electric field generated from the antenna element#1. It is also validated when antenna element#2 is excited, while antenna element#1 is terminated with a 50 Ω load impedance, as seen in Figure 7b. This clearly indicates that the antenna elements are invincibly mutually coupled with each other.

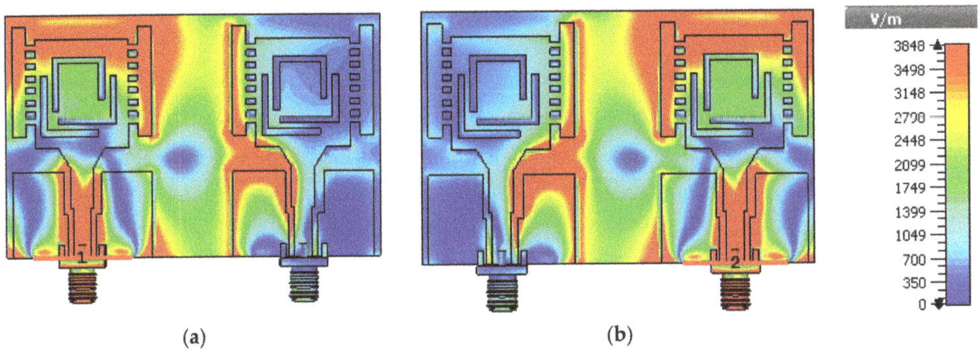

Figure 7. Electric field intensity (V/m) of the proposed two-antenna MIMO without decoupling structure, (**a**) antenna element#1 excited at 5GHz, (**b**) antenna element#2 excited at 5GHz.

4.2. Analysis of the Proposed Two-Antenna MIMO with Decoupling Structure

To reduce the mutual coupling between the two antenna elements, a decoupling structure is loaded between the antenna elements (within the 0.12λ gap distance), as shown in Figure 5b, in which the decoupling structure is comprised of five concentric ring elements. In this case, each ring has an inner and outer radius of 1 mm and 1.5 mm, respectively, and they are printed vertically with a gap distance of 1 mm away from each other. Furthermore, a very thin strip of 0.25 mm in height is added in between the antenna elements in order to connect the ground planes. By further observing the reflection coefficient curves in Figure 8, it is apparent that the deployment of the decoupling structure, as well as connecting the ground plane, does not affect the impedance bandwidth (as seen from the S_{11}/S_{22} curves) of both antenna elements. Notably, the decoupling structure has significantly enhanced the isolation level between the two antenna elements, and a level of >20 dB (seen from S_{12}/S_{21} curves) is observed which was verified by applying two plane waves across the length of the decoupling structure [22]. The isolation of below −20 dB is achieved as the equidistance spaced circles act as reflectors and absorb the surface wave which, therefore, serves as a wide stop band filter and nullifies the surface wave propagation between Ant. 1 and Ant. 2 [23].

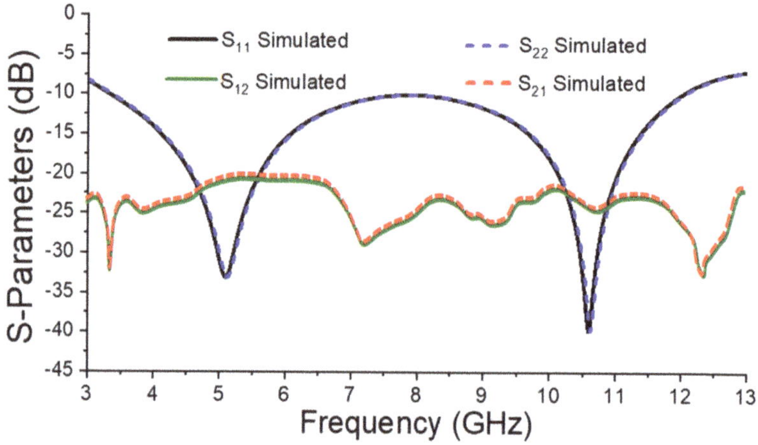

Figure 8. S-parameters of two-port MIMO Antenna with decoupling structure.

To further comprehend the validation of the isolating structure, electric field distribution (V/m) analysis at a resonant frequency of 5GHz is shown in Figure 9, where it can be easily seen that the antenna element#2 is being prevented from the strong electric field of antenna element#1, which is due to the five concentric ring elements that act as a band stop filter to attenuate the correlated signals coming out from antenna element#1. The same phenomenon is also shown when antenna element#2 is excited, while antenna element#1 is terminated with a 50 Ω load impedance.

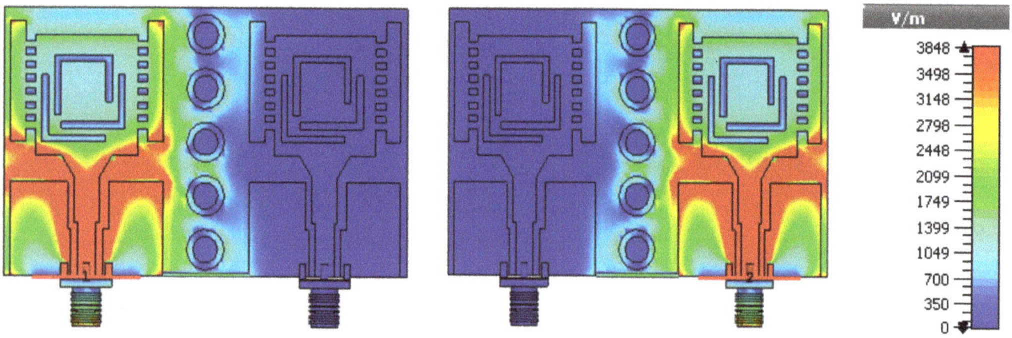

(**a**) (**b**)

Figure 9. Electric field intensity (V/m) of the proposed two-antenna MIMO with decoupling structure, (**a**) antenna element#1 excited at 5GHz, (**b**) antenna element#2 excited at 5 GHz.

5. Results and Discussion of the Proposed Two-Antenna MIMO

To implement the proposed two-antenna MIMO for wireless and satellite applications functioning in Sub-6 GHz 5G NR, 802.11ac/ax, C-band, and X-band, a prototype is manufactured and depicted in Figure 10. The scattering, radiating and diversity performances of the fabricated proposed prototype are verified with simulated performances, and they are discussed in the below sub-sections. Notably, when analysing the performances of the two-antenna MIMO, only one antenna element is excited, whereas the other antenna element is terminated with a 50 Ω load impedance. The Rohde & Schwarz ZNH18 network analyser, having a frequency range of 30 kHz to 18 GHz, was used for measuring S-parameters, whereas the anechoic chamber was used to measure the radiation characteristic of the proposed antenna.

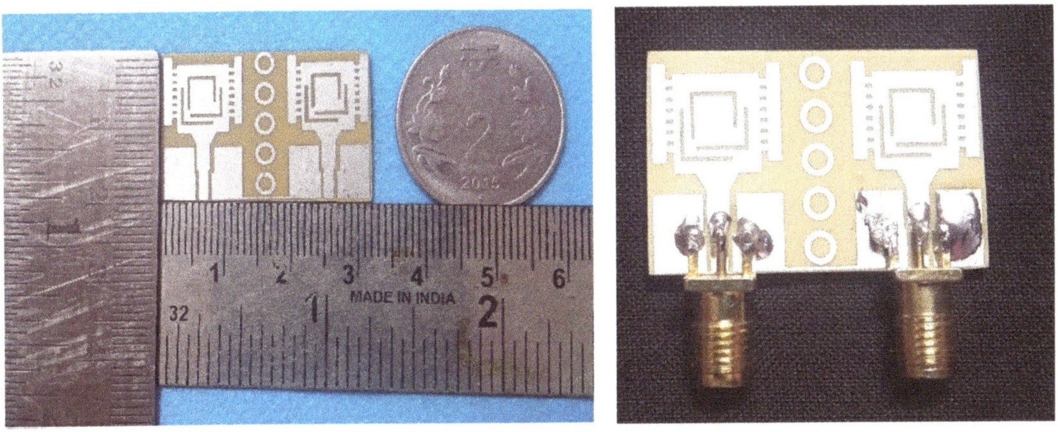

Figure 10. Fabricated prototype of the proposed two-antenna MIMO.

5.1. Simulated and Measured Reflection Coefficient and Isolation

As the reflection coefficient (S_{11}) and isolation (S_{12}) are analogous to S_{22} and S_{21}, respectively, Figure 11 only shows the simulated and measured S_{11} and S_{12} characteristics of the proposed two-antenna MIMO. In this figure, both simulated and measured results are well-validated with each other, and the observed deviation (especially in the high band) may be due to manufacturing tolerances and minor fabricating errors. Nevertheless, a dualband operation is clearly shown in Figure 11a, in which the measured low band and high band operation have exhibited a wide 10-dB impedance bandwidth of 80.54% (3.3–7.75 GHz) and 41.2% (7.9–12.0 GHz), respectively, while isolation larger than 20 dB was achieved across the two operating bands of interest, as seen in Figure 11b.

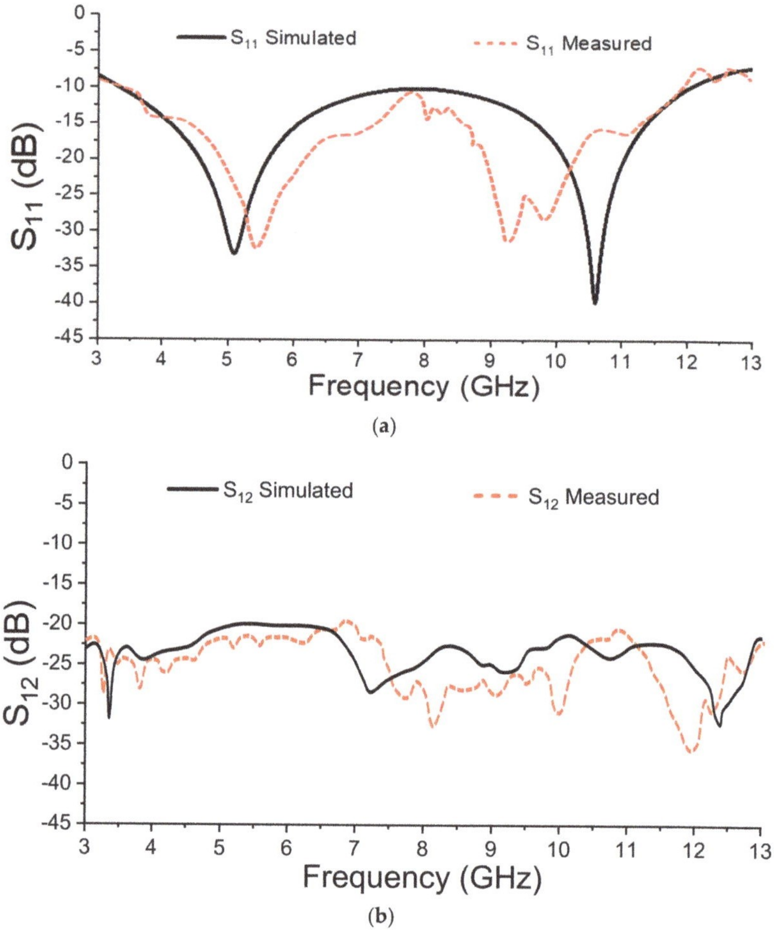

Figure 11. Simulated and measured S-parameters of the proposed two-antenna MIMO, (**a**) reflection coefficient, S_{11}, (**b**) isolation, S_{12}.

5.2. Simulated and Measured Far Field Radiation Patterns

The radiation patterns of the proposed two-antennas MIMO across the E-plane and H-plane are depicted in Figure 12a–d. As shown in Figure 12a,b, across the two bands of interest, 5.0 GHz and 10.5 GHz, the two antenna elements at the E-plane are exhibiting near-omnidirectional patterns and 8-shaped patterns for the co-polarisation (co-polar) and cross-polarisation (cross-pol) radiation, respectively. Furthermore, it is also noteworthy that at both frequencies, the radiation patterns of antenna element#1 are exactly mirror images of the ones shown in antenna element#2. As for its H-plane counterparts, as shown in Figure 12c,d, across the two bands of interest, 5.0 GHz and 10.5 GHz, respectively, the two antenna elements at the H-plane are exhibiting bi-directional patterns (co-polar) and broadside patterns (cross-pol). The above results demonstrated that the proposed two-antenna MIMO has offered acceptable radiation characteristics to meet the desire MIMO diversity performances.

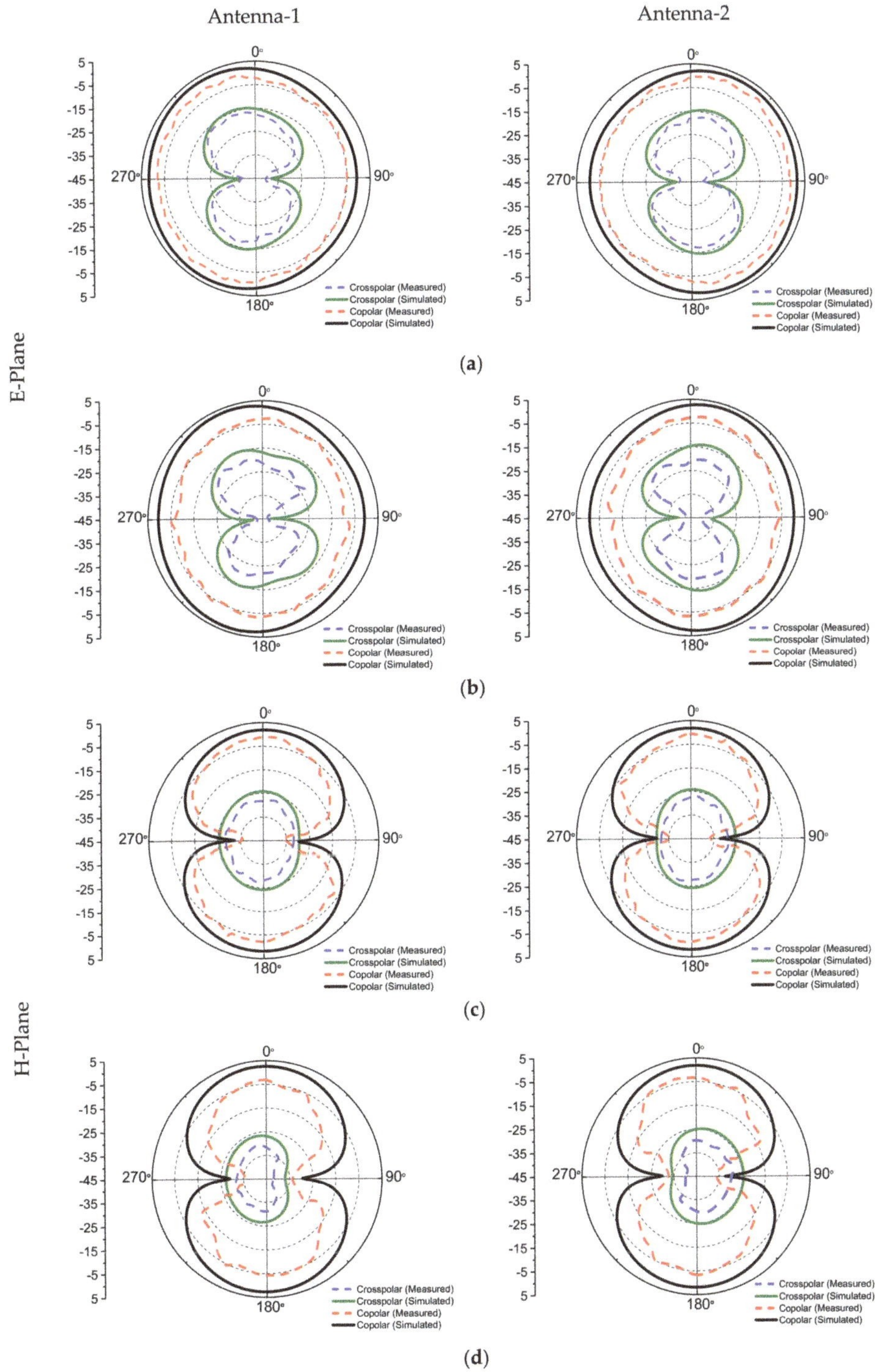

Figure 12. Radiation patterns of antenna element#1 and antenna element#2, (**a**) 5.0 GHz, E-plane, (**b**) 10.5 GHz, E-plane, (**c**) 5.0 GHz, H-plane, (**d**) 10.5 GHz, H-plane.

5.3. Simulated and Measured Realised Gain and Radiation Efficiency

As the two antenna elements are identical to each other, we only plot the gain and efficiency of antenna element#1, and they are illustrated in Figure 13. In this figure, between 3.5 GHz and 12 GHz, the simulated gain was 3.0–4.66 dBi, while the measured one was 2.65–4.00 dBi. As for its corresponding radiation efficiency, the simulated one was ranging from 72.5% to 82%, while the measured one was between 69.64% and 76.8%. Thus, the proposed two-antenna MIMO has exhibited stable gain and efficiency throughout the operating bands of interest. The simulated and measured peak gain and peak efficiency are compared and presented in Table 2, and one can see that these results can ensure good quality of communication.

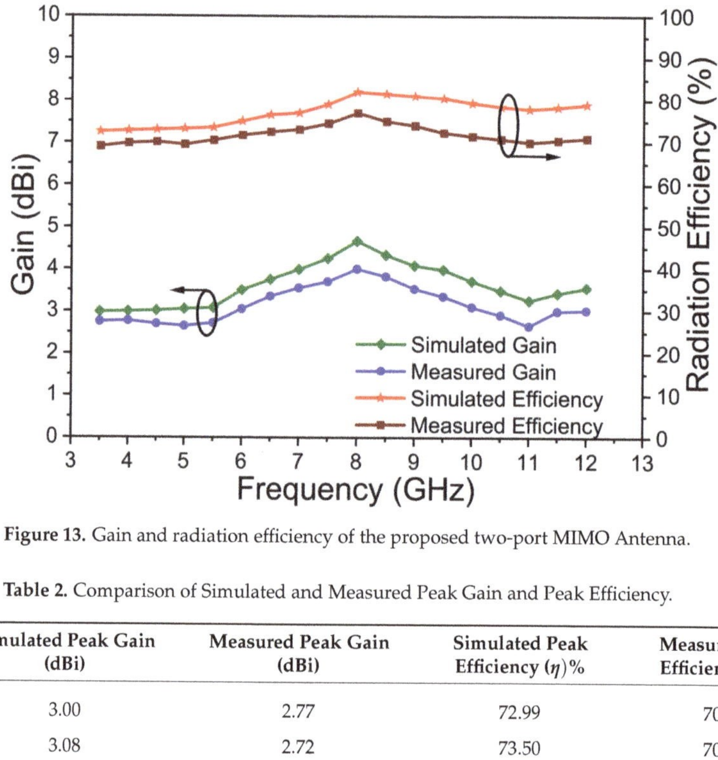

Figure 13. Gain and radiation efficiency of the proposed two-port MIMO Antenna.

Table 2. Comparison of Simulated and Measured Peak Gain and Peak Efficiency.

Functioning Band	Simulated Peak Gain (dBi)	Measured Peak Gain (dBi)	Simulated Peak Efficiency (η)%	Measured Peak Efficiency (η)%
Sub-6 GHz 5G NR & C-band uplink	3.00	2.77	72.99	70.10
IEEE 802.11ac	3.08	2.72	73.50	70.50
IEEE 802.11ax & C-band downlink	3.75	3.35	76.50	72.49
X-band	3.48	2.91	78.48	70.80

6. Diversity Performance Analysis

To prove the potency of the proposed two-antenna MIMO, the diversity performances matrix, such as ECC, DG, MEG, CCL, channel capacity, and TARC, are very essential, and thus they are further verified through simulation as well as measurement.

6.1. Envelope Correlation Coefficient (ECC)

To know how independently antenna elements radiate throughout the operating band, the ECC is a very essential parameter to be investigated. Ideally, the ECC value should be equal to zero, which indicates that the antenna elements radiate independently by producing uncorrelated radiations. However, in a rich fading environment, the value of ECC is not equivalent to zero. Notably, the ECC value of the proposed two-antenna MIMO can be calculated via the S-parameters, as denoted in Equation (1); however, one condition to be noted is that the two antenna elements must have an efficiency of near 100%, so that much accurate ECC can be determined. Thus, it is always better to apply Equation (2) via the far-field radiation patterns to determine the ECC, as the calculation of ECC via this equation is more accurate.

$$\rho_e = \frac{|S_{11} * S_{12} + S_{21} * S_{22}|^2}{\left(1 - |S_{11}|^2 - |S_{21}|^2\right)\left(1 - |S_{22}|^2 - |S_{12}|^2\right)} \tag{1}$$

$$\rho_e = \frac{\left|\iint_{4\pi}\left[\vec{F_1}(\theta,\varnothing) * \vec{F_2}(\theta,\varnothing)\right] d\Omega\right|}{\iint_{4\pi}\left|\vec{F_1}(\theta,\varnothing)\right|^2 d\Omega \iint_{4\pi}\left|\vec{F_2}(\theta,\varnothing)\right|^2 d\Omega} \tag{2}$$

where $\vec{F_i}(\theta,\varnothing)$ is the three-dimensional field pattern of the antenna, when ith port is excited. Ω is solid angle.

Figure 14 shows the ECC values for various frequency ranges. It is seen that the ECC values obtained from Equations (1) and (2) are well below 0.05, which are very close to zero throughout the bands of interest of the proposed two-antenna MIMO. This confirms that both the antenna elements are uncorrelated from each other, and hence contribute to increasing the data rate of the system.

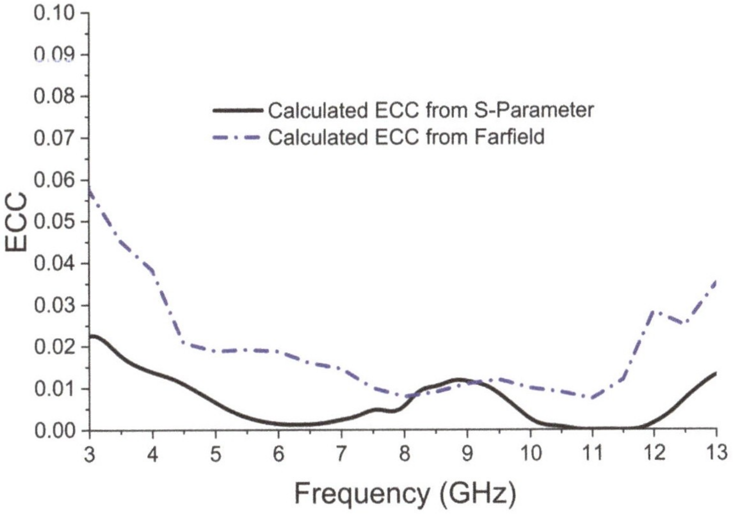

Figure 14. ECC of the proposed two-antenna MIMO.

6.2. Diversity Gain (DG) dB

The DG is another metric of interest as it helps the MIMO antenna to resolve multipath signals, thus improving the receiver's ability to recover intelligent data from multipath signals at a fixed rate of transmission. The *DG* is the parameter used to determine the increment in signal to noise ratio (SNR) magnitude of each path due to the introduction of a spatial diversity scheme, and it is calculated by using the below formula mentioned in Equation (3).

$$DG = 10\sqrt{1 - |\rho_e|^2} \tag{3}$$

Figure 15 shows the DG of the proposed two-antenna MIMO. The calculated *DG* values obtained from the far-field and S-parameters are very close to 10 dB. This confirms that both the antenna elements are strongly uncorrelated with each other and are good candidates for MIMO applications.

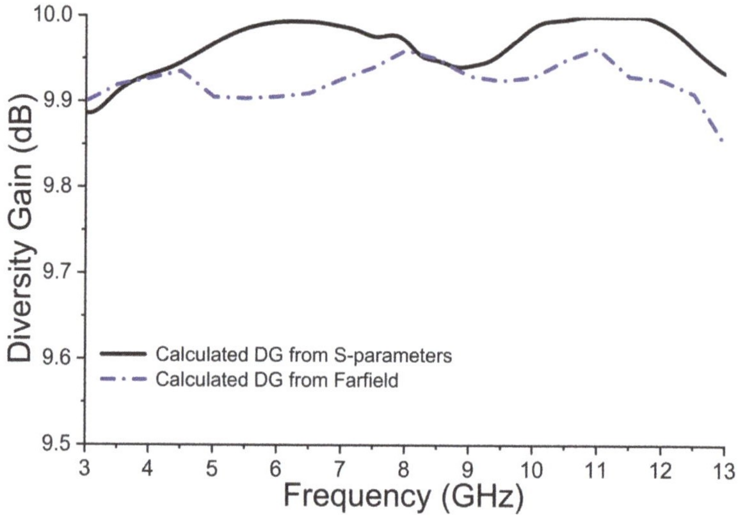

Figure 15. DG of the proposed two-antenna MIMO.

6.3. *Mean Effective Gain (MEG)*

The MEG determines the performance of antenna elements in a rich multipath fading practical environment. *MEG* is the ratio of mean received power to the mean incident power at the antenna element, and it is obtained using the efficiency method as denoted in Equation (4).

$$MEG_i = 0.5\mu_{irad} = 0.5\left(1 - \sum_{j=1}^{K} |S_{ij}|^2\right) \tag{4}$$

where K is the number of antenna elements, 'i' is the excited antenna, and η_{irad} is the radiation efficiency of the ith antenna.

Further Solving Equation (4), the *MEG* of each antenna element can be computed by using Equations (5) and (6):

$$MEG_1 = 0.5\left(1 - |S_{11}|^2 - |S_{12}|^2\right) \tag{5}$$

$$MEG_2 = 0.5\left(1 - |S_{21}|^2 - |S_{22}|^2\right) \tag{6}$$

where, MEG_1 and MEG_2 are the MEG of antenna element#1 and antenna element#2, respectively. By observing Figure 16, one can see that the calculated *MEG* of the two antenna elements is identical (approximately −3 dB) across the bands of interest, thus their corresponding ratio (MEG_1/MEG_2) is one. Table 3 concludes the *MEG* values of the proposed two-antenna MIMO. In this table, the two antenna elements have acquired very good values, which ascertains that the proposed two-antenna MIMO has a maximum *DG*, better isolation across functioning bands, and smaller losses in diversity performances. Moreover, because the *MEG* ratios of the two antenna elements are closer to 1, it also validates better diversity performance from the proposed two-antenna MIMO under a very rich multipath fading environment of wireless channels.

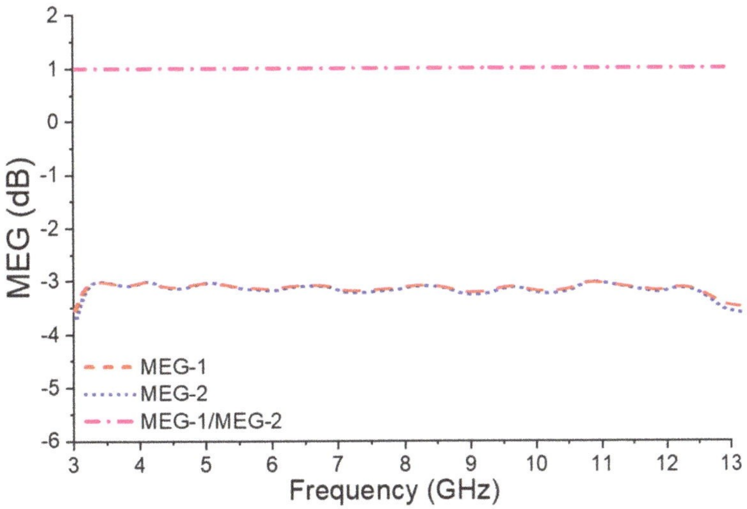

Figure 16. MEG of the proposed two-antenna MIMO.

Table 3. Mean effective gains (MEG) of the proposed two-antenna MIMO.

Frequency (GHz)	MEG (-dB) of Antenna Elements			
	Antenna Element#1	Antenna Element#2	Ratio of Antenna Element#1/Antenna Element#2	Ratio of Antenna Element#2/Antenna Element#1
Sub-6 GHz 5G and C-band uplink	−3.10	−3.11	0.99	1.00
IEEE 802.11ac	−3.21	−3.22	0.99	1.00
IEEE 802.11ax 5G and C-band downlink	−3.15	−3.14	1.00	0.99
X-band	−3.05	−3.04	1.00	0.99

6.4. Total Active Reflection Coefficient (TARC)

The TARC validates the diversity performance of a MIMO antenna and is calculated by using the below Equation (7).

$$\Gamma = \frac{\sqrt{\left(\left|S_{ii} + S_{ij}e^{j\theta}\right|^2\right) + \left(\left|S_{ji} + S_{jj}e^{j\theta}\right|^2\right)}}{\sqrt{2}} \tag{7}$$

where θ is the input phase angle which is changed from 0° to 180° at an interval of 30°, and S_{ii} and S_{jj} are the reflection coefficients (dB) of antenna element#1 and antenna element#2, respectively.

Figure 17 illustrates the measured TARC values which are almost stable and below −10 dB in the entire band. This validates that the proposed two-antenna MIMO has obtained good isolation, as well as serving as a good candidate for the integration with the phase shifter system.

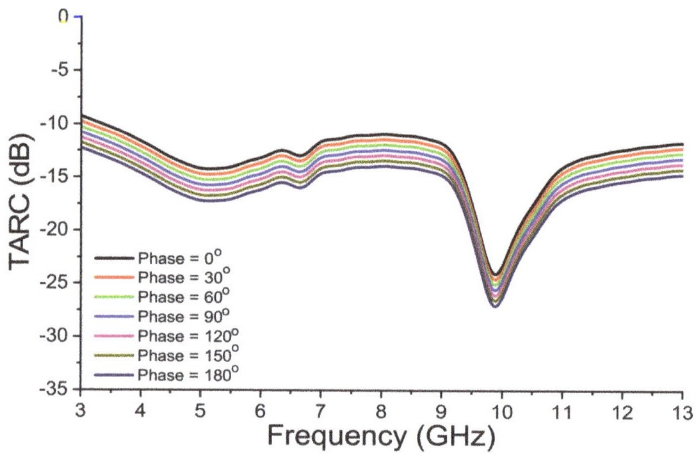

Figure 17. TARC of the proposed two-antenna MIMO.

6.5. Channel Capacity Loss (CCL)

The CCL is an essential metric to characterise the diversity performance of a MIMO antenna, as it confirms the higher bound on the rate of transmission. Thus, to ensure a high data rate, the CCL values must be below 0.5 bits/s/Hz across the operational band, and it is estimated by the following Equations (8) to (13).

$$C_{loss} = -log_2 det\left(\mu^R\right) \tag{8}$$

where, Ψ^R is the correlation matrix and is described as:

$$\mu^R = \begin{bmatrix} \mu_{11} & \mu_{12} \\ \mu_{21} & \mu_{22} \end{bmatrix} \tag{9}$$

where,

$$\mu_{11} = 1 - \left(|S_{11}|^2 + |S_{12}|^2\right) \tag{10}$$

$$\mu_{12} = -(S_{11} * S_{12} + S_{21} * S_{22}) \tag{11}$$

$$\mu_{21} = -(S_{22} * S_{21} + S_{12} * S_{11}) \tag{12}$$

$$\mu_{22} = 1 - \left(|S_{22}|^2 + |S_{21}|^2\right) \tag{13}$$

The simulated and measured CCL values of the proposed two-antenna MIMO using the S-parameters are demonstrated in Figure 18. Here, average values of CCL less than 0.35 bits/s/Hz are observed in the entire band, which ensure a better performance of the MIMO antenna by fulfilling the limits defined by the industry standards. Furthermore, the small discrepancy observed between the simulated and measured CCL values could be because of fabrication inaccuracies.

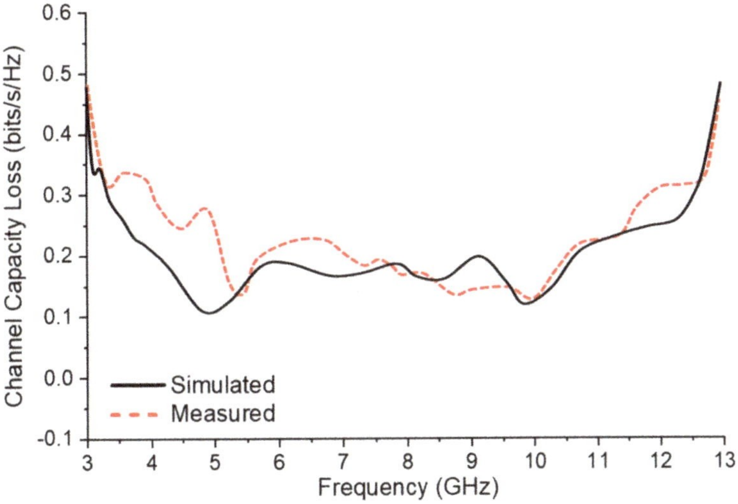

Figure 18. Channel Capacity Loss of the proposed two-antenna MIMO.

6.6. Channel Capacity (bits/s/Hz)

Channel capacity is a vital parameter used to calculate the multiplexing performance of the MIMO antenna, which is calculated using the equation in [20]. Figure 19 depicts the comparison graph of the channel capacity for the SISO and MIMO antenna. From Figure 19 it is easily observed that the channel capacity of the proposed two-antenna MIMO is greater than 10.00 bits/s/Hz in the entire band of interest, that is approximately 1.78 times higher as compared to the maximum limit of an ideal SISO antenna (approx. 5.65 bps/Hz). Further, it is also noticed that the channel capacity values are very near to the maximum limit for an ideal 2 × 2 MIMO system (11.35 bps/Hz).

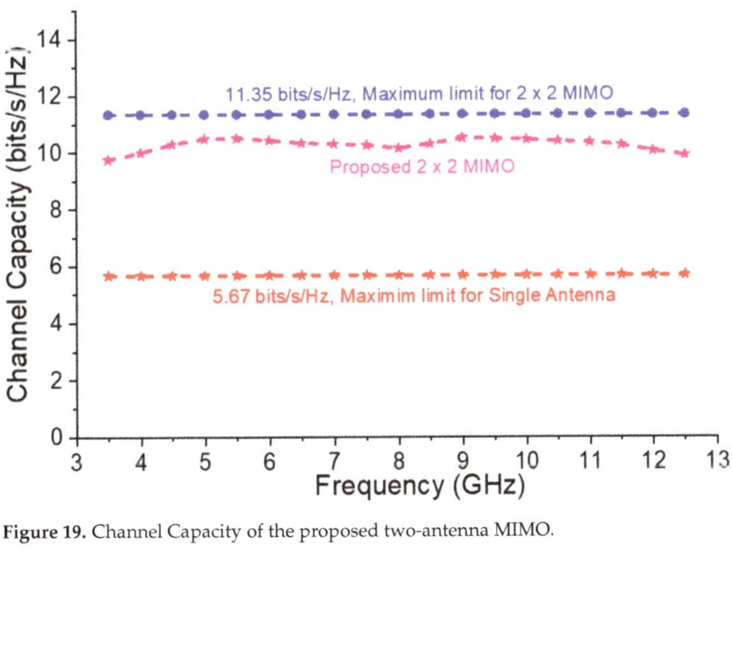

Figure 19. Channel Capacity of the proposed two-antenna MIMO.

7. Performance Comparison of the Proposed Two-Antenna MIMO with Existing State of Arts

The performance comparison of the proposed two-antenna MIMO with other existing two-antenna MIMO types, including bandwidth, dimensions, decoupling structure, gain, efficiency, ECC, and DG are mentioned in Table 4. Here, it is observed that the proposed two-antenna MIMO has exhibited wide dual impedance bandwidths with compact dimensions and good ECC and DG values.

Table 4. Performance comparison of proposed two-antenna MIMO with existing state of art.

Ref	Dimension (mm^3)	Sub	BW (GHz)	Gain (dBi)	ECC	DG	Isol (dB)	Decoupling Structure
[1]	32 × 98 × 1	FR-4	0.6–0.7 1.7–1.9 2.4–2.7 3.2–4.1 5.1–5.9	5.14	0.04	9.8	>15	Not used
[2]	56 × 37 × 1.6	FR-4	2.24–2.50 3.60–3.99 4.40–4.60 5.71–5.90	2	0.08	9.5	>15	Not used
[4]	70 × 52 × 1.6	FR-4	3.10–3.21 6.20–6.33 7.60–7.90	5.84	0.025	9.5	>31	Not used
[9]	90 × 21 × 1.6	FR-4	2.22–2.54 3.14–3.90 5.30–5.90	3.22	0.01	10	>20	Not used
[10]	34 × 34 × 1.44	FR-4	3.50–3.60 5.00–5.40	4.7	0.01	-	>19	U-shape slot in ground plane
[13]	20 × 35 × 0.8	FR-4 epoxy	3.34–3.87	2.5	0.01	-	>20	T-shape ground stub
[15]	59 × 55 × 8.1	FR-4	3.00–7.00	4	0.2	8.94	>20	DGS
[18]	38.2 × 95.94 × 1.6	FR-4	2.43–2.50	4.25	0.008	9.99	>24	Fractal EBG
[19]	55 × 28 × 1.6	FR-4	2.01–3.92	2	0.01	9.8	>15	EBG
[20]	100 × 60 × 1	Jeans	1.34–3.92 4.34–6.34	5	0.04	9.0	>18	Meta-Inspired
[21]	47.5 × 40 × 1.6	FR-4	3.35–3.78	3.5	0.05	-	>15	Metamaterial
This Work	32 × 20 × 0.8	FR-4	3.3–7.8 8.0–12.0	4.0	0.05	9.9	>20	Concentric Rings

From the above table, the following features of the proposed two-antenna MIMO are observed:

1. It has the smallest physical dimensions (volume size) as compared to all the antennas mentioned in Table 1.
2. Unlike [10,13,16–19], the proposed one has exhibited a very wide dual bandwidth operation.
3. It has a higher gain, as compared to [2,9,10,13,15–17,19].
4. Unlike [2,15], the proposed one has higher ECC and DG.
5. The decoupling structure applied in this work is very simple and easy to design, and it can yield high isolation of >20 dB, as compared to antennas reported in [1,2,4,9].

8. Conclusions

A two-antenna MIMO functioning in the Sub-6 GHz 5G NR, IEEE 802.11ac/ax, C-band, and X-band has been successfully studied. Besides showing very wide dual 10-dB impedance bandwidths of 81.08% (3.3–7.8 GHz) and 40% (8.0–12.0 GHz), the proposed two-antenna MIMO has also exhibited a gain >3 dBi and efficiency greater than 69% throughout the two bands of interest. By loading a decoupling structure (five concentric ring elements) between the two adjacent antenna elements, a very desirable isolation above 20 dB is obtained. The MIMO performance metrics, such as ECC (<0.05), DG (>9.9 dB), CCL (<0.35/bits/s/Hz), TARC (<−10 dB), and MEG1/MEG2 ratio (approximately equal to 1) are investigated, and their corresponding values are well within the acceptable practical values. Furthermore, the calculated channel capacity is larger than 10.00 bits/s/Hz. Therefore, because of its compact size, good scattering and radiation characteristics, better diversity performance, the proposed two-antenna MIMO is a potential candidate in futuristic devices for above aforementioned wireless and satellite applications.

Author Contributions: Conceptualization, A.G.A., J.K. and C.-Y.-D.S.; methodology, A.D.; software, A.G.A. and J.K.; validation, J.K., C.-Y.-D.S. and A.P.; formal analysis, A.G.A. and A.D.; investigation, A.G.A. and J.K.; writing—original draft preparation, A.G.A., J.K. and A.D.; writing—review and editing, J.K. and A.D.; visualization, A.G.A.; supervision, C.-Y.-D.S. and A.P.; project administration, A.P. and J.K.; funding acquisition, A.G.A. All authors have read and agreed to the published version of the manuscript.

Funding: This research received no external funding.

Data Availability Statement: All data are included within manuscript.

Conflicts of Interest: The authors declare no conflict of interest.

References

1. Hassan, M.M.; Zahid, Z.; Khan, A.A.; Rashid, I.; Rauf, A.; Maqsood, M.; Bhatti, F.A. Two element MIMO antenna with frequency reconfigurable characteristics utilizing RF MEMS for 5G applications. *J. Electromagn. Waves Appl.* **2020**, *34*, 1210–1224. [CrossRef]
2. Chouhan, S.; Panda, D.K.; Kushwah, V.S.; Singhal, S. Spider-shaped fractal MIMO antenna for WLAN/WiMAX/Wi-Fi/Bluetooth/C-band applications. *AEU Int. J. Electron. Commun.* **2019**, *110*, 1–8. [CrossRef]
3. Soltani, S.; Lotfi, P.; Murch, R.D. A Dual-Band Multiport MIMO Slot Antenna for WLAN Applications. *IEEE Antennas Wirel. Propag. Lett.* **2016**, *16*, 529–532. [CrossRef]
4. Babu, K.V.; Anuradha, B. Design and performance analysis of tri-band Wang shaped MIMO antenna. *Int. J. Inf. Technol.* **2020**, *12*, 559–566. [CrossRef]
5. El Ouahabi, M.; Zakriti, A.; Essaaidi, M.; Dkiouak, A.; Hanae, E. A miniaturized dual-band MIMO antenna with low mutual coupling for wireless applications. *Prog. Electromagn. Res. C* **2019**, *93*, 93–101. [CrossRef]
6. Kulkarni, J.; Sim, C.Y.D.; Deshpande, V. Low-profile, compact, two port MIMO antenna conforming Wi-Fi-5/Wi-Fi-6/V2X/DSRC/INSAT-C for wireless industrial applications. In Proceedings of the IEEE 17th India Council International Conference (INDICON), New Delhi, India, 10–13 December 2020; pp. 1–5.
7. Dkiouak, A.; Zakriti, A.; El Ouahabi, M. Design of a compact dual-band MIMO antenna with high isolation for WLAN and X-band satellite by using orthogonal polarization. *J. Electromagn. Waves Appl.* **2019**, *34*, 1254–1267. [CrossRef]
8. Kumar, M.; Nath, V. Design and development of triple-band compact ACS-fed MIMO antenna for 2.4/3.5/5 GHz WLAN/WiMAX applications. *Analog. Integr. Circ. Sig. Process* **2020**, *103*, 461–470. [CrossRef]
9. Ekrami, H.; Jam, S. A compact triple-band dual-element MIMO antenna with high port-to-port isolation for wireless ap-plications. *Int. J. Electron. Commun.* **2018**, *96*, 219–227. [CrossRef]
10. Islam, S.N.; Das, S. Dual-band CPW fed MIMO antenna with polarization diversity and improved gain. *Int. J. RF Microw. Comput. Eng.* **2020**, *30*, e22128. [CrossRef]
11. Tiwari, R.N.; Singh, P.; Kanaujia, B.K.; Kumar, S.; Gupta, S.K. A low-profile dual band MIMO antenna for LTE/ Blue-tooth/Wi-Fi/WLAN applications. *J. Electromagn. Waves Appl.* **2020**, *34*, 1239–1253. [CrossRef]
12. Kulkarni, J.; Desai, A.; Sim, C. Two port CPW-fed MIMO antenna with wide bandwidth and high isolation for future wireless applications. *Int. J. RF Microw. Comput. Eng.* **2021**, *31*, e22700. [CrossRef]
13. Saurabh, A.K.; Meshram, M.K. Compact sub-6 GHz 5G-multipleinput- multiple-output antenna system with enhanced isolation. *Int. J. RF Microw. Comput. Aided Eng.* **2020**, *30*, e22246. [CrossRef]
14. Khan, A.A.; Jamaluddin, M.H.; Aqeel, S.; Nasir, J.; Kazim, J.U.R.; Owais, O. Dual-band MIMO dielectric resonator antenna for WiMAX/WLAN applications. *IET Microwaves, Antennas Propag.* **2017**, *11*, 113–120. [CrossRef]

15. Kumari, T.; Das, G.; Sharma, A.; Gangwar, R.K. Design approach for dual element hybrid MIMO antenna arrangement for wideband applications. *Int. J. RF Microw. Comput. Eng.* **2018**, *29*, e21486. [CrossRef]
16. Kumar, M.; Nath, V. Analysis of low mutual coupling compact multi-band microstrip patch antenna and its array using defected ground structure. *Eng. Sci. Technol. Int. J.* **2016**, *19*, 866–874. [CrossRef]
17. Pandhare, R.A.; Zade, P.L.; Abegaonkar, M.P. Miniaturized microstrip antenna array using defected ground structure with enhanced performance. *Eng. Sci. Technol. Int. J.* **2016**, *19*, 1360–1367. [CrossRef]
18. Sharma, K.; Pandey, G.P. Two port compact MIMO antenna for ISM band applications. *Prog. Electromagn. Res. C* **2020**, *100*, 173–185. [CrossRef]
19. Biswas, A.K.; Chakraborty, U. Reduced mutual coupling of compact MIMO antenna designed for WLAN and WiMAX applications. *Int. J. RF Microw. Comput. Eng.* **2018**, *29*, e21629. [CrossRef]
20. Roy, S.; Chakraborty, U. Mutual Coupling Reduction in a Multi-band MIMO Antenna Using Meta-Inspired Decoupling Network. *Wirel. Pers. Commun.* **2020**, *114*, 3231–3246. [CrossRef]
21. Wang, C.; Yang, X.; Wang, B. A metamaterial-based compact broadband planar monopole MIMO antenna with high isolation. *Microw. Opt. Technol. Lett.* **2020**, *62*, 2965–2970. [CrossRef]
22. Kulkarni, J.; Desai, A.; Sim, C.Y.D. Wideband four-port MIMO antenna array with high isolation for future wireless sys-tems. *AEU-Int. J. Electron. Commun.* **2021**, *128*, 1–4. [CrossRef]
23. Pirasteh, A.; Roshani, S.; Roshani, S. Compact microstrip lowpass filter with ultrasharp response using a square-loaded modified T-shaped resonator. *Turk. J. Electr. Eng. Comput. Sci.* **2018**, *26*, 1736–1746. [CrossRef]

electronics

MDPI

Article

Design and Analysis of Wideband Flexible Self-Isolating MIMO Antennas for Sub-6 GHz 5G and WLAN Smartphone Terminals

Jayshri Kulkarni [1,*], Abdullah G. Alharbi [2], Arpan Desai [3], Chow-Yen-Desmond Sim [4] and Ajay Poddar [5]

1 Sourayan Technologies Pvt. Ltd., Pune 411037, India
2 Department of Electrical Engineering, Faculty of Engineering, Jouf University, Sakaka 42421, Saudi Arabia; a.g.alharbi@ieee.org
3 Department of Electronics and Communication Engineering, CSPIT, Charotar University of Science and Technology (CHARUSAT), Changa 388421, India; arpandesai.ec@charusat.ac.in
4 Department of Electrical Engineering, Feng Chia University, Taichung 407, Taiwan; cysim@fcu.edu
5 Synergy Microwave Corp., Paterson, NJ 07504, USA; Poddar_ajay@yahoo.com
* Correspondence: jayah2113@gmail.com; Tel.: +91-9552466969

Abstract: A single radiator that is a part of four-port diversity Multiple-Input Multiple-Output (MIMO) antenna design is composed of four octagonal rings embedded between the two opposite sides of a T-shaped conductive layer surrounded by inverted angular edge cut L-shaped and E-shaped structures. The radiators are placed at the four corners with common ground at the center of a smartphone to form a four-element mobile MIMO antenna. The printing of the antenna is carried out on the flexible polyamide substrate (dielectric constant = 3.5 and loss tangent = 0.0027) with dimensions of $70 \times 145 \times 0.2$ mm^3. A wide impedance bandwidth of (84.12%) 2.39 to 5.86 GHz is achieved for all four radiators. The compact size of the radiators along with their placement enables the proposed MIMO antenna to occupy much less area while preserving the space for 2G/3G/4G antennas. The placement of the antennas results in self-isolation between antenna elements by achieving isolation greater than 17.5 dB in the desired operating bands. Furthermore, besides showing a high efficiency of 85% and adequate gain above 4 dBi, good diversity performances such as Envelope Correlation Coefficient (ECC) of less than 0.05, Diversity Gain (DG) of above 9.8 dB, Mean Effective Gain (MEG) of −3.1 dB, Channel Capacity of 21.50 bps/Hz, and Total Active Reflection Coefficient (TARC) of below −10 dB are achieved by the flexible MIMO smartphone antenna. The effect of bending along the X and Y-axis on the performance of the proposed MIMO antenna is also analyzed where decent performance is observed. This makes the proposed flexible four-element MIMO antenna a potential candidate to be deployed in future smartphones.

Keywords: smartphone antennas; 5G; multiple-input multiple-output (MIMO); flexible; four-element

Citation: Kulkarni, J.; Alharbi, A.G.; Desai, A.; Sim, C.-Y.-D.; Poddar, A. Design and Analysis of Wideband Flexible Self-Isolating MIMO Antennas for Sub-6 GHz 5G and WLAN Smartphone Terminals. *Electronics* **2021**, *10*, 3031. https://doi.org/10.3390/electronics10233031

Academic Editors: Faisel Tubbal, Ladislau Matekovits and Raad Raad

Received: 31 October 2021
Accepted: 1 December 2021
Published: 4 December 2021

1. Introduction

The upcoming promising technology for 5G smartphone communication is the use of MIMO, which uses two or more antennas for simultaneous transmission and reception over a communication channel. With the increasing number of mobile users, there is a need to increase channel capacity, which is possible through MIMO, where multiple antennas can be employed in the mobile terminals and base stations [1,2]. Presently, various countries are focusing on sub-6 GHz Long Term Evolution (LTE) bands such as 3.4–3.6 GHz and 3.6–3.8 GHz. In addition, 3.5–4.2 GHz, 3.3–3.4 GHz/4.8–5 GHz, and 3.6–4.2 GHz/4.4–4.9 GHz are also incorporated by the USA, Japan, and China, respectively, for 5G applications [3]. The allocation of the 5G spectrum for countries is different, stretching from 3.3–6 GHz, and for future 5G applications, the multiple licensed and unlicensed bands will be amalgamated. Therefore, a multi-port antenna operating between 3.3 and 6 GHz is essential for smartphone applications [4–6]. Employing multiple antennas with MIMO and 5G capability requires high inter-elemental isolation to ensure smooth communication between the wireless channels along with high data rates in handheld devices. Due to the use of multipath technology, higher data rates using MIMO technology are possible, leading to an increase in the range and reliability without extra bandwidth usage. Thus, spectral efficiency significantly improves and ultimately helps in coping with the necessity of higher data rates for various services [7]. To achieve the desired Electro-Magnetic (EM) signal, the antennas are placed in the smartphones. However, if the antennas are non-conformable, they consume extra space. Therefore, conformal antennas are designed to be accommodated anywhere within the limited space available inside smartphones. Thus, conformal antennas save space, which can then be utilized for increasing the functionality of the smartphones [8].

2. Literature Review

Recently, several researchers have carried out intensive work on MIMO antennas for mobile phones [9–25]. However, the space limitation in mobile phones brings challenges in integrating a greater number of antennas with a good value of isolation and lower ECC. Single-port [9], dual-port [10–16], and quad-port [17–25] MIMO antennas are proposed for smartphone terminals by researchers. The single-port antenna proposed in [9] resonates at a dual-frequency band; however, the efficiency of the antenna is much lower, and it does not cover the 5G band. Dual-port smartphone antennas resonating at dual [10–12], quad [13], wideband [14,15], and ultrawideband [16] frequency bands are also proposed. The antenna proposed in [10–12] has acceptable gain, efficiency, and MIMO performance; however, it resonates at mm-wave 5G band, thus not covering the required sub-6-GHz 5G band. The antenna in [11–13] has a thickness of 5 and 6 mm, respectively, occupying a larger space within the smartphone and thus hindering practical usage. Antennas in [14,15] cover applications such as LTE/Wireless Local Area Network (WLAN) bands; however, the gain in [14] is negative and no efficiency is calculated, whereas the antenna in [15] operates only in the LTE band. The antenna in [16] achieves wide bandwidth, very high isolation, and the lowest thickness (0.508 mm) as compared to other two-port antennas; however, to increase the processing capability while attaining much higher download/upload speeds for applications such as mobile Internet of Things (IoT) and improved data rates, a greater number of antenna elements are needed. To achieve the same, various four-port antennas [17–25] are proposed, out of which the antenna in [18,20–23,25] operates at the sub-6 GHz 5G band while other antennas are resonating at LTE [19] and WLAN [24] bands. The antenna in [18] has a decent frequency response and isolation without using any decoupling techniques; however, the efficiency and SAR analysis of the antenna are not evaluated while the gain achieved is also lower than the proposed antenna. The four-element antenna [20] is printed on the front and back of the substrate, where lack of gain and SAR evaluation with lower efficiency make the antenna less usable in practical smartphone applications. Even though SAR analysis is carried out in [21], the lower bandwidth in the 5G regime with a lack of planar antenna geometry leads to the occupation of more space and thus makes the antenna less practical for slim mobile phones. Even after achieving decent frequency response and MIMO performance, the lack of SAR analysis and lower flexibility make the antennas in [22,23,25] a weak contender as compared to the proposed antenna for WLAN and sub-6 GHz 5G smartphone applications. Moreover, no bending analysis is carried out in any of the dual or four-port antennas discussed.

The novel contributions of the proposed work are as follows:

1. The antenna is a four-port structure with all the four elements connected to a common ground plane.
2. The antenna operates in the entire sub-6 GHz 5G band (3.3 GHz–5 GHz) along with the WLAN band (2.4 GHz and 5 GHz).
3. SAR analysis is carried out at two resonant peaks (3.5 and 5.1 GHz) in order to ensure the safe usage of the antenna in mobile terminals.

4. Bending analysis is performed, which shows decent performance in terms of MIMO diversity, transmission, and scattering parameters that guarantee the practical usage of the antenna for slim and foldable smartphone devices.
5. The antenna has demonstrated high gain (>4 dBi), high inter-elemental isolation (>17.5 dB), and reasonable efficiency (85%) with all the diversity parameters such as ECC, DG, TARC, MEG, and channel capacity meeting the requirements of MIMO antennas. Scattering, transmission, and MIMO parameters matched well under normal as well as bending conditions.

This paper illustrates a planar four-port and four-element flexible MIMO antenna resonating at WLAN and sub-6 GHz 5G bands. Having features such as flexibility, a bi/omnidirectional radiation pattern, planar structure, wide bandwidth covering Sub-6 GHz and WLAN bands, high inter-elemental isolation, and suitable MIMO diversity parameters under normal and bending conditions makes the proposed antenna appropriate for smartphone applications. Each radiator has a single feeding port connected to the central common radiating plane. The antenna achieves the required isolation as the four antenna elements are positioned at the diagonally opposite corners of the ground plane, thus achieving the required spatial diversity. SAR analysis and the flexibility of the antenna prove the practical usage of the antenna for slim and foldable smartphone devices.

3. Mobile Antenna Design and Geometry

Figure 1a shows the layered view of a flexible four-element MIMO antenna geometry with the common conducting ground at the center. A flexible polyamide substrate (dielectric constant of 3.5 and loss tangent of 0.0027) is used with diagonally placed antenna elements at the corner of a common ground plane. Figure 1b shows the top view where it is observed that the dimensions of the antenna are $145 \times 70 \times 0.2$ mm^3. Four conductive antenna elements (diagonally placed) with the common center-ground and fed using a discrete port (A: feeding point and B: grounding point) are all printed on the top side of the substrate.

The detailed view of the single element structure of the antenna is illustrated in Figure 1c, which consists of four octagonal rings embedded between two opposite sides of a T-shaped conductive layer surrounded by inverted angular edge cut L-shaped and E-shaped structures. The inset view of the solid octagonal shape located on a T-shaped branch and four hollow octagonal shapes on the right side of a T-shaped branch is shown. The latter hollow octagonal shapes are connected with two rectangular stubs separated by a distance of 0.8 mm to the vertical side edges. All the edges of the inverted L-shaped strips and E-shaped strip are cut at an angle of 45°. The dimensions are indicated in the figure itself, where the units are in terms of millimeters (mm).

While designing the proposed flexible single-element antenna, a few steps are followed to reach the final design, as shown in Figure 2, and their corresponding reflection coefficient results are shown in Figure 3. Initially, an inner E-shaped strip having angular cuts at the edges (step 1) is adopted. The antenna in this case shows a single band performance at a resonance of 3.8 GHz. In the second step, an additional angular edge cut inverted L-shaped conductive strip along with a vertical strip embedded with a solid octagon is inserted for enhancing the frequency response where dual-band performance is observed; however, the lower band is not in the WLAN/5G regime. Thus, in the third step, a horizontal line is incorporated above the E-shaped strip, which is interfaced with the top edge of the E-shaped strip through three vertical conductive strips. The same leads to improved frequency response; however, the wideband performance is still not achieved. So, in the final stage, four octagonal rings are inserted for achieving bandwidth and reflection coefficient performance to cover the essential WLAN/5G bands. Thus, the final step shows the simulated −10 dB impedance bandwidth spanning from (84.67%) 2.37–5.85 GHz.

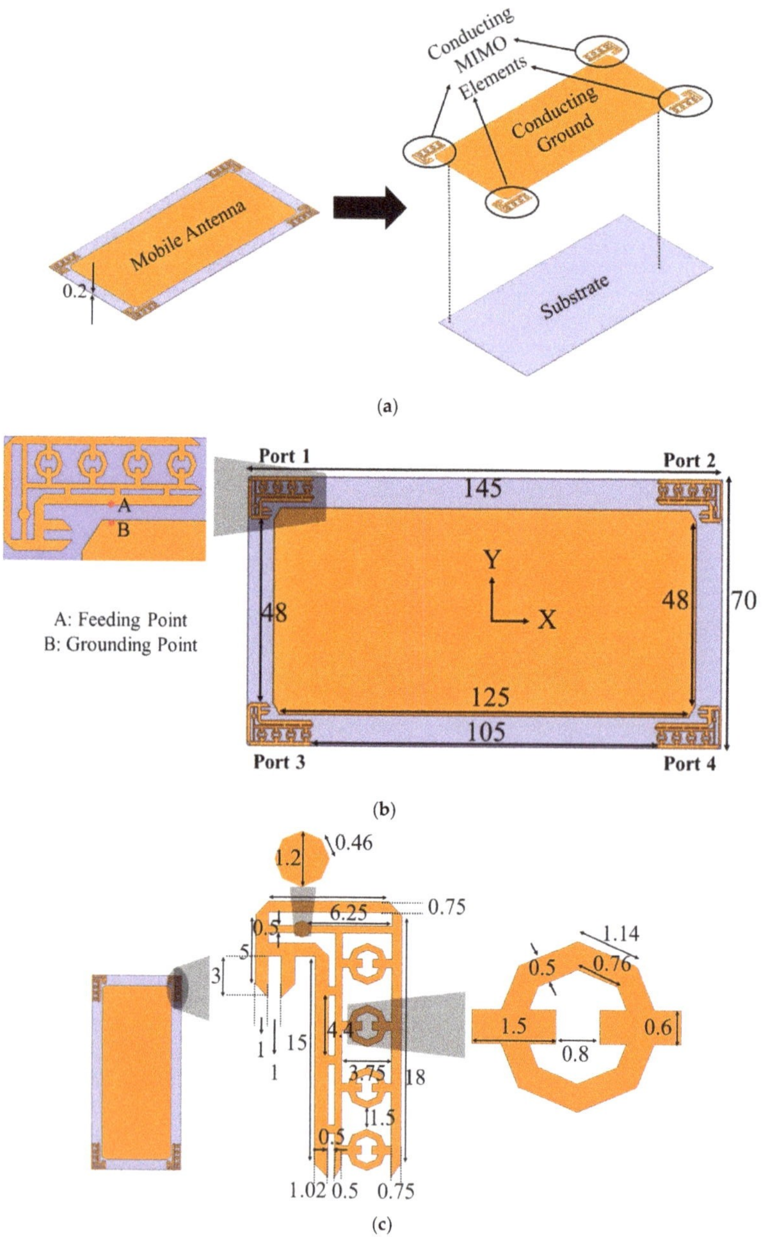

Figure 1. Four-element MIMO antenna geometry: (**a**) Layered view; (**b**) Top view; (**c**) Inset view of antenna element (all dimensions in mm).

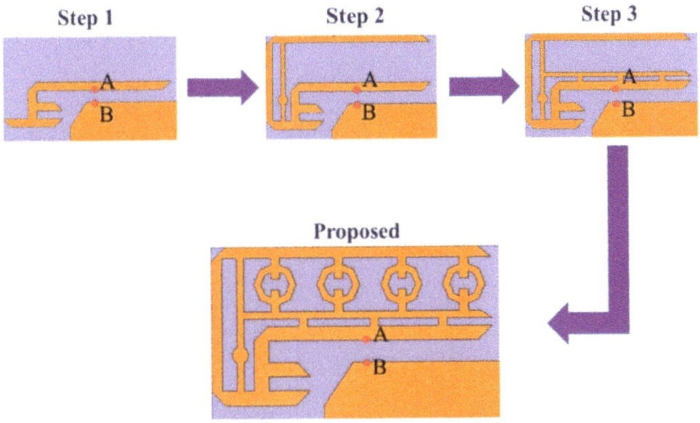

Figure 2. Evolution steps of the proposed mobile MIMO antenna element.

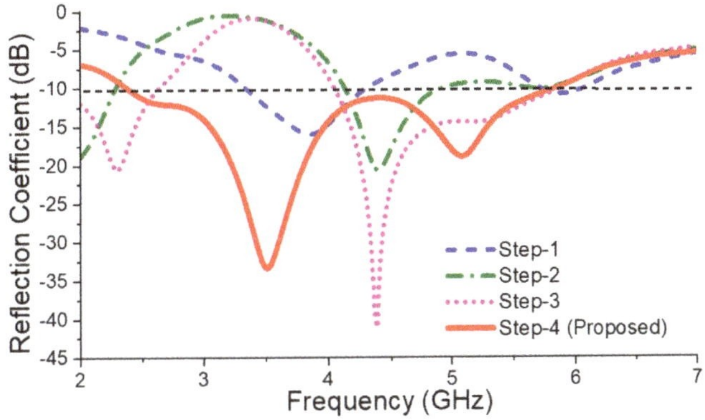

Figure 3. Reflection coefficient S11 (dB) of evolution steps of mobile MIMO antenna.

4. Results and Discussion

The proposed four-port flexible smartphone MIMO antenna is fabricated and measured. The fabricated antenna is demonstrated in Figure 4. Each element is given a separate feed through Sub Miniature version A (SMA) to UFL connector concerning the common ground. The S parameters and radiation patterns are measured using the Keysight vector network analyzer and shielded anechoic chamber. While measuring the S-parameters, when two antenna elements are considered as an active radiator, the other two antenna elements are terminated with a load impedance of 50 Ω, as shown in Figure 4b. The inset view is illustrated in Figure 4c in order to clearly showcase the fabricated antenna geometry.

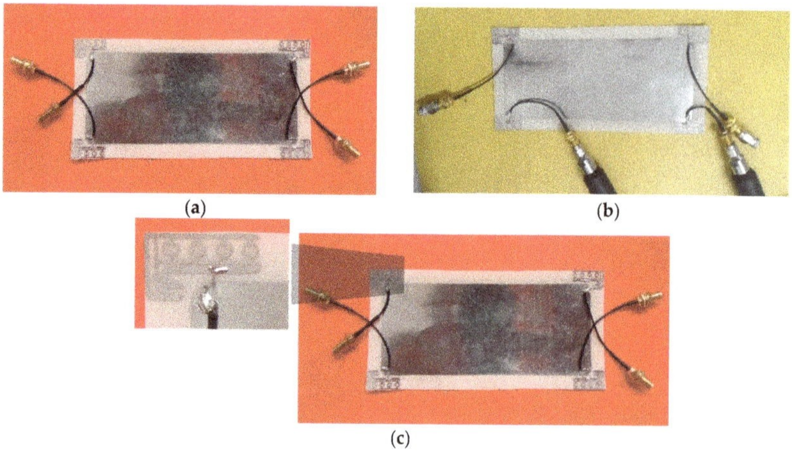

Figure 4. Fabricated antenna geometry of 4-element mobile MIMO antenna: (**a**) Top view; (**b**) Top view with 50 Ω termination; (**c**) Inset view of antenna element.

4.1. Reflection Coefficient (dB)

The simulated reflection coefficient results as seen in Figure 5 depicts that all four antenna elements radiate in the frequency band ranging from (84.67%) 2.37–5.85 GHz. The measured reflection coefficient results illustrate that all four antenna elements radiate in the frequency band ranging from (84.12%) 2.39–5.86 GHz. The measured S parameters agree very well with the simulated ones. The slight disparity may be due to the tolerance in size due to the fabrication process and losses across the UFL to SMA connector.

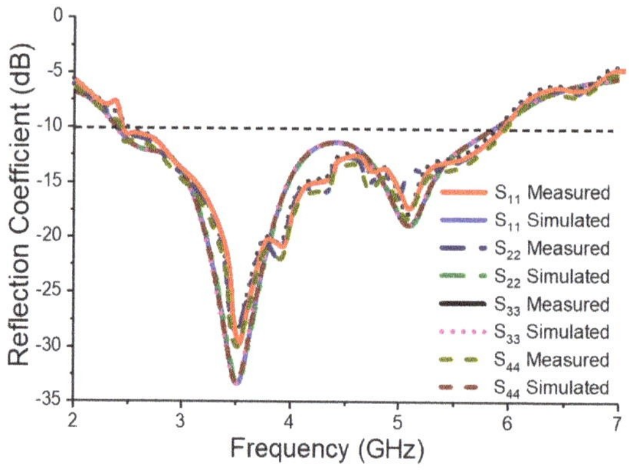

Figure 5. Simulated and measured reflection coefficient of 4-element mobile MIMO antenna.

4.2. Transmission Coefficient (dB)

The simulated and measured transmission coefficient (Sij/Sji) results are illustrated in Figure 6. It is easily seen that the mutual coupling between port 1 and port 3 (S_{13}) exhibits the worst-case mutual coupling of −17.5 dB as they are spaced only 48 mm apart. Further, the antenna elements (port 1 and port 2 (S_{12})) show the minimum coupling of −18 dB as they are spaced at 145 mm apart. Moreover, all the antenna elements exhibit the worst-case mutual coupling of −17.5 dB, which is acceptable for the independent operation of each antenna element in the desired operating band of interest [12,25,26].

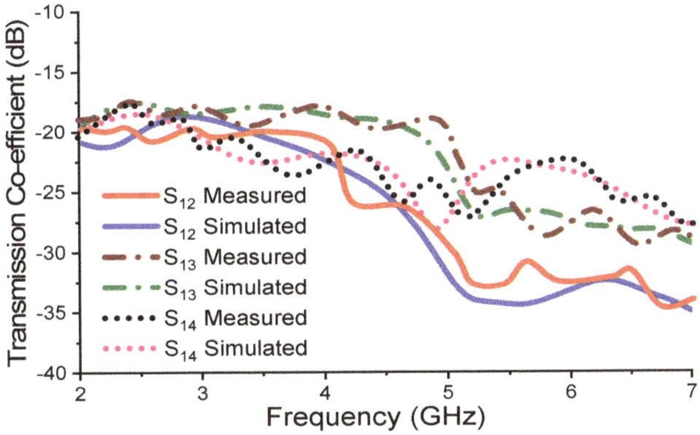

Figure 6. Simulated and measured transmission coefficient of 4-element mobile MIMO antenna.

4.3. Surface Current Disctribution (A/m)

To better understand the working mechanism of the proposed mobile MIMO antenna, the surface current distribution at frequencies 3.5 GHz and 5.1 GHz is presented in Figure 7. As observed from Figure 7a, the maximum current flows in the lower part of the radiator, i.e., the E-shaped structure, which results in the generation of resonance at 3.5 GHz. Similarly, from Figure 7b, it is visualized that the current flows through the octagonal ring as well as from the inverted L-shaped strip, which helps in the generation of resonance at 5.1 GHz. So, these two modes are overlapped with the fundamental mode of 3.5 GHz lying in the frequency range of 2.37–5.85 GHz band.

(a)

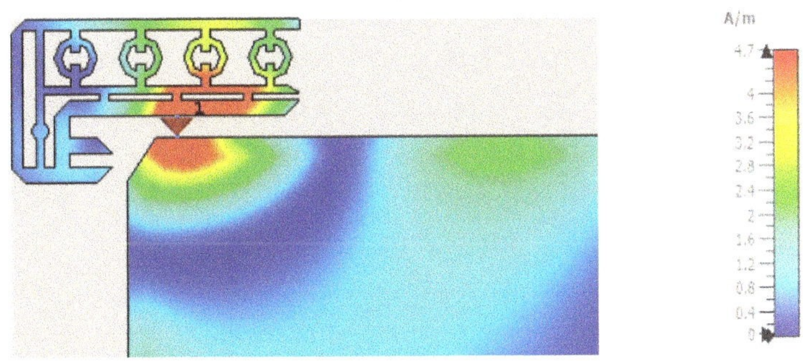

(b)

Figure 7. Current distribution of 4-element mobile MIMO antenna at (a) 3.5 GHz and (b) 5.1 GHz.

4.4. Radiation Pattern (2D)

To evaluate the radiating performance including far-field patterns of the proposed wideband mobile terminal antenna, the simulated and measured radiation patterns in the E-plane and H-plane at 3.5 GHz and 5.1 GHz are depicted in Figure 8. While measuring the radiation patterns of one port, the other ports are terminated with 50 Ω load impedance. From Figure 8a, it is easily observed that in the E-plane, Ant-1 and Ant-3 as well as Ant-2 and Ant-4 exhibit the same radiation patterns, while the radiation patterns of Ant-2 are a mirror image of Ant-1 and radiation patterns of Ant-4 are a mirror image of Ant-3. Likewise, from Figure 8b, it is noted that in the H-plane the radiation patterns of Ant-1 and Ant-2 as well as Ant-3 and Ant-4 are the same, while the radiation patterns of Ant-4 are a mirror image of Ant-1 and the radiation patterns of Ant-3 are a mirror image of Ant-2. However, in both the E-plane and H-plane, all the antenna elements exhibit the radiation patterns with minimum cross-polar magnitude of −15 dB.

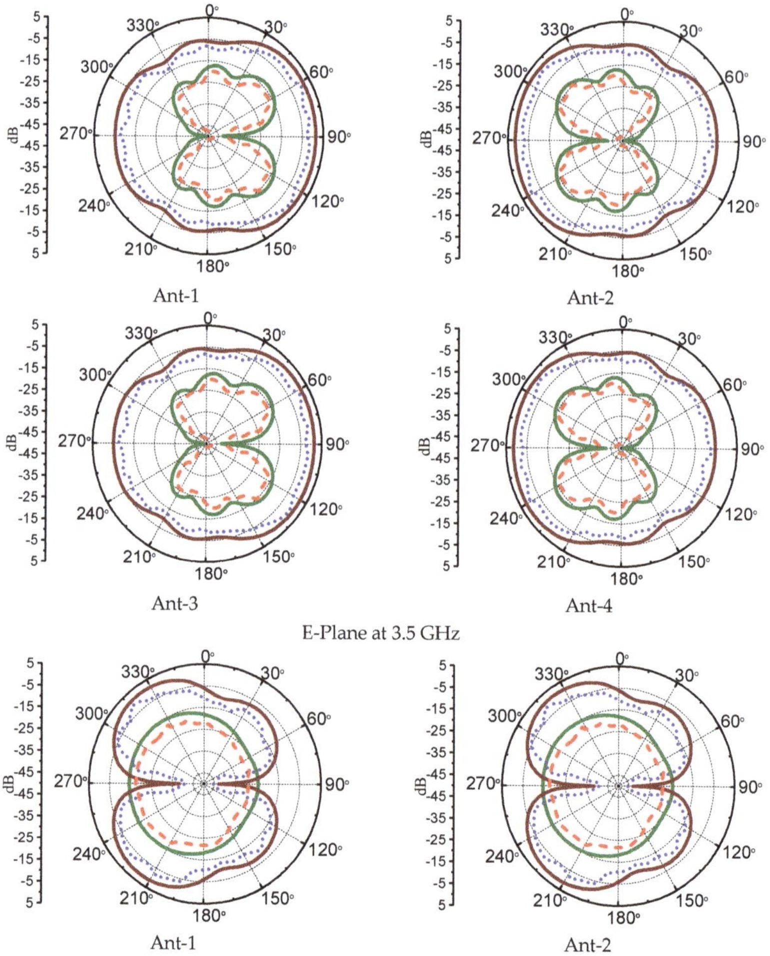

Figure 8. *Cont.*

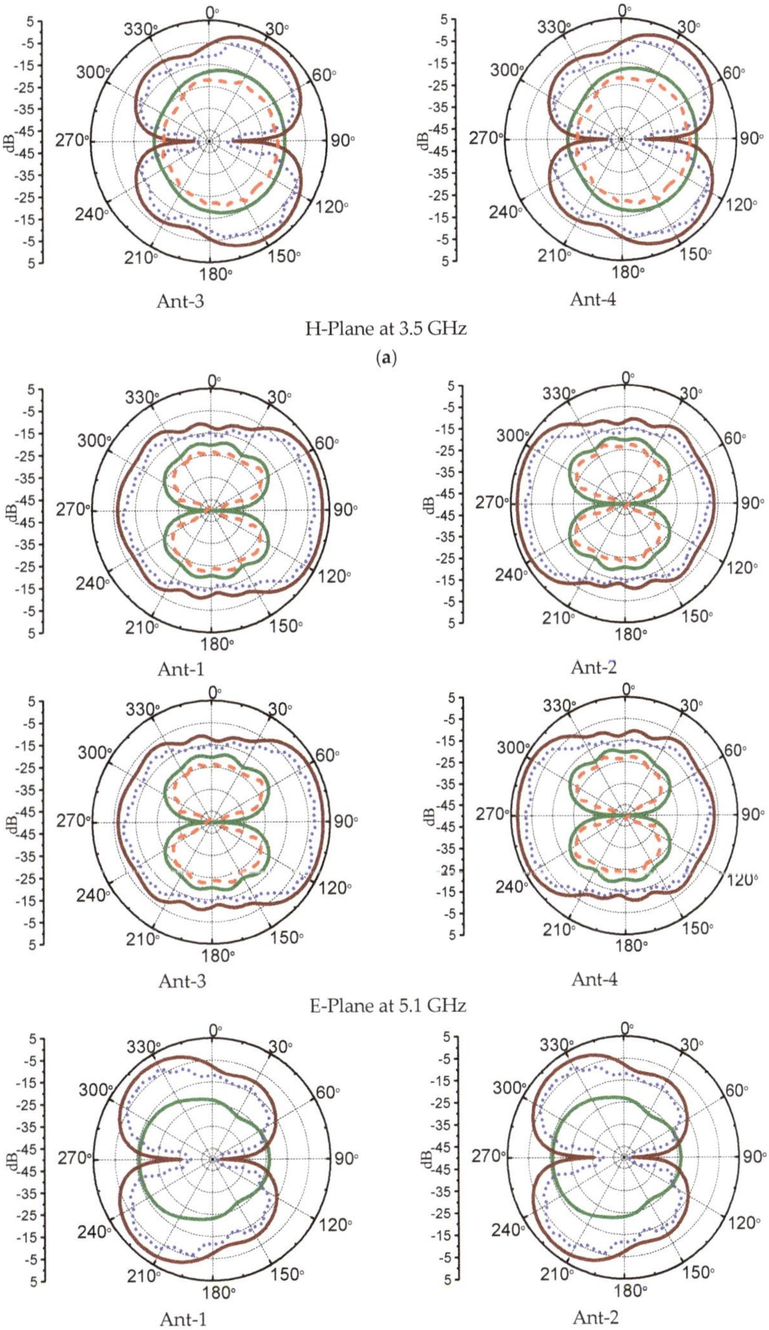

Figure 8. *Cont.*

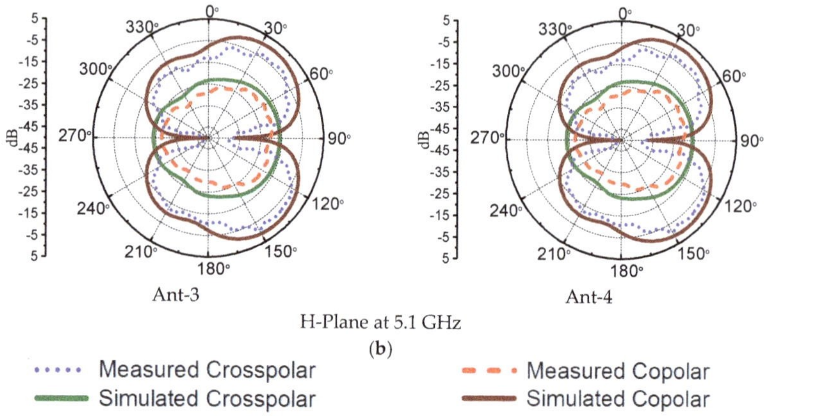

Figure 8. Simulated and measured Co/Cross pol plot of 4-element mobile MIMO antenna at (**a**) 3.5 GHz and (**b**) 5.1 GHz.

4.5. Radiation Pattern (3D)

The three-dimensional (3D) radiation plots for the total radiated power for all the antenna elements at 3.5 GHz and 5.1 GHz are depicted in Figure 9. The 3D patterns are drawn under the condition when one port is excited while the other ports are terminated with 50 Ω load impedance. As seen from Figure 9 at 3.5 GHz, the maximum gain is achieved for all the ports 1, 2, 3 and 4 with the value of 5.56 dBi, which are placed in the vicinity of the cleared ground region at the corners of the back plane. Likewise, as seen in Figure 9b at 5.1 GHz, the slightly lower gain values around 4.63 dBi are obtained since the size of the antenna elements becomes relatively smaller as compared to the guided wavelength λg.

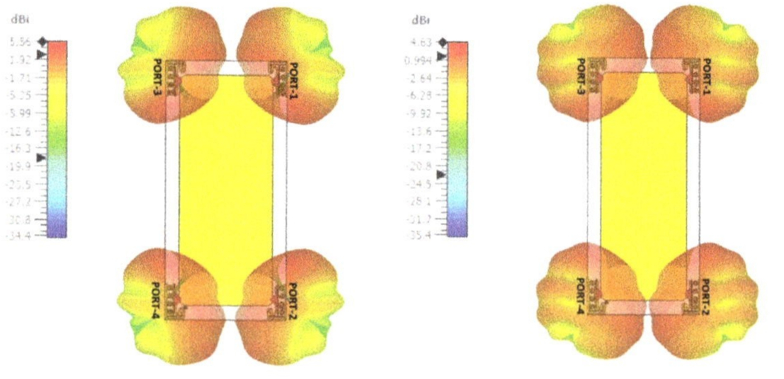

(**a**) (**b**)

Figure 9. Simulated 3D radiation pattern at (**a**) 3.5 GHz (**b**) 5.1 GHz.

4.6. Gain and Efficiency

The simulated and measured values of the gain and efficiency of proposed antenna elements are depicted in Figure 10. Due to the similar performances and deployment of antenna elements in the layout of a smartphone antenna, the gain and efficiency of a single antenna element are shown in Figure 10. A minimal difference between simulated and measured values of gain and efficiency is observed, which may be due to the tolerance of the coaxial cable with the SMA connector, 50 Ω load termination impedance, and the impact of hand soldering. However, the simulated gain and measured gain vary from 4 dBi to 5 dBi in the desired frequency range. Likewise, the simulated and measured efficiency is well above 85% throughout the desired operating band of interest.

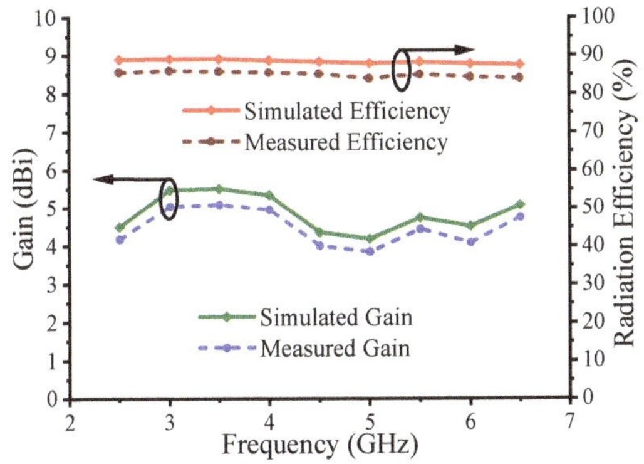

Figure 10. Gain and efficiency of 4-element mobile MIMO antenna.

5. MIMO Diversity Analysis

This section investigates the MIMO performance including ECC, DG, TARC, MEG, and MIMO channel capacity for analyzing the performance of the proposed wideband flexible MIMO antenna for smartphone applications.

5.1. ECC and DG

The ECC is a very essential metric to analyze the performance of a smartphone in the presence of a rich scattering environment as the performance of the smartphone is significantly affected by different hand postures. The ECC values should be <0.5 in the presence of a scattering environment to quantify the uncorrelated operation of each antenna element in the multiple antenna system. The ECC is computed from either S-parameters or from 3D far-field radiation patterns. By considering Equation (1), the ECC values are calculated as [26]:

$$\mathrm{ECC} = \frac{\left| \int \int \left[\vec{F_i}(\theta, \varphi) * \vec{F_j}(\theta, \varphi) \right] d\Omega^2 \right|}{\int \int \left| \vec{F_i}(\theta, \varphi) \right|^2 d\Omega \int \int \left| \vec{F_j}(\theta, \varphi) \right|^2 d\Omega} \quad (1)$$

where, in Equation (1), $\vec{F_i}(\theta, \varphi)$ and $\vec{F_j}(\theta, \varphi)$ are the 3D radiation patterns of the antenna when port i is excited and port j is excited, respectively. Ω represents the solid angle, θ, φ represents spherical angles (elevation, azimuth), and * denotes the Hermitian product.

The calculated ECC values for the proposed flexible MIMO antenna are well below 0.05, as depicted in Figure 11.

The diversity gain is calculated by using Equation (2) [26]:

$$\mathrm{DG} = 10\sqrt{1 - |\mathrm{ECC}|^2} \quad (2)$$

Figure 11 depicts the value of DG obtained using Equation (2) from far-field radiation patterns. It can be noted that the DG value is greater than 9.8 dB throughout the functioning band. This confirms the good diversity performance of the proposed flexible MIMO antenna.

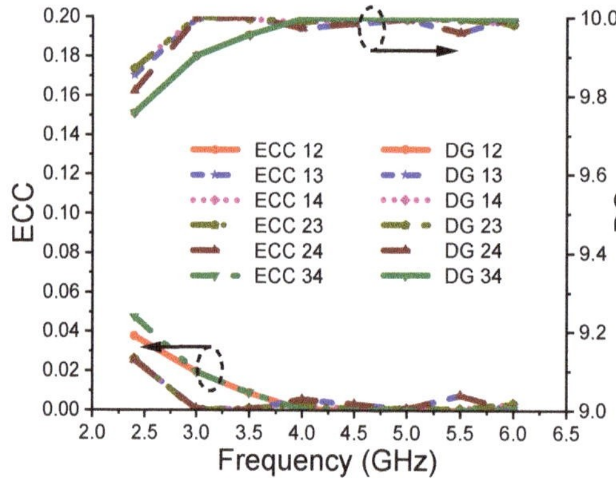

Figure 11. ECC and DG of 4-element mobile MIMO antenna.

5.2. TARC

The TARC is a parameter used to validate the performance of an MIMO antenna in a rich multipath fading environment where the phase angle of every path keeps on changing. TARC can be calculated using Equation (3) [26]:

$$\Gamma = \frac{\sqrt{\left(|S_{ii} + S_{ij}e^{j\theta}|^2\right) + \left(|S_{ji} + S_{jj}e^{j\theta}|^2\right)}}{\sqrt{2}} \tag{3}$$

where θ is the input phase angle that is varied from 0° to 180° with an interval of 30°, S_{ii} and S_{jj} are the reflection coefficients of port one and port two, respectively, and S_{ij} and S_{ji} are the port isolation (dB) between antenna elements.

Figure 12 illustrates the TARC values, which are almost stable throughout the operating bands, which ensures that the proposed four-element mobile MIMO antenna system has achieved high port isolation as well as very stable performance under a multipath fading scattering environment.

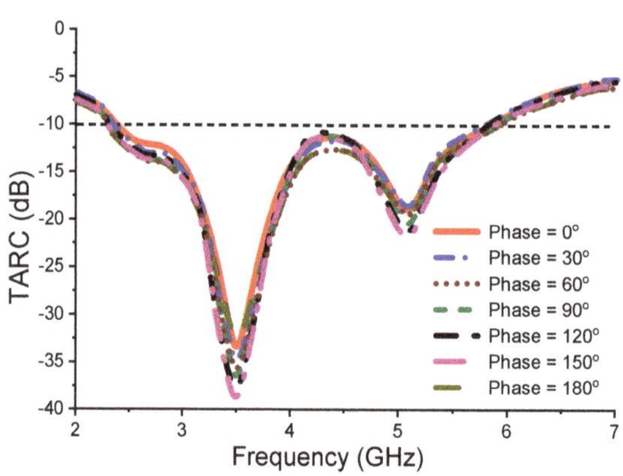

Figure 12. TARC of 4-element mobile MIMO antenna.

5.3. MEG

The MEG values are calculated referring to Equation (4) [26] by considering cross-polarization ratio (XPR) = 0 dB for indoor applications and XPR = 5 dB for outdoor applications under a fading environment.

$$\mathrm{MEG} = \int_{-\pi}^{\pi}\int_{0}^{\pi}[\frac{\mathrm{XPR}}{\mathrm{XPR}+1}G_{\theta}(\theta,\ \phi)P_{\theta}(\theta,\phi) + \frac{1}{1+\mathrm{XPR}}G_{\phi}(\theta,\ \phi)P_{\phi}(\theta,\ \phi) \quad (4)$$

where, in Equation (4), G_{θ}, G_{ϕ} are power gain patterns of antenna elements, and P_{θ}, P_{ϕ} are the available power in vertical and horizontal polarization, respectively. The *XPR* is the cross-polarization ratio and is expressed as in Equation (5):

$$\mathrm{XPR} = 10log_{10}(\frac{P_{vpa}}{P_{hpa}}) \quad (5)$$

where P_{vpa} and P_{hpa} represent the power received by a vertically polarized antenna and horizontally polarized antenna, respectively. The calculation of MEG signifies the effect of total efficiency, gain, and wireless medium to measure the antenna channel mismatch across the desired operating band of interest. For good diversity performance and power balance, the ratio of MEG of antenna i/MEG of antenna j should be below −3 dB. The calculated MEG values are articulated in Table 1, where it is easily confirmed that all the MEG values are well below −3 dB.

Table 1. MEG of the proposed antenna.

Frequency (GHz)	MEGAnt-1/2 (dB)	MEG Ant-1/3 (dB)	MEG Ant-1/4 (dB)	MEG Ant-2/3 (dB)	MEG Ant-2/4 (dB)	MEG Ant-3/4 (dB)
3.5	−3.01	−3.01	−3.01	−3.01	−3.01	−3.01
5.1	−3.01	−3.01	−3.01	−3.01	−3.01	−3.01

5.4. Channel Capacity

The channel capacity is an essential metric to evaluate MIMO antenna performance in a practical environment and is shown in Figure 13. The channel capacity for the proposed four-port flexible MIMO antenna is calculated using Equation (6), as mentioned in [27,28]. From Figure 13, it is observed that the measured efficiency is around 21.50 bps/Hz under the condition of averaging 10,000 Rayleigh fading with a signal-to-noise ratio (SNR) of 20 dB. It is also noted that the channel capacity of the proposed four-port MIMO antenna is much closer to the ideal channel capacity and is almost 1.89 times larger than the maximum limit for a 2 × 2 MIMO antenna.

$$C = k\left\{log_2\left[det([I]+\eta\frac{SNR}{k}[H][H^*])\ \right]\right\} \quad (6)$$

In Equation (6), I is the identity matrix, *SNR* denotes the mean *SNR*, *k* is the rank of the matrix H^H, *H* is the channel matrix, and H^H is the Hermitian Transpose matrix.

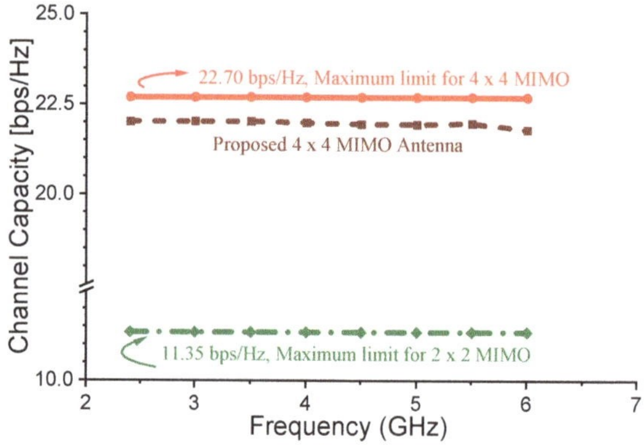

Figure 13. Channel capacity of 4-element mobile MIMO antenna.

6. Bending Analysis

The performance of the proposed four-element flexible MIMO antenna is investigated by varying the operative electrical length under bending conditions. Simulation is carried out in Computer Simulation Technology (CST) Studio Suite®, where the antenna is bent over a cylindrical structure of radius 50 cm, as shown in Figure 14. The antenna validation for flexible applications is performed by plotting the scattering and transmission characteristics as well as gain, efficiency, ECC, and DG [29].

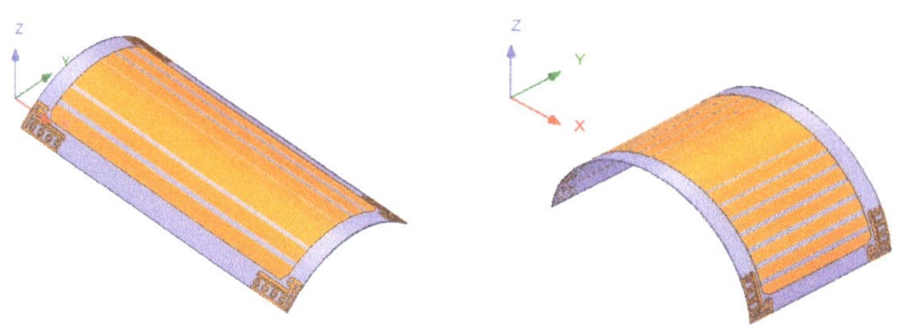

(a) (b)

Figure 14. Geometry of bending the antenna along (**a**) X-axis, (**b**) Y-axis.

6.1. Reflection Coefficient

Figure 15a,b show the measured and simulated reflection coefficient when the antenna is bent along the X and Y-axis, respectively. It can be easily visualized that an almost identical impedance bandwidth is observed by bending the antenna along the X and Y-axis, respectively, and it operates very well in the desired 5G and WLAN bands.

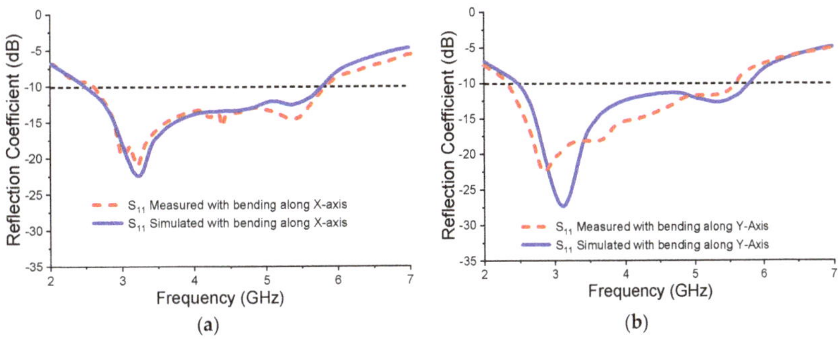

Figure 15. Measured and simulated reflection coefficient of 4-element mobile MIMO antenna by bending along (**a**) X-axis, (**b**) Y-axis.

6.2. Transmission Coefficient

Figure 16a,b show the measured and simulated transmission coefficient when the antenna is bent along the X and Y-axis, respectively. Here, it can be observed that the simulated and measured transmission coefficient is very well above 17.5 dB between all the antenna elements.

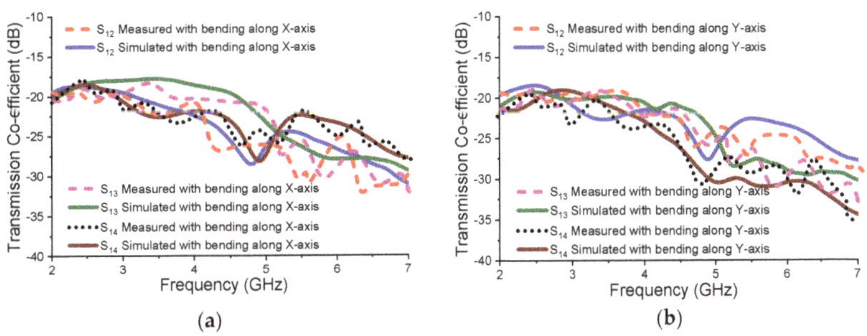

Figure 16. Measured and simulated transmission coefficient of 4-element mobile MIMO antenna by bending along (**a**) X-axis, (**b**) Y-axis.

6.3. Gain and Efficiency

Figure 17 shows the gain and efficiency of the antenna when the antenna is bent along both the X and Y-axis. It can be observed that there is negligible effect when the antenna is bent along a radius of 50 cm in both the X and Y-axis, and both gain and efficiency are above 4 dBi and 85%, respectively.

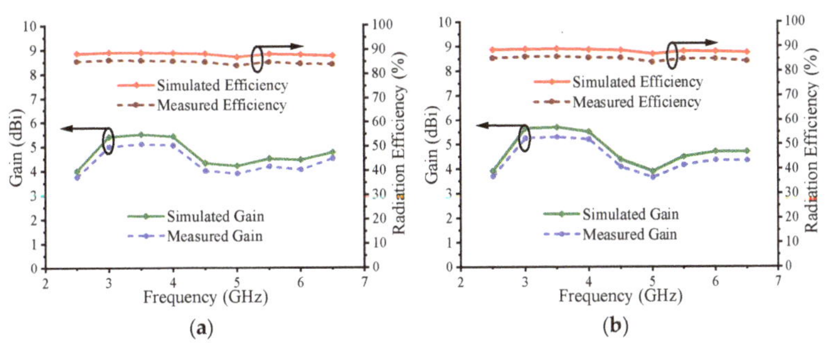

Figure 17. Gain and efficiency of 4-element mobile MIMO antenna by bending along (**a**) X-axis, (**b**) Y-axis.

6.4. ECC and DG

The ECC and DG under bent conditions are shown in Figure 18. The ECC and DG are almost the same as that in the case of unbent conditions. Here, it is observed that the ECC value is less than 0.05, whereas DG is above 9.8 throughout the desired bands.

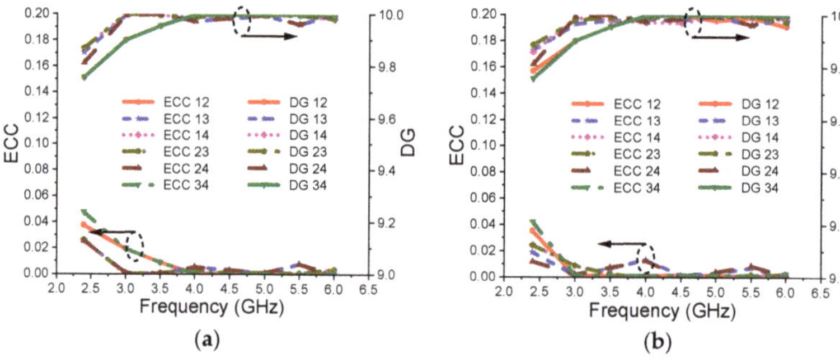

Figure 18. ECC of 4-element mobile MIMO antenna by bending along (**a**) X-axis, (**b**) Y-axis.

7. Impact on User Due to Smartphone Antenna (SAR Analysis)

In this section, the proposed flexible smartphone MIMO antenna is analyzed by interfacing it with the human tissue model. The study on human tissue is essential for assessing its influence on the antenna performance, while it will be interesting to analyze the effect of radiation on a part of the human body close to it.

Figure 19a illustrates the layered view of human tissue underneath the MIMO antenna. The human tissue model simulated for testing with the antenna is shown in Figure 19b, where bone forms the innermost layer covered by muscle, and lastly skin forms the uppermost layer. The characteristics of various layers are depicted in Figure 19c.

7.1. Effect of SAR Analysis on Radiation Pattern (3D)

Figure 20 shows the 3D radiation pattern of the MIMO antenna in the close vicinity of human tissue. As illustrated, good radiation patterns with adequate pattern coverage and gain values are achieved for the MIMO antenna with the human tissue model. Clearly, the human tissue model reduced the antenna characteristics. As seen, the antenna with human tissue in its vicinity still achieves gain values more than 3 dB. The patterns are radiating in different directions, and thus the antenna demonstrated decent spatial diversity at a resonating frequency of 3.5 GHz and 5.1 GHz.

7.2. Effect of SAR on Human Tissue

SAR provides information about the EM absorption by a human body during the transmission and reception of radio frequency signals, which is a grave issue for smartphone systems, and the value of this should be as low as possible. The simulated values of SAR for the proposed MIMO smartphone design in the locality of human tissue model at 3.5 GHz and 5.1 GHz are investigated in Figure 21. The obtained SAR value is lower than the SAR limit in the USA (1.6 W/kg).

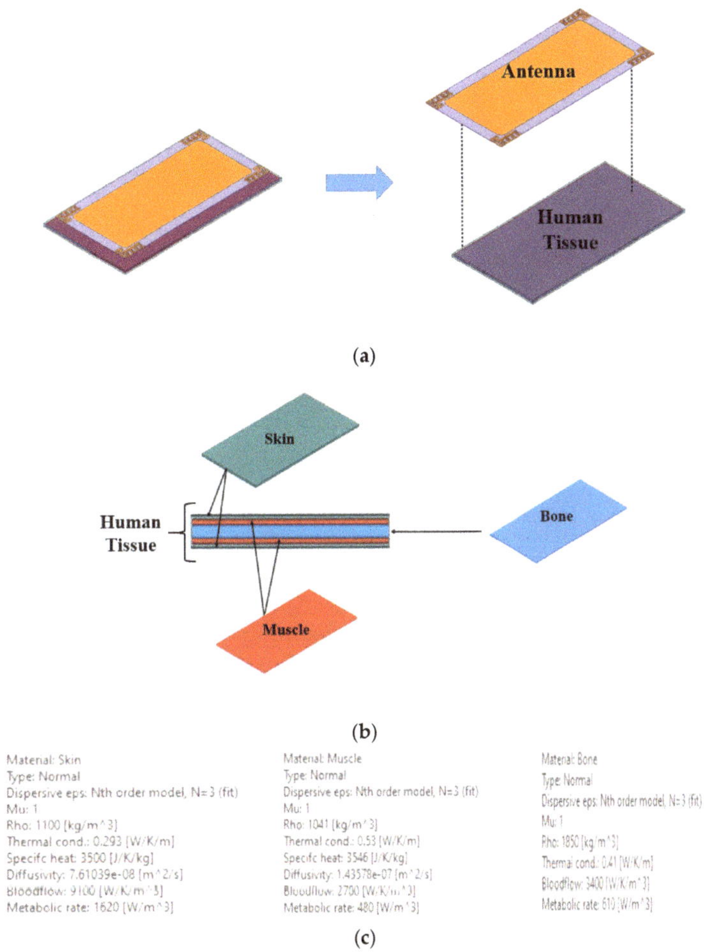

Figure 19. Antenna and human tissue model: (**a**) Layered view; (**b**) Layered view of human tissue; (**c**) Properties of human tissue model.

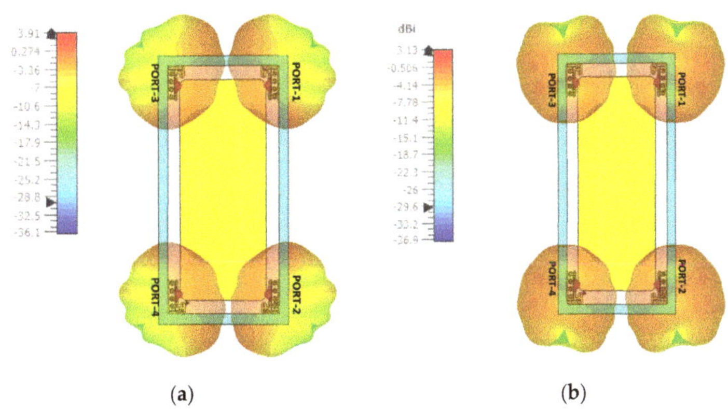

Figure 20. SAR at (**a**) 3.5 GHz, (**b**) 5.1 GHz.

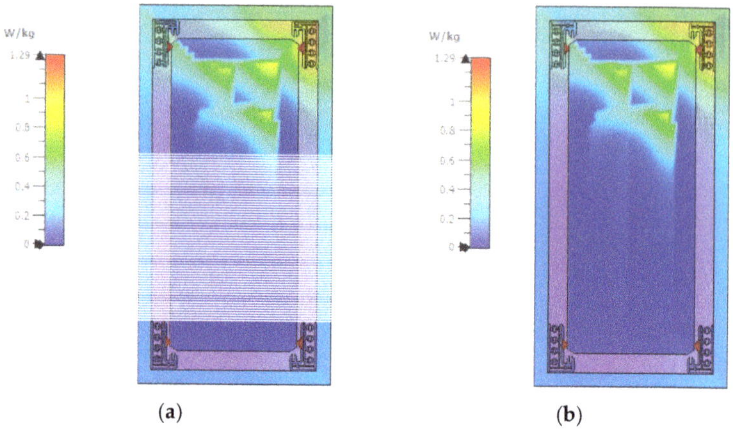

Figure 21. SAR investigation at (**a**) 3.5 GHz, (**b**) 5.1 GHz.

8. Performance Comparison of Proposed Antenna

The performance comparison of the proposed four-element mobile MIMO antenna with other mobile MIMO antennas is presented in Table 2.

From the above table, it is observed that the proposed antenna outperforms the other smartphone MIMO antennas in terms of the following criteria:

1. Four-port structure unlike other antennas [9–16] that have two-port configurations.
2. In comparison to four-port antennas [17–25], the proposed antenna covers the entire sub-6 GHz 5G band along with the WLAN band (2.37–5.85 GHz) with SAR analysis carried out at two resonant peaks (3.5 and 5.1 GHz).
3. The antennas in [18,20–23,25] also cover the sub-6 GHz band; however, in [18,22–25], no SAR analysis is carried out. In [22] as well, the analysis of gain and SAR is missing.
4. The antenna in [21] has a lower bandwidth (3.4–3.6 GHz). Additionally, no bending analysis is carried out for any of the antennas.
5. Flexible structure, high gain, high inter-elemental isolation, and reasonable efficiency with all the diversity parameters such as ECC, DG, TARC, MEG, and channel capacity meeting the requirements of the MIMO antenna make the proposed antenna a good contender for future mobile devices. Scattering, transmission, and MIMO parameters matched well under normal as well as bending conditions.

Table 2. Performance comparison of the proposed antenna.

Ref	Year	Size (mm^3)	Flexible	No. of Ports	Frequency Band (GHz)	Gain (dBi)	Efficiency (%)	Isolation	ECC
[9]	2018	142 × 79 × 7.5	No	1	(698–960 MHz) and (1.710–2.690 GHz)	–	42	–	–
[10]	2017	60 × 100 × 0.965	No	2	1.870–2.530 GHz 28 GHz	4, 8	75	14	~0.18
[11]	2012	110 × 45 × 5	No	2	698 –960 and 1.710–2.690 GHz	0–5/2–7	20–70/40	10	~0.4/0.03
[12]	2017	145 × 72 × 0.8	No	2	824–960 MHz, 1.710–2.690 GHz	−0.32–1.4 i/1.6–4.8	59–72	17	0.02/0.4
[13]	2018	136 × 68 × 6	No	2	3.5–3.9 GHz, 2.3–2.5 GHz, 3.3–3.5 GHz and 4.25–4.45 GHz	–	50	10	~0.2
[14]	2019	60 × 100 × 1	No	2	2.5–3.6 GHz	−4.9– −1.9	–	10	0.16
[15]	2019	65 × 130 × 1	No	2	1710–2690 MHz	–	60–80	19	<0.005
[16]	2018	110 × 120 × 0.508	Semi	2	3 to 10	2.6	–	38	~0.0002
[17]	2016	100 × 60 × 0.8	No	4	2017 and 2265 MHz	4.27	70	10	~0.18
[18]	2020	120 × 65 × 1.6	No	4	N77 (3.3–4.2 GHz), n78 band (3.3–3.8 GHz) and n79 band (4.4–5 GHz),	2,4,4.71	–	18.8	<0.018
[19]	2018	160 × 85 × 0.8	No	4	824–960 MHz/ 1710–2690 MHz) Reconfigurable	–	70	10	~0.2
[20]	2018	153 × 77 × 1	No	4	3.3–3.6 GHz	–	64.2	10	~0.11
[21]	2018	150 × 73 × 0.8	No	4	3.4–3.6	–	51–74	20	<0.06
[22]	2019	140 × 70 × 0.8	No	4	3400–3600 MHz	–	51	11.6	–
[23]	2019	50 × 100 × 4.5	No	4	2.7–3.6 GHz	3	80–90	−25 dB	–
[24]	2020	39 × 30 × 1	No	4	5.15–5.85 GHz	2.8	>70	20	<0.02
[25]	2019	38.3 × 38.3 × 0.8	No	4	3–13.2 GHz	0.5–6.3	72–97	17	0.02
Proposed		70 × 145 × 0.2	Yes	4	2.37–5.85 GHz	4–5.5	85	17.5	<0.05

9. Conclusions

A four-port flexible smartphone antenna with a wideband regime is proposed for sub-6 GHz 5G and WLAN communications. The planar antenna configuration includes four conducting radiators connected to the common ground, which is printed on the flexible substrate. The optimized antenna elements radiate in the range spanning from (84.64%) 2.37–5.85 GHz. The antenna achieves self-isolation greater than 17.5 dB, a decent gain of above 4 dBi, an efficiency of 85%, and bi/omnidirectional radiation patterns with co-cross levels greater than 15 dB. The diversity parameters of MIMO are also analyzed in terms of ECC, TARC, MEG, and channel capacity, where sufficient results meeting the desired limits are achieved. In addition, SAR and bending analyses are also carried out, where safe SAR values are achieved while scattering, transmission, and MIMO parameters match well under normal and bending conditions. The obtained outcomes demonstrate the potential of the proposed flexible MIMO antenna for use in future smart mobile handsets.

Author Contributions: Conceptualization, J.K. and C.-Y.-D.S.; methodology, A.D.; software, J.K.; validation, J.K., C.-Y.-D.S. and A.P.; formal analysis, A.G.A. and A.D.; investigation, J.K. and A.P.; writing—original draft preparation, J.K. and A.D.; writing—review and editing, J.K. and A.D.; visualization, A.G.A.; supervision, C.-Y.-D.S. and A.P.; project administration, J.K.; funding acquisition, A.G.A. All authors have read and agreed to the published version of the manuscript.

Funding: This research received no external funding.

Data Availability Statement: All data are included within the manuscript.

Conflicts of Interest: The authors declare no conflict of interest.

References

1. Sim, C.-Y.-D.; Chen, H.-D.; Kulkarni, J.; Lo, J.-J.; Hsuan, Y.-C. Recent Designs to Achieving Wideband MIMO Antenna for 5G NR Sub-6GHz Smartphone Applications. In Proceedings of the 2020 International Symposium on Antennas and Propagation (ISAP), Osaka, Japan, 25–28 January 2021; pp. 417–418.
2. Javed, I.; Tang, X.; Shaukat, K.; Sarwar, M.U.; Alam, T.M.; Hameed, I.A.; Saleem, M.A. V2X-Based Mobile Localization in 3D Wireless Sensor Network. *Secur. Commun. Netw.* **2021**, 1–13. [CrossRef]
3. Desai, A.; Upadhyaya, T.; Patel, J.; Patel, R.; Palandoken, M. Flexible CPW fed transparent antenna for WLAN and sub-6 GHz 5G applications. *Microw. Opt. Technol. Lett.* **2020**, *62*, 2090–2103. [CrossRef]
4. Ishteyaq, I.; Khalid, M. Multiple input multiple output (MIMO) and fifth generation (5G): An indispensable technology for sub-6 GHz and millimeter wave future generation mobile terminal applications. *Int. J. Microw. Wirel. Technol.* **2021**, 1–17. [CrossRef]
5. Huang, J.; Dong, G.; Cai, Q.; Chen, Z.; Li, L.; Liu, G. Dual-band MIMO antenna for 5G/WLAN mobile terminals. *Micromachines* **2021**, *12*, 489. [CrossRef]
6. Kulkarni, J.; Dhabre, S.; Kulkarni, S.; Sim, C.-Y.D.; Gangwar, R.K.; Cengiz, K. Six-Port Symmetrical CPW-Fed MIMO Antenna for Futuristic Smartphone Devices. In Proceeding of the 2021 6th International Conference for Convergence in Technology (I2CT), Maharashtra, India, 2–4 April 2021; pp. 1–5.
7. Srivastava, K.; Kumar, S.; Kanaujia, B.K.; Dwari, S. Design and packaging of ultra-wideband multiple input-multiple-output/ diversity antenna for wireless applications. *Int. J. RF Microw. Comput. Aided Eng.* **2020**, *30*, e22357. [CrossRef]
8. Jha, K.R.; Jibran, Z.A.P.; Singh, C.; Sharma, S.K. 4-Port MIMO Antenna Using Common Radiator on a Flexible Substrate for Sub-1GHz, Sub-6GHz 5G NR, and Wi-Fi 6 Applications. *IEEE Open J. Antennas Propag.* **2021**, *2*, 689–700. [CrossRef]
9. Huang, D.; Zhengwei, D.; Wang, Y. Eight-band antenna for full-screen metal frame LTE mobile phones. *IEEE Trans. Antennas Propag.* **2018**, *67*, 1527–1534. [CrossRef]
10. Hussain, R.; Alreshaid, A.T.; Podilchak, S.K.; Sharawi, M.S. Compact 4G MIMO Antenna Integrated with a 5G Array for Current and Future Mobile Handsets. *IET Microw. Antennas Propag.* **2017**, *11*, 271–279. [CrossRef]
11. Ren, Y.-J. Ceramic Based Small LTE MIMO Handset Antenna. *IEEE Trans. Antennas Propag.* **2012**, *61*, 934–938. [CrossRef]
12. Xu, Z.-Q.; Sun, Y.-T.; Zhou, Q.-Q.; Ban, Y.-L.; Li, Y.-X.; Ang, S.S. Reconfigurable MIMO antenna for integrated-metal-rimmed smartphone applications. *IEEE Access* **2017**, *5*, 21223–21228. [CrossRef]
13. Shi, X.; Zhang, M.; Wen, H.; Wang, J. Compact quadruple band MIMO antenna for 5G mobile applications. In Proceedings of the 12th European Conference on Antennas and Propagation (EuCAP 2018), London, UK, 9–13 April 2018; pp. 1–5.
14. Lee, W.W.; Jang, B. A tunable MIMO antenna with dual-port structure for mobile phones. *IEEE Access* **2019**, *7*, 34113–34120. [CrossRef]
15. Deng, C.; Xin, L. Wideband MIMO antenna with small ground clearance for mobile terminals. *IET Microw. Antennas Propag.* **2019**, *13*, 1419–1426. [CrossRef]
16. Nie, L.Y.; Lin, X.Q.; Qiang, Q.Y.Z.; Zhang, J.; Wang, B. Structure-shared planar UWB MIMO antenna with high isolation for mobile platform. *IEEE Trans. Antennas Propag.* **2018**, *67*, 2735–2738. [CrossRef]
17. Ikram, M.; Hussain, R.; Hammi, O.; Sharawi, M.S. An L-shaped 4-element monopole MIMO antenna system with enhanced isolation for mobile applications. *Microw. Opt. Technol. Lett.* **2016**, *58*, 2587–2591. [CrossRef]
18. Biswas, A.; Gupta, V.R. Design and development of low-profile MIMO antenna for 5G new radio smartphone applications. *Wirel. Pers. Commun.* **2020**, *111*, 1695–1706. [CrossRef]
19. Zhang, Y.-H.; Yang, S.-R.; Ban, Y.-L.; Qiang, Y.-F.; Guo, J.; Yu, Z.-F. Four-feed reconfigurable MIMO antenna for metal-frame smartphone applications. *IET Microw. Antennas Propag.* **2018**, *12*, 1477–1482. [CrossRef]

20. Sun, L.; Haigang, F.; Yue, L.; Zhijun, Z. Tightly arranged orthogonal mode antenna for 5G MIMO mobile terminal. *Microw. Opt. Technol. Lett.* **2018**, *60*, 1751–1756. [CrossRef]
21. Sun, L.; Haigang, F.; Yue, L.; Zhijun, Z. Compact 5G MIMO mobile phone antennas with tightly arranged orthogonal-mode pairs. *IEEE Trans. Antennas Propag.* **2018**, *66*, 6364–6369. [CrossRef]
22. Deng, C.; Di, L.; Xin, L. Tightly arranged four-element MIMO antennas for 5G mobile terminals. *IEEE Trans. Antennas Propag.* **2019**, *67*, 6353–6361. [CrossRef]
23. Chattha, H.T. 4-port 2-element MIMO antenna for 5G portable applications. *IEEE Access* **2019**, *7*, 96516–96520. [CrossRef]
24. Cheng, Y.; Hui, L.; Sheng, B.Q.; Zhu, L. A compact 4-element MIMO antenna for terminal devices. *Microw. Opt. Technol. Lett.* **2020**, *62*, 2930–2937. [CrossRef]
25. Ricardo, G.-V.; Hildeberto, J.-A. Compact UWB uniplanar four-port MIMO antenna array with rejecting band. *IEEE Antennas Wirel. Propag. Lett.* **2019**, *18*, 2543–2547.
26. Kumar, A.; Ansari, A.Q.; Kanaujia, B.K.; Kishor, J. A novel ITI-shaped isolation structure placed between two-port CPW-fed dual-band MIMO antenna for high isolation. *Int. J. Electron. Commun. AEU* **2019**, *104*, 35–43. [CrossRef]
27. Wen, J.; Liu, B.; Cui, Y.; Hu, W. High-isolation eight-element MIMO array for 5G smartphone applications. *IEEE Access* **2019**, *7*, 34104–34112.
28. Saxena, G.; Jain, P.; Awasthi, Y. High diversity gain super-wideband single band-notch MIMO antenna for multiple wireless applications. *IET Microw. Antennas Propag.* **2020**, *14*, 109–119. [CrossRef]
29. Desai, A.; Palandoken, M.; Kulkarni, J.; Byun, G.; Nguyen, T.K. Wideband Flexible/Transparent Connected-Ground MIMO Antennas for Sub-6 GHz 5G and WLAN Applications. *IEEE Access* **2021**, *9*, 147003–147015. [CrossRef]

electronics

MDPI

Article

Low-Cost, Low-Profile Wide-Band Radar Cross Section Reduction Using Dual-Concentric Phase Gradient Modulated Surface

Yousef Azizi [1,2], Mohammad Soleimani [1], Seyed Hasan Sedighy [3] and Ladislau Matekovits [2,4,5,*]

1 Department of Electrical Engineering, Iran University of Science and Technology, Tehran 1684613114, Iran; yousefazizi@elec.iust.ac.ir (Y.A.); soleimani@iust.ac.ir (M.S.)
2 Department of Electronics and Telecommunications, Politecnico di Torino, Corso Duca degli Abruzzi 24, 10129 Turin, Italy
3 School of New Technologies, Iran University of Science and Technology, Tehran 1684613114, Iran; sedighy@iust.ac.ir
4 Istituto di Elettronica e di Ingegneria dell'Informazione e delle Telecomunicazioni, National Research Council of Italy, 10129 Turin, Italy
5 Department of Measurements and Optical Electronics, University Politehnica Timisoara, 300006 Timisoara, Romania
* Correspondence: ladislau.matekovits@polito.it; Tel.: +39-011-090-4119

Abstract: Design criteria of low-cost, dual-concentric metasurface possessing wideband phase gradient (PG) are introduced. The radar cross-section reduction (RCSR) is explained by anomalous reflection that characterizes the superficial planar. The geometry consists of two single band RCSR modulated surfaces (MSs) that are triggered in each other. Each MS is built-up of square patch (SP) unit cells configured as a modulation structure to realize PG that causes anomalous reflection and monostatic RCSR behavior. Applying sinusoidal modulation to the sequence of the SP unit cells leads to the formation of PG along the surface and hence the intensity of the reflected wave is reduced for the broadside direction ($\theta_r = 0°$). The proposed structure fabricated on a 0.8 mm thin FR-4 substrate extends over 249 × 249 mm^2. It achieves a wide RCSR bandwidth from 20.9 GHz to 45.7 GHz (i.e., relative bandwidth of 75%) as designed in Dassault Systèmes (CST) Microwave Studio as a full-wave simulator and confirmed by the measurement results.

Keywords: phase gradient; radar cross section reduction; square patch unit cell; modulated surface

Citation: Azizi, Y.; Soleimani, M.; Sedighy, S.H.; Matekovits, L. Low-Cost, Low-Profile Wide-Band Radar Cross Section Reduction Using Dual-Concentric Phase Gradient Modulated Surface. *Electronics* **2021**, *10*, 1552. https://doi.org/10.3390/electronics10131552

Academic Editor: J.-C. Chiao

Received: 10 May 2021
Accepted: 25 June 2021
Published: 26 June 2021

1. Introduction

RCSR by phase gradient metasurface (PGM) has attracted considerable attention in recent years. Unlike the phase cancellation method that uses two unit cells with phase variation of ±37° around 180° [1–3], in the PGM design several unit cells with phase difference less than | 180° ± 37° | between them are considered. The operational principal of PGM is based on two different phenomena: (i) anomalous reflection and (ii) surface wave conversion, achieving the ability to reduce the reflected power [4]. By using large number of unit cells in the metasurface structure, phase differences between adjacent unit cells were decreased and 74% RCSR bandwidth has been achieved by thick (2.5 mm) substrate [5]. By controlling the anomalous and diffuse reflection parameters in modified unit cells, the PGM metasurface can also be designed as an antenna superstrate [6]. Although the RCSR bandwidth increased up to 82.4% without degradation of the antenna performance, the high thickness substrate (3 mm) limits its practical applications. By using hybrid resonant structure, a dual band PGM with anomalous reflection mechanism presented both theoretically and experimentally (with substrate thickness of 0.5 mm) is able to reduce the backscattering RCS of the surface for two frequencies, namely for 8.9 and 11.4 GHz [7]. Narrow bandwidth performance, i.e., lower than 2% at each band is considered as a limitation on practical applications. In [8], the chessboard structure with 77% RCSR bandwidth using holographic surface (HS) and MS redirector is introduced such that each of the MS and HS reduce the RCS in a single band; by placing them in a chessboard configuration, wideband RCSR is achieved. Although the structure is simple and low-cost, its substrate thickness is still relatively high (1.6 mm). It is also possible to create a PGM with coded PGM (CPGM). Using holographic surfaces to convert the impinging wave into surface one, as well as anomalous or diffuse reflection of the incidence, RCSR can be achieved as introduced, for example [9], that suffer from low bandwidth. In 2017 Zheng et al. introduced a concept of CPGM that it is more flexible to manipulate the reflected wave by changes of the coded configuration [10]. Although wide angle RCSR performance highlighted the use of this CPGM structure, RCSR bandwidth (lower than 64%) and 3 mm substrate thickness are their issues. In another effort by applying the modulation to the coding element and coding sequence simultaneously, two dimensional CPGM has been developed that presents a more flexible pattern manipulation [11]. However, the low bandwidth (46%) is a disadvantage of this structure. By using 4 bit optimized coded metasurface, RCSR bandwidth improved up to 91% [12]. Even though the RCSR bandwidth is enough for practical application, it uses a high cost substrate (F4B). In [13] by using amplitude and phase gradient modulated surface (APGM), RCSR improvement was achieved with 128% bandwidth. Using amplitude gradient as an auxiliary variable caused the RCSR bandwidth improvement simultaneously with improvement of RCS depth. Even though the structure has an ultra-wideband RCSR performance, their dual layer thick substrate (3.2 mm) is considered as a disadvantage.

In this letter, a new dual concentric phase gradient modulated surface (PGMS) is introduced for RCS reduction at wideband frequency range, 20.9–45.7 GHz (75%). In this concentric surface, the anomalous reflection is employed as the main RCS reduction mechanism. The introduced structure consists of two single band concentric modulated surfaces, MS1 and MS2, which reduce RCS from 20.5–32 GHz (45%) for MS1 and 31.8–46.5 GHz (40%) for MS2, consequently. In both MS1 and MS2, RCSR formation is based on gradual phase difference between adjacent unit cells. The square patches (SPs) unit cell is used with sinusoidal pattern for both MS1 and MS2 by changing cell gap size (contrarily to the usual change in the patch size). By placing MS2 in the central part of MS1, a concentric configuration is obtained that exhibits wideband 75% (20.9–45.7 GHz) RCSR performance. The proposed surface has significant advantages with respect to the state of the art references such as low-cost, very low profile (0.8 mm thin), simple manufacturing process (by standard microstrip technology) and wideband monostatic RCSR performance.

2. PGM Design

Based on the generalized Snell's law, the existence of PG along the x and/or y directions in a given coordinate system (see Figure 1) leads to the reflection of an incident wave with incidence angle θ_i (with respect to the normal $\hat{z}$) in the direction θ_r (reflection angle of the scattered wave), where the two angles are related by Equation (1) [14]:

$$\sin(\theta_r) - \sin(\theta_i) = \frac{1}{k_0}\frac{d\Psi_{x,y}}{d_{x,y}} \tag{1}$$

where k_0 is the free space wave number and $\frac{d\Psi_{x,y}}{d_{x,y}}$ defines the PG along the x,y coordinate.

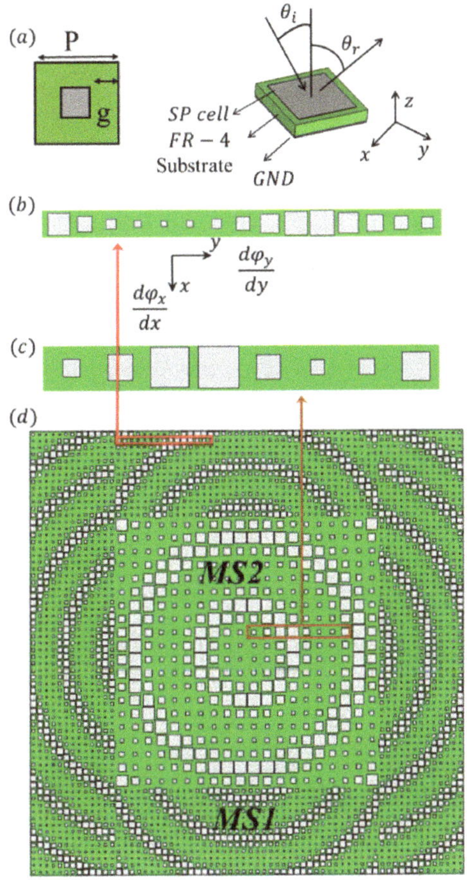

Figure 1. (**a**) SP unit cell, (**b**) PG formation by sinusoidal modulation applied to the MS1 and (**c**) MS2, (**d**) CAD model of the dual-concentric configuration.

According to Equation (1), considering a normally incident plane wave ($\theta_i = 0$) that illuminates the MS with PG characterized by a sinusoidal modulation leads to $\theta_r > 0°$. In the case of constant PG value, constant reflection phase ($\theta_r =$ const.) occurs which can be useful in some practical applications. However, in the case of monostatic RCSR, not only constant PG but variable PG along the x,y directions could make arbitrary reflections that decrease the backscattering of the incident waves. In fact, by assuming $\theta_i = 0$ and $\frac{d\Psi_{x,y}}{d_{x,y}} \neq 0$ according to Equation (1) the reflection angle can be expressed as:

$$\theta_r = \arcsin\left(\frac{1}{k_0}\frac{d\Psi_{x,y}}{d_{x,y}}\right) > 0 \tag{2}$$

Based on Relation 2, the design of the metasurface with wideband non-zero PG causes the reflected signal being in opposite direction with the incident one and this is enough for monostatic RCSR achievement. Therefore, sinusoidal modulation is applied into SP unit cell to form a PG and reflection on arbitrary direction, consequently as depicted in Figure 1. The introduced PGM consists of dual concentric MSs (MS1 and MS2) which is plotted in Figure 1d. For the arrangement of the final structure (Figure 1d), Matlab and Dassault Systèmes (CST) Microwave studio was used. By linking Microwave Studio with Matlab the dual concentrate structure was built in the simulation environment and analyzed in full-wave structure that was discussed in [8–13]. Resizing the SP gaps in Figure 1 is based on the formation of the appropriate PG and redirecting the input wave. In fact, the tradeoff between maximum RCSR bandwidth, wave redirection and structure dimensions is done by an optimization process. In each MSs, the sinusoidal modulation is applied to the SP unit cell to reduce the RCS in a certain bandwidth, indicated by Δf_1 and Δf_2, respectively. SP unit cells are placed on the grounded FR-4 substrate with $\varepsilon_r = 4.4$, $\tan\delta = 0.025$ and of 0.8 mm thickness. Due to the reduced thickness of the supporting dielectric and presence of the ground plane, the unit cell reflects the total incident power. Therefore, the reflection phase can be considered as the main degree of freedom in the unit cell design that gradually changes with the gap size between patches. As depicted in Figure 1a, the following relation $g = (P - d)/2$ between cell parameters is considered, where d and g correspond to the length of the patch and gap size, respectively. The period (P) of the unit cells in MS1, MS2 is 3.5 mm and 7.1 mm, respectively. Two MS1 and MS2 dimensions were extracted by full-wave (CST Microwave Studio) simulations that lead to the gradual phase difference between SP unit cells. Because of the SP symmetry with respect to x, y, and using sinusoidal modulation, MS1 and MS2 PG values are the same. Reflection phase of the MS1 and MS2 are plotted in Figure 2, with the gaps size of 0.2, 0.5, 0.9, 1.4, 1.6 mm and 0.4, 0.8, 1.4, 2, 2.5 mm, respectively. It is observed that MS1 (solid lines) in the 20–33 GHz band has a non-constant PG versus radial distance which can be helpful for monostatic RCSR. In a similar manner, MS2 (dashed lines) exhibits similar behavior in 33–45 GHz band. Based on Equation (2), in case of non-constant PG ($\frac{d\Psi_{x,y}}{d_{x,y}} \neq 0/const.$) along the x,y direction it is expected that the reflected wave can be redirected in different directions with respect to the incident one. While constant PG leads redirecting of the reflected wave into a specific direction, the non-constant PG cause the redirection of the incidence wave in all directions and of effective monostatic RCSR, consequently.

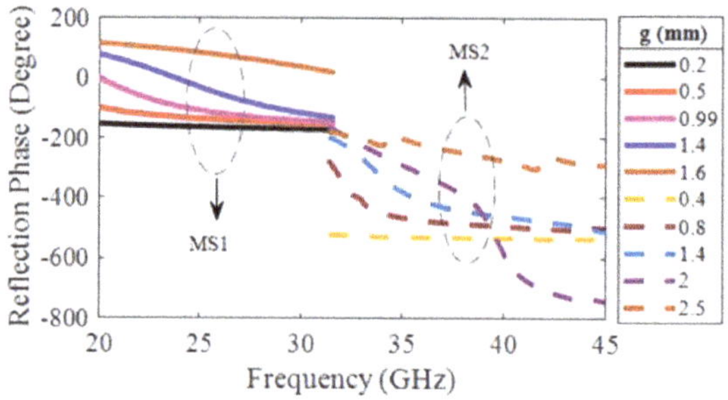

Figure 2. Reflection phase of MS1 (Solid Lines) and MS2 (dashed lines) versus gap variation of the cells.

To prove the idea, two MS1 and MS2 RCS simulation results are plotted in Figure 3a versus frequency: both MS1 and MS2 exhibit the expected behavior for PG with the variation of unit cell gap size in the $\Delta f_1 = 31.8 - 46.5$ GHz (40%) and $\Delta f_2 = 20.5 - 32$ GHz (45%) frequency bands, respectively. In both MS1 and MS2, there are certain numbers of unit cells with constant gap size and reflection phase at each constant radius (with respect to the center of the surface). By gradual change along the radius, the gap size of the cells changes in a sinusoidal fashion which leads to the PG formation that increases the monostatic RCSR by benefiting of an anomalous reflection mechanism that in turn leads to wideband monostatic RCSR. The RCS patterns of MS1 and MS2 at 28 GHz and 35 GHz for a normally incident plane wave are shown in Figure 3b,c, respectively. Based on Figure 3b,c it is observed that the maximum reflections happen in the $\varphi = 45^{\circ}$ plane for both MS1 and MS2. It can be seen that both MS1 and MS2 redirect the incident plane wave to the angle of 30° and 16° (which was extracted from two-dimensional RCS simulation results of MS1 and MS2 in $\varphi = 45^{\circ}$ plane), which proves the formation of anomalous reflection.

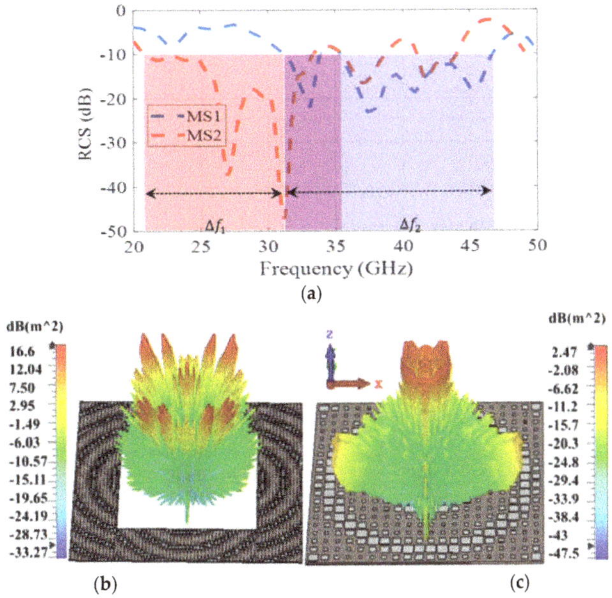

Figure 3. RCS of MS1 and MS2: (**a**) Monostatic RCS of MS1 and MS2; (**b**) RCS pattern of MS1 at 28 GHz; (**c**) RCS pattern of MS2 at 35 GHz.

3. Measurement and Results

The final dimensions of MS1 and MS2 require some optimization due to their location in the concentric composition and the total size of the structure (249 × 249 mm^2). As a result of the optimization aiming to obtain a better RCS bandwidth, the final MS1 and MS2 unit cells periods are changed to 3.8 and 7.5 mm, respectively. Additionally, the gap sizes of the MS1 and MS2 are 0.2, 0.6, 1, 1.3, 1.6 and 0.4, 0.9, 1.5, 2.1, 2.6 mm, respectively. It is noteworthy that the external dimension of MS1 and MS2 has been selected by a tradeoff between their surface area and wideband RCSR performance to obtain the maximum RCSR bandwidth. In Figure 4a–c the fabricated prototypes of MS1/MS2 and Tx/Rx antenna for monostatic RCS measurements are shown, respectively. In fact, MS1 and MS2 were fabricated together, and separated in a second moment by cutting the general structure, and the monostatic RCSR of each metasurface has been measured separately. The measured result in Figure 4d depicts good agreement with the simulation one. RCS measurements have been carried out using N5227A PNA network analyzer as a transmitter and receiver that cover frequency range of 10 MHz-67 GHz. The test frequency range is from 22 GHz to 42 GHz (20 GHz band). The limitation of the frequency range is due to the limitation of instruments such as waveguide feeder, horn antenna and LNA. The 20 GHz test bandwidth is divided into 1601 frequency points and the resolution bandwidth of the PNA chosen as 100 Hz to obtain the proper results in a time gating process extraction. The anechoic chamber supported the RCS measurement based on IEEE standard 1502TM and was suitable for high frequency RCSR measurement [15]. Tx/Rx antenna and dual concentrate MS have a 1.5 m height and the distance between them is about 2.5 m.

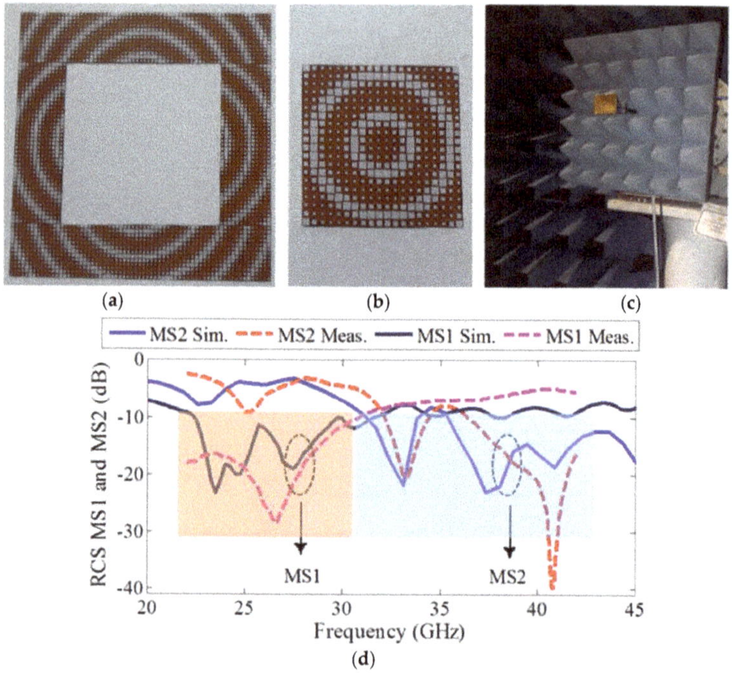

Figure 4. MS1, MS2 and their monostatic performance (**a**), (**b**) Prototype of MS1 and MS2; (**c**) wideband horn antenna as a Tx/Rx antenna; (**d**) RCS Simulation and measurement results of the MS1 and MS2.

The RCS test method based on time gating has consisted of two steps: (i) measurement of monostatic RCS of the MS1, MS2 and the dual-concentric MS structure, which is done by using a Tx/Rx antenna simultaneously and using a directional coupler to separate the forward/backward waves, and (ii) RCS pattern of the dual concentric MS is performed at 30 GHz by using Marconi set up (Figure 5a). In the Marconi test set-up, two antennas are used: the Tx antenna is placed in front of the introduced MS and the Rx antenna, which is located on the rotating and calibrated arm, and is moved manually to measure the RCS at different angles. Figure 4b shows the simulation and measurement monostatic RCSR results of MS1 and MS2.

It is noteworthy that there is some difference in the measurement results with respect to the simulated one, especially at Δf_1 frequency band, due to the existence of fabrication and measurement errors, as well as extracting the results from the time gating method and applying time filters is acceptable. Figure 5a depicts the photograph of the realized prototype that is mounted in anechoic chamber for monostatic RCS measurement. Combining the two single-band RCSR surface (MS1 and MS2) in the form of a concentric MS effectively increases the 10 dB RCS bandwidths to 20.9–45.7 GHz (75% fractional bandwidth) verified by measurement as depicted in Figure 5b. The RCS measurement (bi-static at 30 GHz) and simulation results of dual concentric MS sample shown in Figure 5c, which indicates the good agreement between simulation and measurement results. The anomalous reflection resulting from PG or gradual phase difference between the adjacent unit cells present in the MS1 and MS2 reduces the RCS of the introduced dual concentrated MS. The bi-static RCS results in the introduced structure and PEC with equal size with 1° angular step was performed at 30 GHz and plotted in Figure 5c. It can be seen that at 30 GHz the introduced dual-concentric MS achieves 23 dB RCSR compared with the PEC surface with same dimensions. For a better presentation of the results allowing an easier comparison, the measurement and simulation results of Figure 5c are normalized with respect to 0 dB. According to the results of Figure 5c, there is good agreement between measurement and simulation for angles of 0°–15°. The measurement and simulation results have some differences of few dB between the angles of 60°–90°; however, these results are acceptable due to the low amplitude of the reflection signal at these angles relative to the main lobe.

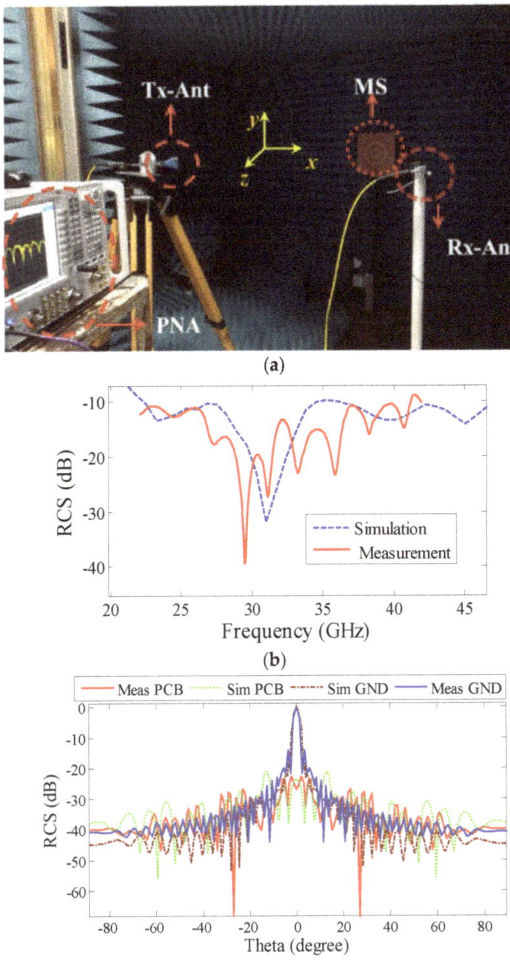

(a)

(b)

(c)

Figure 5. RCS results and test set-up: (**a**) fabricated dual concentric MS in anechoic chamber; (**b**) measurement and simulation results of monostatic RCS; (**c**) RCS of the structure vs. similar PEC surface that measured by Marconi set-up.

In order to express the features of the introduced structure, a comparison with available data from the literature is given in Table 1. Low-cost, ease of construction (thin single-layer structure with 0.8 mm thickness and no need for lump elements) and wideband RCSR performance (75%) are the main advantages of the introduced structure. In the references of Table 1 there is a thin substrate with 0.5 mm thickness. This structure has a very narrowband RCSR performance (about 2% of fractional bandwidth). Therefore, it concludes that there is a trade-off between substrate thickness and RCSR performance. The design of the dual concentrated MS consisting of two MS1 and MS2 (MS1 with RCSR bandwidth of Δf_1 and MS2 with the RCSR bandwidth of Δf_2) using a gradual phase difference between adjacent cells on a single-layer thin substrate (with 0.8 mm thickness) is one of the main advantages of this structure compared to the state of the art. In simple words, nonconstant PG performance with wideband monostatic RCSR is the main concept of this introduced structure that is certified by measurement result.

Table 1. Comparison between this work and the state-of-the-art.

Structures	Thickness (mm)	BW (%)/ Freq. Rang (GHz)	No of Layers	Substrates
[1]	2.28	85/9.4–23.38	2	RO4003
[2]	5	109/6.4–21.7	3	F4B/Air/F4B
[3]	6.35	60/4.2–7.8	1	RT5880
[5]	2.5	74/7.8–17	1	FR4
[6]	3	82.4/7–16.8	1	RO5880
[7]	0.5	<2/8.9–11.4	1	RO4350
[10]	3	64/9.83–19.12	1	F4B
[11]	3	46/12.1–18.8	1	F4B
This work	0.8	75/20.9–45.7	1	FR4

4. Conclusions

In this paper a new, low-cost, commercially available, simple structure with one layer substrate, thin and wideband dual concentrated MS RCS reducer is introduced. The proposed structure consists of two single bands RCS MS1 and MS2 with RCSR fractional bandwidth of 46% and 40%, respectively. Gradual phase difference between adjacent unit cells on radially modulated surface structure is the main concept behind the MS1 and MS2 design that causes nonconstant PG along the surface that efficiently redirects the incident wave. Placing two PG MS1 and MS2 in the concentrated configuration with one-layer thin substrate (0.8 mm) increases the monostatic RCSR bandwidth up to 75% (20.9–45.7 GHz), which is verified by measurement results.

Author Contributions: Conceptualization, Y.A. and S.H.S.; methodology, Y.A. and L.M.; software, Y.A.; validation, Y.A., S.H.S. and L.M.; formal analysis, Y.A.; investigation, Y.A.; resources, Y.A., L.M.; data curation, M.S., S.H.S., L.M.; writing—original draft preparation, Y.A.; writing—review and editing, Y.A., S.H.S., L.M.; visualization, M.S.; supervision, S.H.S., L.M., M.S.; project administration, M.S., L.M.; funding acquisition. All authors have read and agreed to the published version of the manuscript.

Funding: This research received no external funding.

Acknowledgments: The authors are willing to express their gratitude to Gianluca Dassano and Gianfranco Albis for their cooperation and assistance in mounting RCS measurement test set up and extraction of results with time gating method.

Conflicts of Interest: The authors declare no conflict of interest.

References

1. Esmaeli, S.; Sedighy, S. Wideband Radar Cross-Section Reduction by AMC. *Electron. Lett.* **2016**, *52*, 70–71. [CrossRef]
2. Xue, J.; Jiang, W.; Gong, S. Wideband RCS Reduction of Slot-Coupled Patch Antenna by AMC Structure. *Electron. Lett.* **2017**, *53*, 1454–1456. [CrossRef]
3. Chen, W.; Balanis, C.A.; Birtcher, C.R. Checkerboard EBG Surfaces for Wideband Radar Cross Section Reduction. *IEEE Trans. Antennas Propag.* **2015**, *63*, 2636–2645. [CrossRef]
4. Shi, H.; Li, J.; Zhang, A.; Jiang, Y.; Wang, J.; Xu, Z.; Xia, S. Gradient Metasurface with Both Polarization-Controlled Directional Surface Wave Coupling and Anomalous Reflection. *IEEE Antennas Wirel. Propag. Lett.* **2015**, *14*, 104–107. [CrossRef]
5. Li, Y.; Zhang, J.; Qu, S.; Wang, J.; Chen, H.; Xu, Z.; Zhang, A. Wideband Radar Cross Section Reduction Using Two-Dimensional Phase Gradient Metasurfaces. *Appl. Phys. Lett.* **2014**, *104*, 221110. [CrossRef]
6. Zhang, W.; Liu, Y.; Gong, S.; Wang, J.; Jiang, Y. Wideband RCS Reduction of a Slot Array Antenna Using Phase Gradient Metasurface. *IEEE Antennas Wirel. Propag. Lett.* **2018**, *17*, 2193–2197. [CrossRef]

7. Cheng, Y. An Ultra-Thin Dual-Band Phase-Gradient Metasurface Using Hybrid Resonant Structures for Backward RCS Reduction. *Appl. Phys. A* **2017**, *123*, 143. [CrossRef]
8. Azizi, Y.; Soleimani, M.; Sedighy, S.H. Low Cost, Simple and Broad Band Radar Cross Section Reduction by Modulated and Holography Metasurfaces. *J. Phys. D Appl. Phys.* **2019**, *52*, 435003. [CrossRef]
9. Liu, Y.; Hao, Y.; Li, K.; Gong, S. Wideband and Polarization-Independent Radar Cross Section Reduction Using Holographic Metasurface. *IEEE Antennas Wirel. Propag. Lett.* **2015**, *15*, 1028–1031. [CrossRef]
10. Zheng, Q.; Li, Y.; Zhang, J.; Ma, H.; Wang, J.; Pang, Y.; Han, Y.; Sui, S.; Shen, Y.; Chen, H.; et al. Wideband, Wide-Angle Coding Phase Gradient Metasurfaces Based on Pancharatnam-Berry Phase. *Sci. Rep.* **2017**, *7*, srep43543. [CrossRef]
11. Feng, M.; Li, Y.; Zheng, Q.; Zhang, J.; Han, Y.; Wang, J.; Chen, H.; Sai, S.; Ma, H.; Qu, S.; et al. Two-Dimensional Coding Phase Gradient Metasurface for RCS Reduction. *J. Phys. D Appl. Phys.* **2018**, *51*, 375103. [CrossRef]
12. Saifullah, Y.; Waqas, A.B.; Yang, G.M.; Zhang, F.; Xu, F. 4-Bit Optimized Coding Metasurface for Wideband RCS Reduction. *IEEE Access* **2019**, *7*, 122378–122386. [CrossRef]
13. Azizi, Y.; Soleimani, M.; Sedighy, S.H. Ultra-wideband radar cross section reduction using amplitude and phase gradient modulated surface. *J. Appl. Phys.* **2020**, *128*, 205301. [CrossRef]
14. Yu, N.; Capasso, F. Flat optics with designer metasurfaces. *Nat. Mater.* **2014**, *13*, 139–150. [CrossRef] [PubMed]
15. Monebhurrun, V. IEEE Standard 1502-2020: IEEE Recommended Practice for Radar Cross-Section Test Procedures [Stand on Standards]. *IEEE Antennas Propag. Mag.* **2021**, *63*, 106. [CrossRef]

electronics

Article

A Quad-Port Dual-Band MIMO Antenna Array for 5G Smartphone Applications

Jianlin Huang, Guiting Dong, Jing Cai, Han Li and Gui Liu *

College of Electrical and Electronics Engineering, Wenzhou University, Wenzhou 325035, China; jlhuang2021@163.com (J.H.); gtdong@yeah.net (G.D.); jcai2018@163.com (J.C.); lihan@wzu.edu.cn (H.L.)
* Correspondence: gliu@wzu.edu.cn

Abstract: A quad-port antenna array operating in 3.5 GHz band (3.4–3.6 GHz) and 5 GHz band (4.8–5 GHz) for fifth-generation (5G) smartphone applications is presented in this paper. The single antenna element consists of an L-shaped strip, a parasitic rectangle strip, and a modified Z-shaped strip. To reserve space for 2G/3G/4G antennas, the quad-port antenna array is printed along the two long frames of the smartphone. The evolution design and the analysis of the optimal parameters of a single antenna element are derived to investigate the principle of the antenna. The prototype of the presented antenna is tested and the measured results agree well with the simulation. The measured total efficiency is better than 70% and the isolation is larger than 16.5 dB.

Keywords: 5G; sub-6 GHz; MIMO; dual-band antenna; smartphone

Citation: Huang, J.; Dong, G.; Cai, J.; Li, H.; Liu, G. A Quad-Port Dual-Band MIMO Antenna Array for 5G Smartphone Applications. *Electronics* **2021**, *10*, 542. https://doi.org/10.3390/electronics10050542

Academic Editor: Faisel Tubbal

Received: 7 February 2021
Accepted: 23 February 2021
Published: 25 February 2021

1. Introduction

With the increasing demand for the quality of wireless communication, the fifth-generation (5G) mobile communication technology provides a promising solution to higher transmission rates, shorter latency, more connection density, and larger communication capacity. Recently, many 5G sub-6 GHz Multiple-Input Multiple-Output (MIMO) antennas have been developed for mobile terminals and base stations [1–5]. However, with the limited internal space of mobile phones, it is a challenging task to integrate many antennas with high isolation and low envelope correlation coefficient. Dual-band four-port MIMO antenna can expand system capacity and realize multi-mode communication. To minimize the unwanted mutual coupling, various isolation techniques have been reported in recent years [3–14]. The rectangle slots of the defected ground plane and two rectangular microstrip lines are employed to decrease the mutual coupling in [3]. In [4], the vertical patch is used to offset the coupling between the antenna and nearby components. A ground-connected T-shaped decoupling stub and an additional modified T-shaped decoupling stub is inserted between the two mirrored dual-antenna arrays to reduce the coupling in [5]. By introducing two short T-shaped strips and a π-shaped strip, the isolation between two antenna elements is enhanced [6]. An inverted-F antenna is used to reduce the mutual coupling of antenna elements [7]. Other decoupling techniques have also been presented, such as using a short neutral line [8,9], high-pass filter [10], pattern diversity [11], self-decoupling [12], defected ground structure [13], and metamaterial structure [14].

In the World Radiocommunication Conference 2015, the 3.5 GHz (3.4–3.6 GHz) frequency band became one of the 5G mobile phone spectrums. The Ministry of Industry and Information Technology of China approved 3.3–3.6 GHz and 4.8–5 GHz as the sub-6 GHz 5G frequency bands. Moreover, except for the sub-6-GHz spectrum, 5G smartphones are also expected to be phased array millimeter-wave antenna at 28 GHz frequency band [15].

In this paper, a novel dual-band antenna operating at 3.5 GHz band (3.4–3.6 GHz) and 5 GHz band (4.8–5 GHz) is proposed for the applications of a 5G MIMO antenna in smartphones. The proposed antenna shows high isolation, high gain and low ECC, which can be used to achieve high transmission rates and large channel capacity. The proposed antenna array consists of an L-shaped feeding strip, a parasitic rectangle strip, and a modified Z-shaped radiating strip. Different evaluative structures and optimal parameter analysis are carried out for the proposed compact quad-port dual-band MIMO antenna.

2. Antenna Geometry

The geometry of the proposed quad-port dual-band MIMO antenna array is shown in Figure 1. There are two types of printed circuit boards (PCBs) including a main board and two side boards, as depicted in Figure 1a. The size of the main board is 150 mm × 75 mm × 0.8 mm, and the size of the side board is 150 mm × 6.2 mm × 0.8 mm. Four antenna elements are printed on two side boards which are positioned vertically to the main board. Both the side boards and the main board are printed on an FR4 substrate with $\varepsilon_r = 4.4$ and tan $\delta = 0.02$. The side boards are bonded with the main boards by metal adhesive. Other spaces of both frames (side boards) are reserved for 2G/3G/4G or other wireless communication antennas in the mobile handsets. Figure 1b shows the detailed dimensions of single antenna element. The single antenna element includes a L-shaped feed strip, a parasitic rectangular strip and a modified Z-shaped radiation strip which is connected to the ground plane. The total size of a single antenna element is 14.9 mm × 7 mm × 0.8 mm.

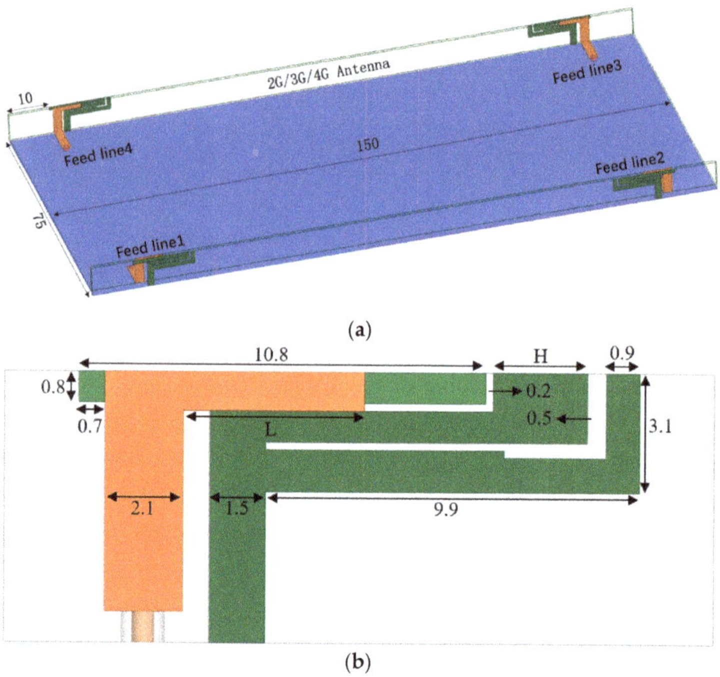

Figure 1. Geometry of the proposed quad-port MIMO antenna (in millimeters). (**a**) Prospective view; (**b**) Single antenna element.

3. Antenna Analysis

In order to understand the mechanism of the proposed MIMO antenna, the design evolution of the antenna element and the optimal parameter analysis are both carried out. Since the structure of the four antenna elements is identical, the parameters of a single element are used for analysis. The 3D full-wave electromagnetic simulator HFSS is used for design and optimization.

3.1. Analysis of the Design Evolution of the Proposed Antenna Element

The proposed antenna is generated by the evolution demonstrated in Figure 2. Ant. 1 is a structure composed of an L-shaped feeding strip and a Z-shaped radiation strip. By adding a parasitic rectangle strip to Ant. 1, Ant. 2 is obtained. Compared with Ant. 2, Ant. 3 has an additional L-shaped strip. The gaps between the L-shaped and Z-shaped strip are 0.5 mm and 0.2 mm, respectively. Compared with Ant. 3, Ant. 4 cuts a rectangular slit on the additional L-shape strip.

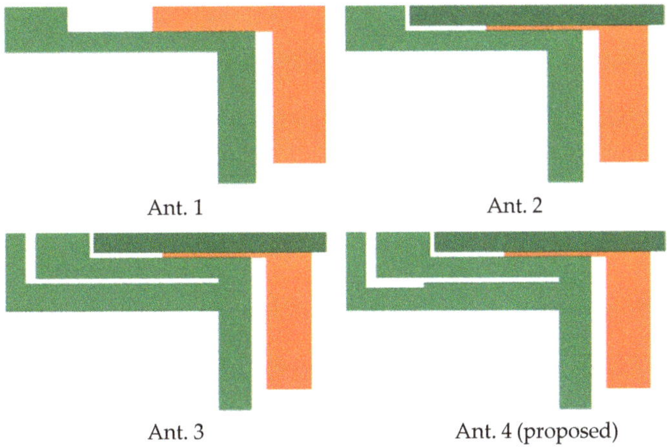

Figure 2. Design evolution of a single MIMO antenna element.

Figure 3 depicts the simulated S_{11} parameters of the evolution antennas. It can be seen that parasitic rectangular strip can increase the bandwidth, and L-shaped strip can create a 3.4 GHz resonant point. Furthermore, the modified Z-shaped strip can shift the frequency band to the higher frequencies which can cover 4.9 GHz (4.8–5 GHz) frequency band.

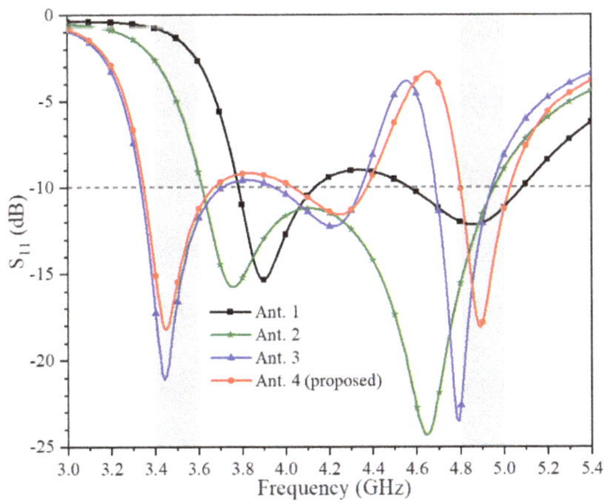

Figure 3. Simulated S_{11} of various antennas.

3.2. Analysis of the Variables of the Proposed Antenna Element

Figures 4 and 5 show simulated S_{11} of a single antenna element as a function of L and H, respectively. The value of L can be effectively used to change the resonant frequency of the lower frequency band. The higher resonance of a single element can be optimized by tuning the value of H. Eventually, the antenna can operate at frequencies ranging from 3.4 GHz to 3.6 GHz and 4.8 GHz to 5 GHz.

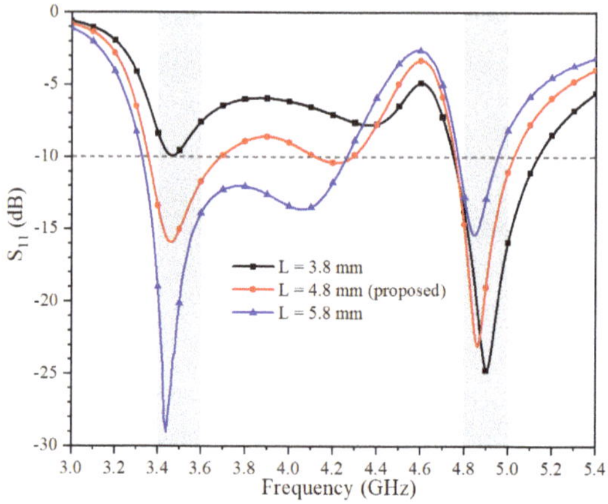

Figure 4. Simulated S_{11} of the proposed antenna with different values of L.

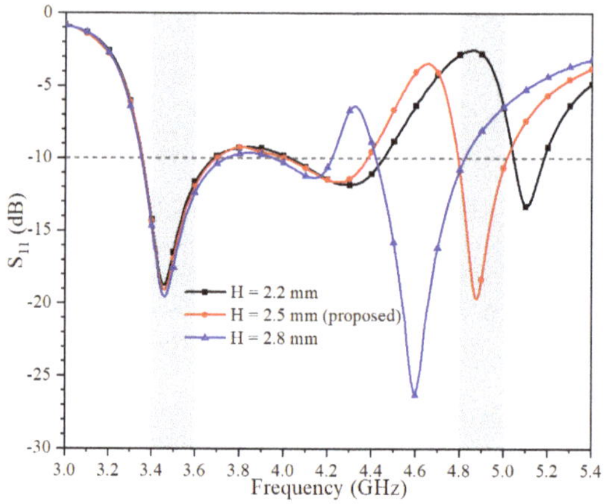

Figure 5. Simulated S_{11} of the proposed antenna with different values of H.

4. Experimental Results and Discussion

To verify the proposed design, an antenna prototype was fabricated using the optimized dimensions listed in Figure 1. Figure 6 is the photograph of the prototype antenna.

Figure 6. Photograph of the fabricated antenna prototype.

Figure 7 shows the S-parameters which are measured by Keysight Vector Network Analyzer N5224A. It can be seen that the measured S_{11} can cover both 3.5 GHz (3.4–3.6 GHz) and 4.9 GHz (4.8–5 GHz) frequency bands, and the isolation is higher than 16.5 dB. The measured isolation between various ports agrees well with the simulated results. The slight frequency offset is mainly due to SMA connective errors.

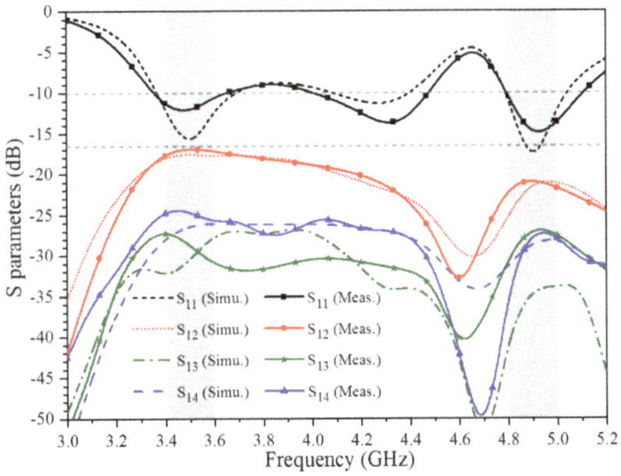

Figure 7. Simulated and measured S-parameters of the proposed antenna.

In Figure 8, the simulated surface current at 3.5 GHz is mainly distributed at the gap between the rectangular strip and the Z-shaped strip. At 4.9 GHz, strong current intensity is observed at the gap between the L-shaped strip and the modified Z-shaped strip. The distribution of the electric field is mostly the same as the distribution of electric current.

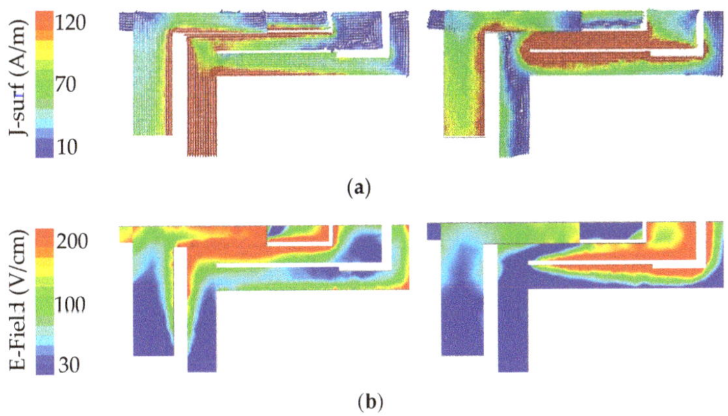

Figure 8. The surface current and electric field distributions of the proposed antenna element at (**a**) 3.5 GHz. (**b**) 4.9 GHz.

The radiation patterns of the proposed antenna element at 3.5 GHz and 4.9 GHz are shown in Figure 9, respectively. The measured co-pol and cross-pol are represented by different symbol lines. E-plane represents the direction in which the feed current flows, and H-plane is perpendicular to that direction. As shown in the E-plane radiation patterns of Figure 9b, the radiation between the feeding strip and the ground is taken into account, resulting in a similar gain between the co-pol and cross-pol.

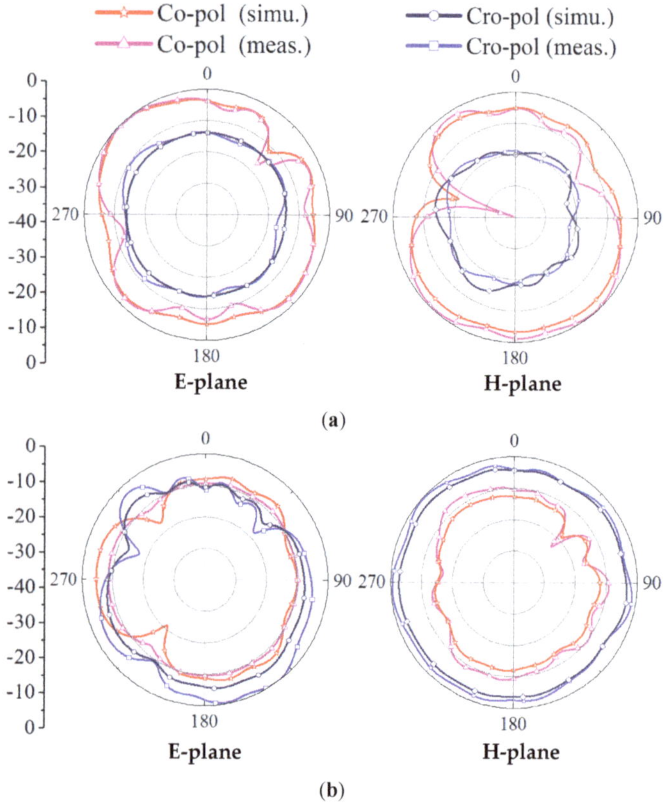

Figure 9. Radiation patterns of the proposed antenna. (**a**) 3.5 GHz. (**b**) 4.9 GHz.

In Figure 10, the 3D radiation patterns of a single antenna element as seen from the front side and back side of the substrate are presented, respectively. The maximum gain at 3.5 GHz and 4.9 GHz are 4.9 dBi and 5.1 dBi, respectively. The gains are high enough to meet the requirements of most mobile phone antennas.

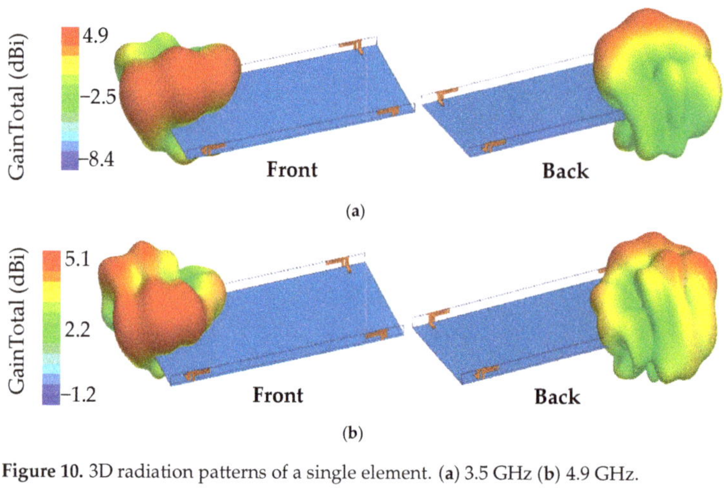

Figure 10. 3D radiation patterns of a single element. (**a**) 3.5 GHz (**b**) 4.9 GHz.

The measured total efficiency and peak gain, as shown in Figure 11, are better than 70% and 4 dBi, respectively. The 3.5 GHz and 4.9 GHz frequency points obtain the maximum total efficiency about 85% and 82%, respectively, while the 3.6 GHz and 5 GHz frequency points have the maximum peak gain about 4.7 dBi and 5 dBi, respectively. The measured efficiency is generally lower than that of simulation, which is mainly due to the current skin effect caused by excess solder and the loss of the chamber. The measured total efficiency and peak gain are in good agreement with the simulation results.

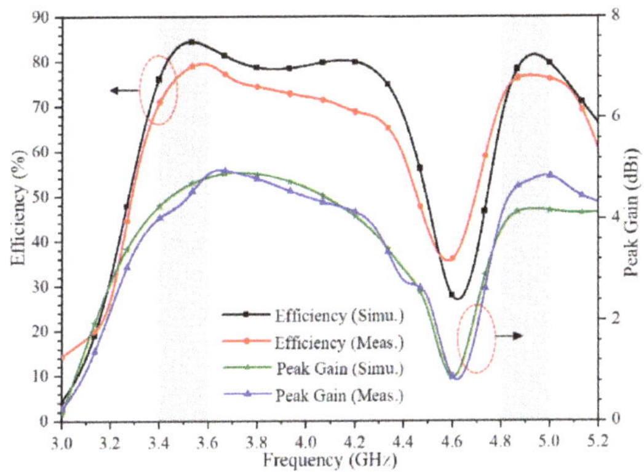

Figure 11. Simulated and measured peak gain and total efficiencies of the proposed antenna.

The envelope correlation coefficient (ECC) between radiating elements is a critical parameter of the MIMO antenna. The ECC can be calculated by the measured S-parameters according to the below equation.

$$ECC = \frac{|S^*{}_{mm}S_{mn} + S^*{}_{mm}S_{mn}|^2}{\left(1-|S_{mm}|2-|S_{nm}|^2\right)\left(1-|S_{nn}|2-|S_{mn}|^2\right)} \quad (1)$$

The calculated ECC of the proposed MIMO antenna is shown in Figure 12. The values of ECC in the working frequency bands are all smaller than 0.01, which means that the proposed quad-port MIMO antenna achieves good MIMO diversity performance.

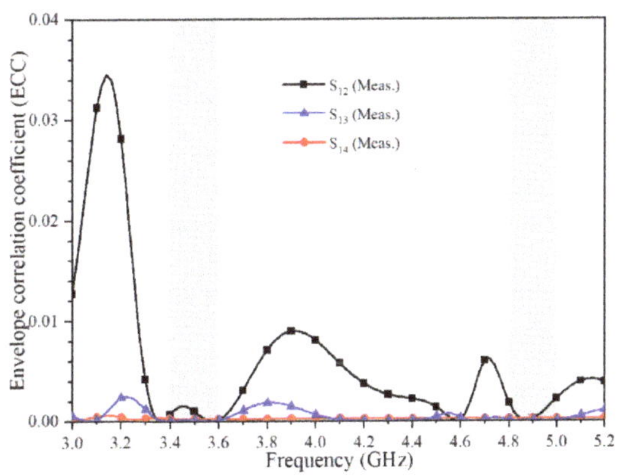

Figure 12. Calculated ECC of the proposed antenna.

Table 1 shows the performance comparison between the proposed antenna and the previously reported smartphone MIMO antennas. It could be concluded from Table 1 that the proposed MIMO antenna had high isolation, high efficiency and low ECC performance.

Table 1. Performance comparison with smartphone MIMO antennas.

Reference	Bandwidth (GHz)	Isolation (dB)	Efficiency (%)	ECC	Size (mm³)
[1]	3.3–3.6 (−6 dB)	20	33–47	0.4	150 × 75 × 5.3
[2]	3.4–3.6 (−6 dB)	13	50–60	0.15	136 × 68 × 1
[5]	3.3–4.2 (−6 dB) 4.8–5.0 (−6 dB)	11.5 15	53.8–76.5 62.6–79.1	0.1 0.12	150 × 75 × 5.5
[8]	3.1–3.85 (−10 dB) 4.8–6 (−10 dB)	17 18	65–75 60–71	0.06 0.06	150 × 70 × 6
[9]	3.4–3.6 (−6 dB) 4.8–5.1 (−6 dB)	11.5	41–72 40–85	0.08 0.05	150 × 75 × 7
[12]	3.3–4.2 (−6 dB)	11.5	63.1–85.1	0.2	150 × 75 × 7.5
[15]	2.45–2.65 (−10 dB) 3.4–3.75 (−10 dB) 5.6–6.0 (−10 dB)	11	40–65 50–70 60–80	0.01	150 × 75 × 1.6
This work	3.4–3.6 (−10 dB) 4.8–5.0 (−10 dB)	16.5	85 82	0.01	150 × 75 × 6.2

5. Conclusions

In this paper, a quad-port MIMO antenna covering 3.5 GHz (3.4–3.6 GHz) and 4.9 GHz (4.8–5 GHz) frequency bands is proposed. The proposed antenna is printed on two long frames of the smartphone, which reserves some space for other wireless communication antennas. The antenna is verified by both simulation and measurement. The size of a single antenna element is only 14.9 mm × 7 mm × 0.8 mm. The measured maximum peak gain at 3.6 GHz and 5 GHz are 4.7 dBi and 5 dBi, respectively. The measured total efficiency is greater than 70% and the isolation is better than 16.5 dB. The proposed MIMO antenna is a good candidate for 5G mobile handsets.

Author Contributions: Conceptualization, J.H.; methodology, G.L.; investigation, G.D., J.C.; writing—original draft preparation, J.H.; writing—review and editing, H.L., G.L.; supervision and funding acquisition, G.L. All authors have read and agreed to the published version of the manuscript.

Funding: This work was supported in part by National Natural Science Foundation of China under Grant No. 61671330and No. 61340049, the Science and Technology Department of Zhejiang Province under Grant No. LGG19F010009, and Wenzhou Municipal Science and Technology Program under Grant No. C20170005 and No.2018ZG019.

Data Availability Statement: The data presented in this study are available on request from the corresponding author.

Conflicts of Interest: The authors declare no conflict of interest.

References

1. Li, R.; Mo, Z.; Sun, H.; Sun, X.; Du, G. A low-profile and high-isolated MIMO antenna for 5G mobile terminal. *Micromachines* **2020**, *11*, 360. [CrossRef] [PubMed]
2. Abdullah, M.; Kiani, S.H.; Abdulrazak, L.F.; Iqbal, A.; Bashir, M.A.; Khan, S.; Kim, S. High-performance multiple-input multiple-output antenna system for 5G mobile terminals. *Electronics* **2019**, *8*, 1090. [CrossRef]
3. Barani, I.R.R.; Wong, K. Integrated inverted-F and open-slot antennas in the metal-framed smartphone for 2 × 2 LTE LB and 4 × 4 LTE M/HB MIMO operations. *IEEE Trans. Antennas Propag.* **2018**, *66*, 5004–5012. [CrossRef]
4. Liu, D.Q.; Zhang, M.; Luo, H.J.; Wen, H.L.; Wang, J. Dual-band platform-free PIFA for 5G MIMO application of mobile devices. *IEEE Trans. Antennas Propag.* **2018**, *66*, 6328–6333. [CrossRef]

5. Cui, L.; Guo, J.; Liu, Y.; Sim, C.Y.D. An 8-element dual-band MIMO antenna with decoupling stub for 5G smartphone applications. *IEEE Antennas Wirel. Propag Lett.* **2019**, *18*, 2095–2099. [CrossRef]
6. Bai, J.; Zhi, R.; Wu, W.; Shangguan, M.; Wei, B.; Liu, G. A Novel Multiband MIMO Antenna for TD-LTE and WLAN Applications. *Prog. Electromagn. Res.* **2018**, *74*, 131–136. [CrossRef]
7. Wu, W.; Zhi, R.; Chen, Y.; Li, H.; Liu, G. A Compact Multiband MIMO Antenna for IEEE 802.11 a/b/g/n Applications. *Prog. Electromagn. Res.* **2019**, *84*, 59–65. [CrossRef]
8. Serghiou, D.; Khalily, M.; Singh, V.; Araghi, A.; Tafazolli, R. Sub-6 GHz dual-band 8 × 8 MIMO antenna for 5G smartphones. *IEEE Antennas Wirel. Propag. Lett.* **2020**, *19*, 1546–1550. [CrossRef]
9. Guo, J.; Cui, L.; Li, C.; Sun, B. Side-edge frame printed eight-port dual-band antenna array for 5G smartphone applications. *IEEE Trans. Antennas Propag.* **2018**, *66*, 7412–7417. [CrossRef]
10. Xiao, F.; Lin, X.; Su, Y. Dual-band structure-shared antenna with large frequency ratio for 5G communication applications. *IEEE Antennas Wirel. Propag Lett.* **2020**, *19*, 2339–2343. [CrossRef]
11. Xu, Z.; Deng, C. High-isolated MIMO antenna design based on pattern diversity for 5G Mobile terminals. *IEEE Antennas Wirel. Propag. Lett.* **2020**, *19*, 467–471. [CrossRef]
12. Sun, L.; Li, Y.; Zhang, Z.; Wang, H. Self-decoupled MIMO antenna pair with shared radiator for 5G smartphones. *IEEE Trans. Antennas Propag.* **2020**, *68*, 3423–3432. [CrossRef]
13. Xing, H.; Wang, X.; Gao, Z.; An, X.; Zheng, H.; Wang, M.; Li, E. Efficient isolation of an MIMO antenna using defected ground structure. *Electronics* **2020**, *9*, 1265. [CrossRef]
14. Iqbal, A.; Saraereh, O.A.; Bouazizi, A.; Basir, A. Metamaterial-based highly isolated MIMO antenna for portable wireless applications. *Electronics* **2018**, *7*, 267. [CrossRef]
15. Parchin, N.O.; Basherlou, H.J.; Abd-Alhameed, R.A. Design of multi-mode antenna array for use in next-generation mobile handsets. *Sensors* **2020**, *20*, 2447. [CrossRef] [PubMed]

MDPI
St. Alban-Anlage 66
4052 Basel
Switzerland
Tel. +41 61 683 77 34
Fax +41 61 302 89 18
www.mdpi.com

Electronics Editorial Office
E-mail: electronics@mdpi.com
www.mdpi.com/journal/electronics

www.ingramcontent.com/pod-product-compliance
Lightning Source LLC
Chambersburg PA
CBHW061838210926
51850CB00006B/921
9783036551517